Handbook of Earth Science

Handbook of Earth Science

Editor: Jasper O'Brien

R CALLISTO
REFERENCE
www.callistoreference.com

Callisto Reference,
118-35 Queens Blvd., Suite 400,
Forest Hills, NY 11375, USA

Visit us on the World Wide Web at:
www.callistoreference.com

ISBN: 978-1-64116-141-1 (Hardback)

Cataloging-in-Publication Data

Handbook of earth science / edited by Jasper O'Brien.
    p. cm.
Includes bibliographical references and index.
ISBN 978-1-64116-141-1
1. Earth sciences. 2. Geology. I. O'Brien, Jasper.
QE26.3 .H36 2019
550--dc23

Table of Contents

Permissions

List of Contributors

Index

Preface

This book was inspired by the evolution of our times; to answer the curiosity of inquisitive minds. Many developments have occurred across the globe in the recent past which has transformed the progress in the field.

Earth science is a branch of science that studies the physical constitution of the Earth, its atmosphere and its various processes. Comprehension of the Earth and its physical characteristics is achieved by developing an understanding of all the Earth's spheres. It includes the lithosphere, atmosphere, hydrosphere, pedosphere and biosphere, as well as the Earth's interior and its magnetic field. Quantitative understanding is developed by an integration of the principles of geography, chronology, physics, chemistry and biology. The fields of science within the umbrella discipline of Earth science include physical geography, geophysics, soil science, ecology, atmospheric sciences, etc. This book provides significant information of this discipline to help develop a good understanding of Earth science and related fields. It brings forth some of the most innovative concepts and elucidates the unexplored aspects of this discipline. The extensive content of this book provides the readers with a thorough understanding of Earth Science.

This book was developed from a mere concept to drafts to chapters and finally compiled together as a complete text to benefit the readers across all nations. To ensure the quality of the content we instilled two significant steps in our procedure. The first was to appoint an editorial team that would verify the data and statistics provided in the book and also select the most appropriate and valuable contributions from the plentiful contributions we received from authors worldwide. The next step was to appoint an expert of the topic as the Editor-in-Chief, who would head the project and finally make the necessary amendments and modifications to make the text reader-friendly. I was then commissioned to examine all the material to present the topics in the most comprehensible and productive format.

I would like to take this opportunity to thank all the contributing authors who were supportive enough to contribute their time and knowledge to this project. I also wish to convey my regards to my family who have been extremely supportive during the entire project.

Editor

Hydrologic controls on aperiodic spatial organization of the ridge–slough patterned landscape

Stephen T. Casey[1], **Matthew J. Cohen**[1], **Subodh Acharya**[1], **David A. Kaplan**[2], and **James W. Jawitz**[3]

[1]School of Forest Resources and Conservation, University of Florida, Gainesville, FL, USA
[2]Engineering School of Sustainable Infrastructure and Environment, Environmental Engineering Sciences Department, University of Florida, Gainesville, FL, USA
[3]Soil and Water Science Department, University of Florida, Gainesville, FL, USA

Correspondence to: Matthew J. Cohen (mjc@ufl.edu)

Abstract. A century of hydrologic modification has altered the physical and biological drivers of landscape processes in the Everglades (Florida, USA). Restoring the ridge–slough patterned landscape, a dominant feature of the historical system, is a priority but requires an understanding of pattern genesis and degradation mechanisms. Physical experiments to evaluate alternative pattern formation mechanisms are limited by the long timescales of peat accumulation and loss, necessitating model-based comparisons, where support for a particular mechanism is based on model replication of extant patterning and trajectories of degradation. However, multiple mechanisms yield a central feature of ridge-slough patterning (patch elongation in the direction of historical flow), limiting the utility of that characteristic for discriminating among alternatives. Using data from vegetation maps, we investigated the statistical features of ridge-slough spatial patterning (ridge density, patch perimeter, elongation, patch size distributions, and spatial periodicity) to establish more rigorous criteria for evaluating model performance and to inform controls on pattern variation across the contemporary system. Mean water depth explained significant variation in ridge density, total perimeter, and length : width ratios, illustrating an important pattern response to existing hydrologic gradients. Two independent analyses (2-D periodograms and patch size distributions) provide strong evidence against regular patterning, with the landscape exhibiting neither a characteristic wavelength nor a characteristic patch size, both of which are expected under conditions that produce regular patterns. Rather, landscape properties suggest robust scale-free patterning, indicating genesis from the coupled effects of local facilitation and a global negative feedback operating uniformly at the landscape scale. Critically, this challenges widespread invocation of scale-dependent negative feedbacks for explaining ridge-slough pattern origins. These results help discern among genesis mechanisms and provide an improved statistical description of the landscape that can be used to compare among model outputs, as well as to assess the success of future restoration projects.

1 Introduction

The coupling of ecosystem processes operating at different scales can cause vegetation communities to form a wide variety of spatial patterns (Borgogno et al., 2009), ranging from highly regular striping, stippling, or maze-like patterns in woodland landscapes (Ludwig et al., 1999), tidal mud flats (Weerman et al., 2012), and boreal peatlands (Eppinga et al., 2010) to scale-free patterning in semi-arid landscapes (Kéfi et al., 2007; Scanlon et al., 2007). The mechanisms that produce these patterns are integral to understanding landscape origins and, thus, for predicting appropriate remedies where patterns and underlying processes have been degraded and require restoration. The spatial arrangement of vegetation on the landscape has long been viewed as a manifestation of the dominant interactions and drivers (Hutchinson, 1957; Levin, 1992) and the scales at which they operate. By quantifying this spatial arrangement we can make process-based inferences about the underlying mechanisms (Gardner et al., 1987; Turner, 2005).

The ridge-slough landscape comprised ~ 55 % of the pre-development Everglades in southern Florida (McVoy et al., 2011). However, processes that created, and in some places still maintain, the characteristic ridge-slough patterning are only partially understood (SCT, 2003; Larsen et al., 2011; Cohen et al., 2011). The landscape pattern consists of flow-parallel bands of higher-elevation ridges dominated by emergent sedge saw grass (*Cladium jamaicense*), interspersed within a matrix of lower-elevation sloughs (~ 25 cm lower in the best conserved portions of the landscape; Watts et al., 2010), which contain a variety of submerged and emergent herbaceous macrophytes. The Everglades has undergone massive hydrologic modification through the construction of a system of levees and canals over the past century (Light and Dineen, 1994), and ensuing ecological degradation has prompted a complex, expensive, and ambitious restoration effort. Because the ridge-slough landscape was so prevalent in the pre-development system, pattern restoration is a central priority (SCT, 2003; McVoy et al., 2011). The mechanisms that control the emergence of patterning and explain variation in pattern geometry are thus integral to specifying hydrologic restoration objectives.

To understand the landscape processes that produce patterning, and by extension gain insight into how to restore them (Pickett and Cadenasso, 1995), requires a testable mechanistic framework for pattern genesis and maintenance. However, experiments to test alternative mechanisms are constrained by the spatial extent and timescales of peat accumulation responses. Paradoxically, compartmentalization by the extensive canal and levee system has created artificial gradients that are informative for assessing trajectories of landscape pattern degradation. Here we focus on Water Conservation Area 3 (WCA-3), located in the central Everglades, an area historically dominated by the ridge-slough landscape (Fig. 1), and where the best conserved patterning is found. The hydrologic gradient in WCA-3 spans from relatively dry (i.e., short hydroperiod) conditions in the north due to major canals that drain water to the southeast, to extended inundation (i.e., long hydroperiod) in the south and southeast due to impoundment caused by US41/Tamiami Trail (which runs orthogonal to flow) and the L-67 levee. The best conserved patterning (SCT, 2003; Watts et al., 2010) is found between these hydrologic extremes.

Several alternative hypotheses have been proposed to explain ridge–slough patterns, and all have been evaluated using process-based models. The mechanisms invoked vary and include evaporative nutrient redistribution (Ross et al., 2006; Cheng et al., 2011), flow-driven sediment redistribution from sloughs to ridges (Larsen et al., 2007; Larsen and Harvey, 2011; Lago et al., 2010), self-optimization of patterning for discharge and hydroperiod (Cohen et al., 2011; Kaplan et al., 2012; Heffernan et al., 2013), and a suite of mechanisms that couple pattern-hydroperiod effects with directional local facilitation processes (i.e., where patches expand more rapidly in one direction than another; Acharya et

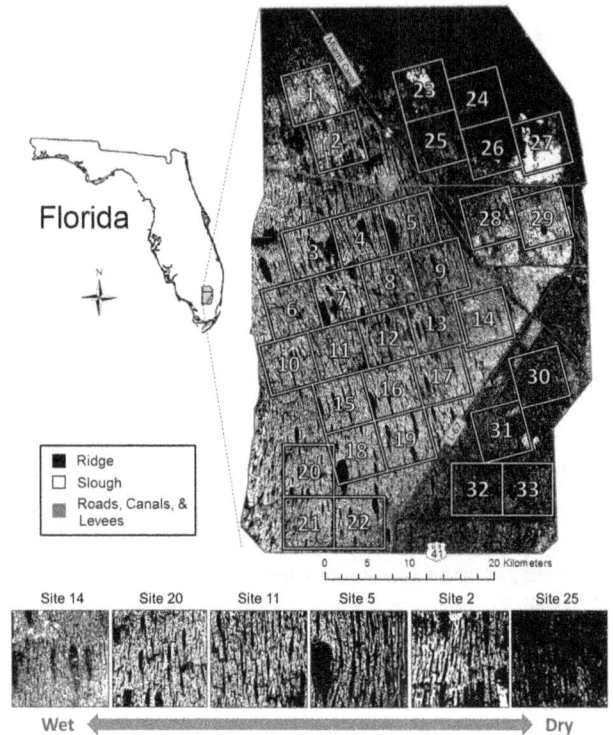

Figure 1. Study area and site locations, including major roads, canals, and levees for the primary map (M1). Sites spanning the pattern gradient in WCA3 are shown in the bottom panel. Two additional maps (Supplement) were used to corroborate the primary results.

al., 2015). Clearly, these mechanisms are not mutually exclusive, so process models have sought to explore the sufficiency of each alternative, while acknowledging the potential that multiple processes may overlap. One central criterion used to evaluate the models has been whether simulations can produce morphologies qualitatively consistent with the extant landscape (principally replicating the elongation of patches in the flow direction). To date, however, almost all models either accomplish (Ross et al., 2006; Larsen and Harvey, 2010; Lago et al., 2010; Cheng et al., 2011; Acharya et al., 2015) or strongly imply (Heffernan et al., 2013) patch elongation (albeit sometimes under conditions markedly different than those observed in the Everglades), limiting discrimination among pattern genesis mechanisms and highlighting the need for a more rigorous and quantitative characterization of landscape pattern.

To better characterize patterns in both the best conserved state and spanning a gradient of degradation requires spatial analyses that yield quantitative properties against which model outputs can be compared. Although numerous metrics have been developed to quantify different pattern attributes (Wu et al., 2006; Yuan et al., 2015), significant gaps in our understanding of how to interpret these metrics remain (Turner, 2001; Remmel and Csillag, 2003). Real landscapes clearly depart from regular Euclidean geometry, making characterization problematic in some cases (Mandelbrot, 1983). Likewise, changes in mapping procedures (e.g., grain size, extent, classification schemes) can yield significantly different metric values for the same landscape (Li and Wu, 2004). To remedy some of these issues, we focused on a set of relatively direct and easily interpreted metrics of fundamental aspects of the pattern and used multiple maps produced with varying methods to rule out mapping-related artifacts. We were interested in three aspects of landscape patterning: density and shape statistics, patch size distributions, and spectral (i.e., pattern wavelength) characteristics. For each aspect, we explored the magnitude of site-to-site variation and the support for hydrologic control of that variation.

Density and shape statistics focus on the most basic and intuitive geometric properties of the landscape: areal coverage of the patch types (density), landscape pattern complexity (perimeter), and the degree of elongation. While inundation has been shown to control species composition (Givnish et al., 2008; Zweig and Kitchens, 2008; Todd et al., 2010), the relationship between hydrologic drivers and other aspects of landscape pattern remain relatively unknown, so this effort also serves as an inventory of hydrologic controls on pattern geometry.

Patch size distributions (i.e., frequency of different patch sizes) have been used in many systems to identify underlying landscape processes (e.g., Manor and Shnerb, 2008a; Kéfi et al., 2011; Bowker and Maestre, 2012; Weerman et al., 2012). For example, regular patterning is associated with a characteristic patch size (Rietkerk and van de Koppel, 2008; von Hardenberg et al., 2010), arising in response to an inhibitory feedback operating at a particular spatial scale (van de Koppel and Crain, 2006) that limits patch expansion. Under these conditions, there should be a distinct mode in patch area distribution, or at least the absence of very large patches (Manor and Shnerb, 2008a; von Hardenberg et al., 2010; Kéfi et al., 2014). In contrast, patch size distributions that follow a power law (i.e., $y = x^{\alpha}$, where α is a scaling parameter) lack a characteristic spatial scale (e.g., Scanlon et al., 2007) and may suggest genesis mechanisms that operate equally across scales. Correspondingly, power law distributions are often referred to as scale-free, in that the distribution form remains the same regardless of the measurement scale.

Scale-free distributions can arise via a number of mechanisms (Newman, 2005). In a landscape where grid cells are randomly occupied, patch distributions show relatively few large patches, up to a critical density (~ 0.59; known as the percolation threshold) at which patches span the domain, yielding power-law area scaling. At densities slightly above and below the percolation threshold, area distributions depart from power laws. The narrow range of density space over which scale-free area distributions emerge would seem to suggest that this mechanism is rare. However, some systems can endogenously maintain themselves near this critical point in a phenomenon referred to as self-organized criticality (Bak et al., 1989). This is accomplished through disturbance processes that propagate via patch contiguity (e.g., forest fires, see Drossel and Schwabl, 1992), maintaining patterns near the percolation threshold through a cycle of large-scale disturbance and slow recovery (Pascual and Guichard, 2005).

Alternatively, power-law scaling of patch areas can arise from the coupled action of local facilitation, which causes patches to expand, and competition for a global resource (Pascual et al., 2002; Scanlon et al., 2007) that ultimately limits the density of that patch type at the landscape scale. In contrast to regular patterning mechanisms, these feedback processes limit landscape-level patch density but not the size of individual patches, leading to the creation, via local facilitation, of very large patches. This is known as robust criticality because power-law scaling in response can occur over a wide range of external conditions and patch densities, including densities well below the percolation threshold. Robust criticality has been noted in Everglades vegetation distributions (Foti et al., 2012), as well as in a variety of dryland vegetation patterns (Kéfi et al., 2011). Widespread occurrence of both local facilitation and global resource competition in ecological systems suggests this process may operate in a multitude of landscapes.

Finally, spectral characteristics provide insights on the presence and wavelength of regular landscape pattern. Useful information about the scale at which spatial feedbacks operate in self-organized systems has been obtained by evaluating two-dimensional pattern periodicity (Couteron, 2002; Kéfi et al., 2014). This is particularly important in the Everglades because the prevailing conceptual model for ridge-slough pattern genesis invokes interactions between spatial feedbacks operating on different characteristic scales, resulting in a pattern wavelength of approximately 150 m in the direction perpendicular to historical flow (SCT, 2003; Watts et al., 2010). Several models (e.g., Ross et al., 2006; Lago et al., 2010; Cheng et al., 2011) produce distinctly periodic landscapes, which arise from the action of local facilitation feedbacks and, crucially, negative feedbacks on patch expansion that operate at a characteristic scale. In contrast, the feedback between hydroperiod and landscape geometry suggested by Cohen et al. (2011), enumerated by Heffernan et al. (2013), and tested at the landscape scale in Kaplan et al. (2012) operates at the global scale, implying no characteristic spatial scale. To that end, we tested the hypothesis that the ridge-slough landscape is regularly patterned (i.e., exhibits a characteristic wavelength), consistent with scale-specific negative feedbacks, or whether the landscape lacks periodicity, consistent with scale-free feedbacks.

Together, these spatial analyses encompass a novel and rigorous set of metrics for improved quantification of observed and modeled landscape pattern. While developed to improve descriptions of the ridge-slough pattern, these metrics may also be useful for identifying pattern and discriminating genesis mechanisms in other patterned landscapes.

2 Methods

2.1 Vegetation and hydrologic data

We used multiple vegetation maps of the central Everglades, which vary in scale, extent, mapping schemes, and time frame. For all maps, we aggregated vegetation types into binary classes (reclassification scheme in Table S2 in the Supplement) of ridges (value $= 1$) and sloughs (value $= 0$). Our primary map (M1) was produced by the South Florida Water Management District (SFWMD) using 1 : 24 000-scale color infrared photos from September 1994 (Rutchey et al., 2005). This map was chosen due to its large, continuous spatial extent and fine mapping detail. The presence of small ($< 25 \, \text{m}^2$) landscape features permitted us to select raster representation of dominant vegetation at high (i.e., 1×1 m cells) resolution. While the presence of small features does not imply map accuracy at that fine scale, it does imply loss of patch geometric detail with larger cells. Features at this scale can be subject to mapping error and artifacts, likely under-representing their prevalence. As such, patches below $100 \, \text{m}^2$ were omitted from patch-level analyses.

We selected 33 6×6 km sites to span the range of current hydrological conditions (i.e., dry in northern areas to wet in southern areas; Fig. 1). We sought to maximize the number of sites with minimal overlap while avoiding roads and canals. All sites except 20–22 and 32–33 were rotated to align with the prevailing direction of patch elongation ($15°$ counterclockwise). Ridge cells were grouped into patches if they shared at least one edge with an adjacent ridge (i.e., a von Neumann neighborhood).

Within each site, *point-specific* daily average water depths at a grid spacing of 200 m were obtained from the Everglades Depth Estimation Network (EDEN) xyLocator (http://sofia.usgs.gov/eden/edenapps/xylocator.php). We note these water depths are spatially interpolated from a network of water elevation monitoring stations and, as such, represent only an estimate of actual conditions. *Site-specific* mean water depth (MWD) values were obtained by averaging all point-specific values in each site over the period of record from 1991 to 2010.

We used two additional maps (M2 and M3), which vary in spatial extent, resolution, and sampling date, to corroborate M1 analyses and test map resolution effects and temporal changes. M2 was generated from 1 : 24 000 scale aerial photographs taken in 2004 (RECOVER, 2014) and rasterized at 50 m resolution. M3 was generated from 1 m resolution digital orthophotos and rasterized at 1 m (Nungesser, 2011). Methodological details for both M2 and M3 are given as Supplement.

2.2 Shape and density

We compared ridge density, edge density, and elongation across sites. Ridge density is the proportion of ridge area to site area, while edge density is total patch perimeter divided by site area. In order to measure elongation, E, we first identify individual lengths and widths (l and w, respectively) as any group of contiguous ridge cells (i.e., unbroken by slough cells) along a row or column. Elongation is the ratio of the mean of these contiguous row and column sections:

$$E = \frac{\frac{1}{n_c}\sum l}{\frac{1}{n_r}\sum w} = \frac{n_r}{n_c}, \tag{1}$$

where n_r and n_c represent the number of contiguous rows and columns. Elongation simplifies to their ratio since the summation terms both yield the total number of ridge cells. Elongation metrics are sensitive to orientation differences between the grid and landscape features. Sites with tortuous flow paths or a poorly aligned grid will underestimate E. We provide estimates of grid alignment with feature orientation as a mean patch angle, $\overline{A}_p$, where A_p is the angle between the grid y axis and the major axis of an ellipse with the same second moment as the patch.

Hydrologic trends were identified by regressing MWD against site-level metrics and were considered statistically significant at $p < 0.05$. For analyses that are highly dependent on mapping resolution (i.e., edge density), we omit M1 sites north of Interstate 75, as these were mapped using significantly lower resolution than those to the south (Rutchey et al., 2005). Because elongation values are dominated by the domain shape at very high ridge densities, we omitted sites where ridge density exceeded 0.8.

2.3 Patch size distributions

Patch size scaling properties were evaluated by comparing empirical distributions to several candidate models. Patch size distributions can be described in terms of their complementary cumulative distribution function (CCDF), which gives the probability that the area of an observed patch is greater than or equal to a given area, x. Preliminary analyses showed that empirical CCDFs exhibited extremely heavy tails consistent with power laws, but only above a minimum cutoff, below which patches were less abundant and the CCDFs were rounded. This form is in relative agreement with both the generalized Pareto (GP) and truncated lognormal distributions. The GP is given by its CCDF as

$$P(x) = \begin{cases} \left(1 + \dfrac{k(x - x_{\min})}{\delta}\right)^{-\frac{1}{k}} & \text{for } k \neq 0 \\[2ex] \exp\left(-\dfrac{x - x_{\min}}{\delta}\right) & \text{for } k = 0, \end{cases} \tag{2}$$

for $x \geq x_{\min}$ when $k \geq 0$, and for $x_{\min} \leq x \leq (x_{\min} - \delta/k)$ when $k < 0$. The GP reduces to the exponential distribution when $k = 0$ and $x_{\min} = 0$ and reduces to a power function when $k > 0$ and $x_{\min} = \delta/k$. For $k > 0$ and $x_{\min} < \delta/k$ the GP shows exponential-like behavior for low values of x, while the tail asymptotically approaches a power law for $x \gg x_{\min}$. Within this range of parameters, δ indicates the curvature in the upper end of the distribution (higher values correspond to greater curvature and, hence, relatively fewer small patches), while k indicates the scaling properties of the tail, such that for $x \gg x_{\min}$, the power-law scaling exponent α approaches $\alpha^* = (1 + 1/k)$ (Pisarenko and Sornette, 2003). Where the GP fits the data well, we can use the estimated parameters as general information about patch size scaling properties. The CCDF for a truncated lognormal distribution uses the mean ($\mu_{\ln_x}$) and standard deviation ($\sigma_{\ln_x}$) of $\ln(x)$.

$$P(x) = \frac{\text{erf}\left(\frac{\sqrt{2}[\mu_{\ln_x} - \ln(x)]}{2\sigma_{\ln_x}}\right) + 1}{\text{erf}\left(\frac{\sqrt{2}[\mu_{\ln_x} - \ln(x_{\min})]}{2\sigma_{\ln_x}}\right) + 1} \qquad x \geq x_{\min} \qquad (3)$$

We compared empirical distributions to synthetic data sets from Monte Carlo simulations ($n = 20\,000$ per model) and compared candidate distributions based on log-likelihood ratios and significance values (Clauset et al., 2009). Distribution testing details are given in SI.

2.4 Spectral characteristics

Spectral characteristics of the ridge-slough landscape were evaluated from 2-D periodograms generated following the methods of Mugglestone and Renshaw (1998). In brief, we constructed a discrete 2-D Fourier transform (available in most computational software packages) for each binary vegetation map (Kéfi et al., 2014) and then took the absolute value to obtain the real number component. The resulting 2-D periodogram (i.e., spectral density) is a grid representing the magnitude of cosine and sine waves of possible wavenumbers (i.e., spatial frequencies) and orientations to the spectrum. Values were averaged across all orientations in equally spaced wavenumber bins to generate radial spectra (r spectra), which indicate the relative spectral density for each corresponding wavenumber bin. Local maxima indicate dominant wavelengths and, thus, suggest the presence of spatial periodicity (Couteron, 2002; Kéfi et al., 2014) or regular patterning. The absence of local maxima indicates an aperiodic landscape. Because the ridge-slough pattern has been described as regular in the direction orthogonal to flow, we generated both lateral and longitudinal r spectra derived from the spectral densities observed within $\pm 10°$ perpendicular and parallel to the main axis of pattern elongation. For both directions, we noted the wavelength at which either clear spectral peaks (i.e., for periodic patterns) or locations of spectral shouldering (i.e., slope breaks), which may indicate a secondary scale-dependent feedback mechanism, were evident. Since smaller features are underrepresented in low-resolution maps, we omitted wavelengths < 10 m from our analyses.

3 Results

3.1 Visual comparisons

Visual inspection of the vegetation maps reveals a remarkable range of pattern morphology (Fig. 1). Ridges in northwestern sites (1–5) show pronounced striping, which is less apparent in southern sites (18–22), where ridges appear more elliptical. Eastern sites located below I-75 (5, 9, 13, 14, 17, 28–33) show fine-scale speckling and disaggregation, with sites 14, 28, and 29 appearing random, with faint outlines of historic pattern.

Individual ridges exhibit numerous connections between adjacent elongated portions, with larger patches forming complex webs composed of multiple individual elements. Although this behavior is apparent in all sites, it appears to be density dependent, with most of the landscape spanned by one large patch in denser sites (e.g., 2, 5, 8, 9, 11–13, 23–28, 30–33). Within sites, large patches are always more web-like than smaller ones, which appear more distinctly separated.

3.2 Density and shape

Ridge density was negatively correlated to MWD (Fig. 2a; $R^2 = 0.38$, $p = 0.0002$). Deviation from this association was similar across maps and related to geographic position. Specifically, ridge densities in the eastern half of the domain (sites 9, 13, 14, 17, 23–33) were consistently higher than in the west, suggesting a strong east–west control on density. The correlation between MWD and ridge density increased markedly when sites were partitioned into east and west blocks (east: $R^2 = 0.81$, $p < 0.0001$; west: $R^2 = 0.61$, $p = 0.0004$). Based on recent aerial imagery, low ridge density in site 1 is a misclassification of sparse saw-grass prairies as slough; that site was omitted from regression analyses.

Site-level elongation was also strongly correlated to MWD (Fig. 2b; $R^2 = 0.65$, $p < 0.0001$). Sites with ridge densities greater than 0.8 showed elongation values much lower than this trend. Average patch orientations ($\overline{A}_p$) indicate consistency between the grid and feature elongation (i.e., $\overline{A}_p$ values close to zero; Table S1). Sites with values of $|\overline{A}_p| \geq 5°$ (e.g., 1, 22) may be underestimated due to mismatch between patch orientation and map orientation. Finally, edge density was strongly correlated to MWD, indicating greater perimeter at deeper sites (Fig. 2c; $R^2 = 0.79$, $p < 0.0001$).

3.3 Patch size distributions

Patch area distributions were consistent with the GP distribution (Fig. 3c), with 16 of 25 sites passing GP Monte Carlo tests for M1 and 4 of 9 passing for M3 (Table S1). The majority of sites that were not significant contained extremely large patches but had little deviation in the rest of the distribution; in some cases (e.g., sites 2, 5, 8, 9, 11, 12, 13, 28, 31) the largest patch was over an order of magnitude larger than

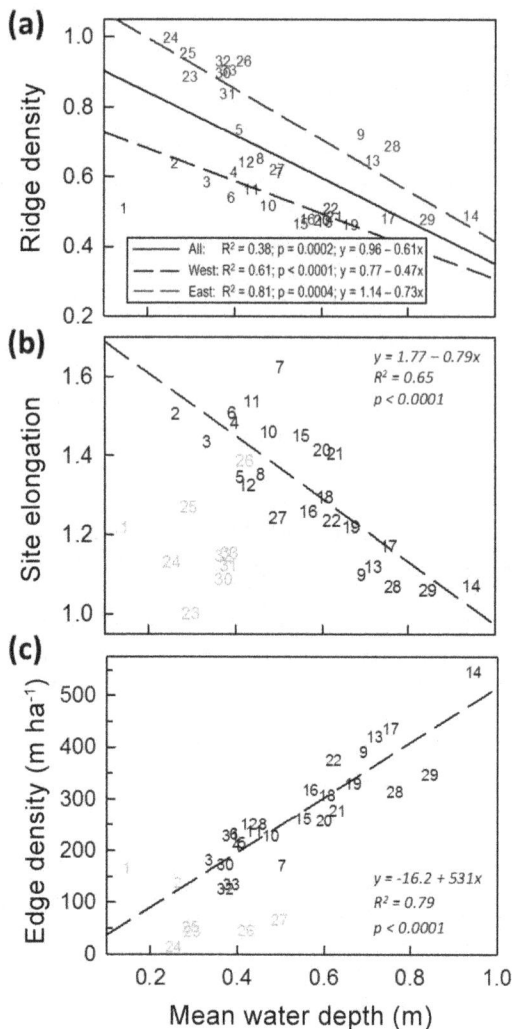

Figure 2. (a) Ridge density is negatively correlated with mean water depth. Eastern sites (blue) show consistently higher ridge densities than those in the west (black). Trends associated with east–west segregation (dashed lines) show much stronger relationships than the composite trend (solid line). Site 1 was omitted due to possible misclassification. **(b)** Site elongation shows a strong negative relationship with mean water depth. Sites with ridge densities greater than 0.8 (indicated in grey) were omitted from regressions and show elongation values lower than expected from this trend. **(c)** Edge density is positively correlated to mean water depth indicating higher perimeters in deeper sites. Sites indicated in grey were mapped at lower resolution and were omitted from regressions. The relationships observed for site elongation and edge density are both consistent with patches becoming disaggregated with increased water depth.

predicted based on the GP distribution. All these sites with extremely large patches have ridge densities above or very close to the percolation threshold of a square lattice (~ 0.59, Stauffer and Aharony, 1991). Above this percolation threshold, the largest patch becomes "over-connected", suggesting that failure of Monte Carlo tests within this group may be

density driven rather than a result of an underlying patterning mechanism. Note that these sites are largely located in the north and eastern sections of the study area, a region typified by high ridge densities. The presence of tree islands, a third landform modality distinct from ridges and sloughs, may also affect patch scaling relationships; however, we neglect these effects here because across all blocks tree islands represent less than 10 % of the area, and often much less. The log-normal distribution was significant in only 4 of 25 sites for M1 and 2 of 9 sites for M3. Although these sites (15, 16, 19, 21) showed slight rounding in the extreme tail, log-likelihood ratios were not different enough to distinguish between the two candidate distributions (Table S1).

Within each map, GP parameters were remarkably consistent across sites, with almost constant estimates of k and δ for sites that passed Monte Carlo tests (Table S1). Area scaling in the tail of the distribution is illustrated by α^* (analogous to the scaling exponent of a power-law distribution) $= 1.77 \pm 0.06$ for M1 and 1.87 ± 0.13 for M3. The δ parameter indicates how sharply the distribution head deviates from a power law, with larger values indicating that smaller patch areas are exceedingly rare. For M1 and M3, $\delta = 474 \pm 88$ and 1490 ± 219; these differences are likely due to map resolution, with M3 under-representing smaller patches.

3.4   Spectral characteristics

We found no evidence of periodicity in either the lateral or longitudinal r spectra. The absence of peak values other than the smallest wavenumber indicates that no dominant pattern wavelength exists, a finding consistent across hydrologic conditions and pattern morphologies (Fig. 3a, Fig. S7 in the Supplement). Spearman correlations, ρ, show the r spectra are nearly perfectly approximated by a monotonic function across all sites (Table S1), with $\rho < -0.99$ for both lateral and longitudinal r spectra. As with patch-scaling relationships, tree islands may introduce some noise in the observed r spectra; however, this effect is likely to be small, given that they constitute less than 10 % of the landscape.

For both lateral and longitudinal directions, the form of the r spectra appeared to contain a mix of both power law and exponential scaling. Lateral r spectra largely appear linear in log–log space (i.e., power law form) at higher wavenumbers while rounding towards an exponential (i.e., curved) at lower wavenumbers. This curvature appears over a wider range of wavenumbers for longitudinal spectra, but the morphology and mean transition location are same laterally and longitudinally. Sites with the best-conserved patterning (e.g., sites 2, 5, 11, 20) show more localized curvature in lateral r spectra compared to degraded sites, potentially signifying the action of a secondary, scale-dependent patterning mechanism. We note, however, that this finding is inconsistent with proposed patterning mechanisms that invoke a characteristic wavelength in the lateral direction but include no mecha-

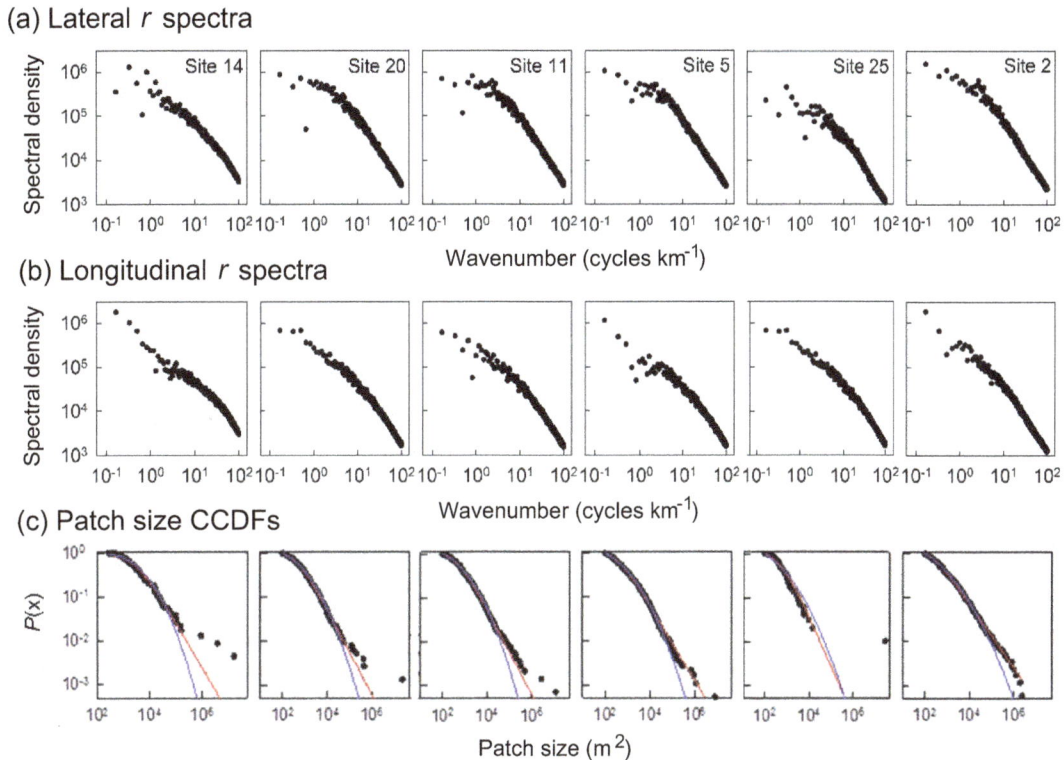

Figure 3. (a) Lateral r spectra (limited to $\pm 10°$ perpendicular to the pattern) monotonically decreased with no evidence of peaks, indicating aperiodic behavior in the direction of presumed regularity. **(b)** Longitudinal r spectra (limited to $\pm 10°$ in the direction parallel to the pattern) show similar monotonic behavior. The form for both lateral and longitudinal directions is similar, with both exhibiting a mixture of power-law and exponential behavior. The location of the exponential-like curvature appears to be influenced by both orientation and pattern condition, suggesting a weak-acting scale-dependent mechanism. **(c)** Patch size distributions across sites are well described by the generalized Pareto distribution (red lines). Sites with high ridge densities (e.g., sites 2, 5 and 25) have maximum patch sizes much greater than expected from the GP distribution. Conversely, sites in excessively inundated sections (e.g., site 20) show slightly steeper tails, consistent with a lognormal distribution (blue lines), though not enough to rule out the GP.

nism to generate regular patterning in the longitudinal direction. Alternatively, this shouldering may result from undersampling large features at low wavenumbers due to a limited domain size.

4 Discussion and conclusions

4.1 Water depth controls pattern attributes

Our results provide strong observational support for water depth as a dominant control on several key shape and density properties of the ridge-slough landscape. Although these findings are correlative and not necessarily mechanistic, they align with current understanding about the mechanisms that create, maintain, and degrade the landscape. The observed decline in ridge abundance with MWD is consistent with conceptual models that predict that changes in water levels precipitate transitions between ridge and slough by modifying production and respiration dynamics (Givnish et al.,

2008; Watts et al., 2010) and inducing state changes in vegetation composition (Zweig et al., 2008). The implication that these dynamics differ in eastern and western sections of the study area was unexpected and points to unexplained controls on ridge expansion. The largest difference between the east and west trends occurs at low water depths, indicating that this control is most pronounced in drier sites. In short, the deviation seen in eastern sites represents a shifting of the relationship to favor saw-grass expansion in extremely dry sites rather than a general reduction of the hydrologic limitation (since deep sites remain the least affected).

Mean water depth also exerted strong control on ridge-slough pattern shape. The most salient features of the pattern, elongation and perimeter, both showed strong dependence on MWD, with maximum elongation observed at low to intermediate water depths and minimum edge density values at low water depths. This is consistent with ridge features fragmenting into smaller, less elongated patches under deeper water conditions, a finding previously observed anecdotally (McVoy et al., 2011) and in the spatial statistics of soil elevation (Watts et al., 2010). Likewise, sites with very low MWD show a significant loss of pattern, with ridge densities approaching unity and elongation values that are largely isotropic. The coherent response of these pattern features to hydrologic modification suggests promise for their use as restoration performance measures (Yuan et al., 2015).

In this work we provide support for hydrological controls on ridge-slough pattern shape; however landscape patterning (specifically ridge density and elongation) has also been shown to exert reciprocal control on regional hydrology (Kaplan et al., 2012). Loss of sloughs in sites with very low MWD alters drainage characteristics. Coupled to observations of patch fragmentation in sites with higher water depths, these results strongly reinforce the commanding role of hydrology in maintaining landscape pattern, indicating that reversal of modern hydrologic modification is paramount for ongoing restoration.

4.2 The ridge-slough landscape is aperiodic and scale-free

Both spatial periodogram results and patch size distributions strongly suggest the ridge-slough landscape pattern is aperiodic, a marked departure from extensive literature qualitatively describing the pattern as periodic (SCT, 2003; Wetzel et al., 2005; Ross et al., 2006; Larsen et al., 2007; Givnish et al., 2008; Larsen and Harvey, 2010; Lago et al., 2010; Watts et al., 2010; Cheng et al., 2011; Nungesser, 2011; Sullivan et al., 2014). Because negative feedbacks operating at a characteristic spatial scale result in regular patterning (Rietkerk and Van de Koppel, 2008), aperiodic patterning in the ridge-slough landscape implies the absence, or least secondary importance, of such feedbacks, ruling them out as the dominant control on patterning many of the mechanisms invoked to explain pattern formation (Borgogno et al., 2009).

While our results clearly support the primacy of aperiodic patterning mechanisms, the r spectra in both lateral and longitudinal directions do exhibit persistent curvature, whose location and degree appears dependant on both orientation and pattern condition. This suggests ridge-slough patterning is secondarily influenced by scale-dependent (but omnidirectional) feedbacks, possibly suggesting links with vegetative propagation or fire behavior. Additional investigation and modeling, requiring higher-resolution mapping, would be necessary to better understand the mode and scale of these secondary feedbacks.

The observation that patch size distributions uniformly follow power-law scaling suggests a scale-free patterning process. While power-law scaling can be produced via several mechanisms (Newman, 2005), our results can be used to rule out some alternatives. For example, power-law scaling of patch areas can arise in systems near the percolation threshold (i.e., at criticality), which occurs within a relatively narrow region of patch density. Observed patch area scaling in our study occurs across a wide range of patch densities, suggesting robust criticality that comports with Foti et al. (2012), who observed similar power-law scaling behavior over a wide range of vegetation types and densities.

Caution is warranted when using contemporary aerial imagery to infer pre-drainage landscape conditions; the first aerials were taken $\sim$65 years after Everglades drainage began. Several pattern attributes (e.g., density, perimeter) may adjust readily with hydrologic modification, and while some areas remain largely unchanged since initial imagery was obtained, pattern in many other areas has degraded, sometimes entirely (Wu et al., 2006; Nungesser 2011). However, pattern properties that are relatively invariant with hydrologic modification (e.g., the general forms of the r spectrum and patch area distributions) are more likely to reflect pre-drainage conditions. In contrast, while measures that vary with hydrologic modification are correlative, they remain useful for understanding landscape responses to hydrologic forcing but may be less informative for inferring pre-drainage conditions and long-term processes such as landscape formation.

Self-organized criticality can also produce power-law scaling at varying densities (i.e., far from the percolation threshold) but requires large temporal variation in ridge density as the system endogenously readjusts towards criticality following disturbances (Pascual and Guichard, 2005). Recent paleoecological evidence (Bernhardt and Willard, 2009) suggests that ridge-slough configurations and densities have remained relatively stable since initial formation 2700 years before present, though temporal variation in density (e.g., during the Medieval Warm Period) may have been sufficient to modestly alter landscape pattern metrics. Moreover, no documented disturbance regime exhibits the characteristic separation of timescales between growth and disturbance associated with self-organized criticality. While peat fires could be invoked, there is little evidence for widespread incidence and large-scale impacts of these prior to modern hydrologic modification (McVoy et al., 2011).

Rather, power-law scaling in patch areas over a range of densities along environmental gradients is consistent with robust criticality, wherein local facilitation induces clustering (i.e., patch growth) while a global limitation maintains landscape heterogeneity (Pascual and Guichard, 2005). Although robust criticality is typically suggested in isotropic landscapes, Acharya et al. (2015) recently showed that anisotropy in the local facilitation kernel of a robust criticality model can produce directional banding without periodicity, yielding simulated ridge-slough patterns with high statistical and visual fidelity to the observed landscape. Local facilitation may take the form of autogenic peat accretion (Larsen et al., 2007), clonal propagation of saw grass (Brewer, 1996), nutrient accumulation dynamics (Cohen et al., 2009; Larsen et al. 2015), or local seed dispersal, although the relative importance and directionality of these mechanisms remains unknown (Acharya et al., 2015). Screening possible mechanisms for anisotropic local facilitation emerges from our analysis as a priority for future investigations.

Several candidate processes could limit patch expansion in the ridge–slough landscape. Each implies a distinct spatial pattern geometry, and we can use the extant scale-free and aperiodic geometry to evaluate their respective plausibilities. A key distinction between limiting processes that produce periodic versus scale-free patterning is the spatial range over which the limiting factor acts (Manor and Shnerb, 2008a; von Hardenberg et al., 2010) the landscape, the effect is considered global or uniform. Conversely, when the limiting effect acts in a more localized manner, limitation gradients can develop and produce periodic patterning.

Phosphorus limitation and sediment transport mechanisms are both potentially important feedbacks on patch expansion. While phosphorus is strongly limiting of primary production in the Everglades (Noe et al., 2001) and can be dramatically enriched in tree islands (Wetzel et al., 2009) and ridges (Ross et al., 2006) via multiple mechanisms, this process of local enrichment and depletion is inconsistent with robust criticality. Indeed, the presence of strong local phosphorus gradients indicates that limitation feedbacks are distinctly local and not spread uniformly across the landscape. If phosphorus limitation were the dominant control, the result would be regular patterning. Similarly, sediment transport mechanisms (Larsen et al., 2007; Lago et al., 2010) yield a balance between entrainment and deposition governed by focused flow in sloughs, the velocity of which is controlled by cross-sectional occlusion of flow by ridges. Because patch geometry is controlled by local heterogeneity in flow velocity, this suggests an inhibitory feedback operating at a limited spatial scale, as the velocity field responds most strongly to local flow occlusion.

Water level (and hydroperiod) is another potential feedback on patch expansion. Our observations of water depth control on ridge density comport with numerous studies (Givnish et al., 2008; Zweig and Kitchens, 2008; Todd et al., 2010) suggesting ridges are significantly impacted by water depths. Moreover, pattern geometry strongly influences landscape hydrology (Kaplan et al., 2012; Acharya et al., 2015). As ridges expand into adjacent sloughs, they displace water and alter landscape flow capacity, causing regional water levels to increase (Kaplan et al., 2012), and creating a negative feedback that likely limits further ridge expansion (Cohen et al., 2011). Indeed, the RASCAL model of ridge-slough development (Larsen and Harvey, 2011) represents this feedback, though in that model velocity-field feedbacks alone could not impose elongation and regular patterning; disentangling sediment transport and water-level feedbacks in that model, and interrogating pattern output, may enable tests of the relative importance of overlapping feedbacks at different scales. We note here that because water depths equilibrate quickly, local patch expansion effects on water level are distributed rapidly and evenly across the landscape. This expansion is consistent with the global limitation necessary to create observed aperiodic and scale-free pattern. Therefore, water depth effects are strong candidates for the requisite global feedback to induce ridge-slough formation.

Our results also indicate that elongated landscape features do not necessarily require pattern periodicity, suggesting that spatial structures in numerous ecosystems may have been misclassified as regularly patterned and that aperiodic banding may be more prevalent than the literature suggests. Invoking robust criticality and directional facilitation, as in Acharya et al. (2015), may be of general value for explaining aperiodic banding in other settings.

The ridge-slough landscape pattern has emerged as a key measure of restoration performance in one of the largest and most ambitious ecosystem management endeavors ever. Enumeration of spatial pattern statistical features is a prerequisite for assessing landscape condition and for comparing models with alternative landscape genesis mechanisms. Our results inform about the metrics for comparison between real and simulated landscape patterns and provide insights into the controls on pattern variation across the contemporary system. Given the potentially significant differences in water management implied by comparative genesis explanations, these metrics of real and simulated landscapes are important for restoration planning and assessment.

Acknowledgements. We gratefully acknowledge funding support from the US Army Corps of Engineers through the Everglades Monitoring and Assessment Program. We also acknowledge the contributions of Danielle Watts, Jim Heffernan and Mark Brown in this work. We also benefited greatly from deep and constructive feedback from Laurel Larsen on an earlier version of the manuscript.

Edited by: R. Moussa

References

Acharya, S., Kaplan, D. A., Casey, S., Cohen, M. J., and Jawitz, J. W.: Coupled local facilitation and global hydrologic inhibition drive landscape geometry in a patterned peatland, Hydrol. Earth Syst. Sci., 19, 2133–2144, doi:10.5194/hess-19-2133-2015, 2015.

Bak, P., Tang, C., and Wiesenfeld, K.: Self-organized criticality, Phys. Rev. A, 38, 364–374, 1989.

Bernhardt, C. E. and Willard, D. A.: Response of the Everglades ridge and slough landscape to climate variability and 20th-century water management, Ecol. Appl., 19, 1723–1738, doi:10.1890/08-0779.1, 2009.

Borgogno, F., D'Odorico, P., Laio, F., and Ridolfi, L.: Mathematical models of vegetation pattern formation in ecohydrology, Rev. Geophys., 47, RG1005, doi:10.1029/2007RG000256, 2009.

Bowker, M. A. and Maestre, F. T.: Inferring local competition intensity from patch size distributions: a test using biological soil crusts, Oikos, 121, 1914–1922, doi:10.1111/j.1600-0706.2012.20192.x, 2012

Brewer, J. S.: Site differences in the clone structure of an emergent sedge, *Cladium jamaicense*, Aquat. Bot., 55, 79–91, 1996.

Cheng, Y., Stieglitz, M., Turk, G., and Engel, V.: Effects of anisotropy on pattern formation in wetland ecosystems, Geophys. Res. Lett., 38, L04402, doi:10.1029/2010GL046091, 2011.

Clauset, A., Shalizi, C. R., and Newman, M. E.: Power-law distributions in empirical data, SIAM Rev., 51, 661–703, 2009.

Cohen, M. J., Osborne, T. Z., Lamsal, S. J., and Clark, M. W.: Regional Distribution of Soil Nutrients-Hierarchical Soil Nutrient Mapping for Improved Ecosystem Change Detection, South Florida Water Management District, West Palm Beach, Florida, USA, 91 pp., 2009.

Cohen, M. J., Watts, D. L., Heffernan, J. B., and Osborne, T. Z.: Reciprocal biotic control on hydrology, nutrient gradients and landform in the Greater Everglades, Crit. Rev. Environ. Sci. Technol., 41, 395–429, 2011.

Couteron, P.: Quantifying change in patterned semi-arid vegetation by Fourier analysis of digitized aerial photographs, Int. J. Remote Sens., 23, 3407–3425, 2002.

Drossel, B. and Schwabl, F.: Self-organized critical forest-fire model, Phys. Rev. Lett., 69, 1629, doi:10.1103/PhysRevLett.69.1629, 1992.

Eppinga, M. B., Rietkerk, M., Belyea, L., Nilsson, M., Ruiter, P., and Wassen, M.: Resource contrast in patterned peatlands increases along a climatic gradient, Ecology, 91, 2344–2355, 2010.

Foti, R., del Jesus, M., Rinaldo, A., and Rodriguez-Iturbe, I.: Hydroperiod regime controls the organization of plant species in wetlands, P. Natl. Acad. Sci. USA, 109, 19596–19600, 2012.

Gardner, R. H., Milne, B. T., Turnei, M. G., and O'Neill, R. V.: Neutral models for the analysis of broad-scale landscape pattern, Landscape Ecol., 1, 19–28, doi:10.1007/BF02275262, 1987.

Givnish, T. J., Volin, J. C., Owen, V. D., Volin, V. C., Muss, J. D., and Glaser, P. H.: Vegetation differentiation in the patterned landscape of the central Everglades: importance of local and landscape drivers, Global Ecol. Biogeogr., 17, 384–402, 2008.

Heffernan, J. B., Watts, D. L., and Cohen, M. J.: Discharge competence and pattern formation in peatlands: a meta-ecosystem model of the everglades ridge-slough landscape, PloS One, 8, e64174, doi:10.1371/journal.pone.0064174, 2013.

Hutchinson, G. E.: A treatise on Limnology, vol. 1, Geography, Physics, and Chemistry, Wiley, New York, 1957

Kaplan, D. A., Paudel, R., Cohen, M. J., and Jawitz, J. W.: Orientation matters: patch anisotropy controls discharge competence and hydroperiod in a patterned peatland, Geophys. Res. Lett., 39, L17401, doi:10.1029/2012GL052754, 2012.

Kéfi, S., Rietkerk, M., Alados, C. L., Pueyo, Y., Papanastasis, V. P., ElAich, A., and De Ruiter, P. C.: Spatial vegetation patterns and imminent desertification in Mediterranean arid ecosystems, Nature, 449, 213–217, 2007.

Kéfi, S., Rietkerk, M., Roy, M., Franc, A., De Ruiter, P. C., and Pascual, M.: Robust scaling in ecosystems and the meltdown of patch size distributions before extinction, Ecol. Lett., 14, 29–35, 2011.

Kéfi, S., Guttal, V., Brock, W. A., Carpenter, S. R., Ellison, A. M., Livina, V. N., Seekell, D. A., Scheffer, M., van Nes, E. H., and Dakos, V.: Early warning signals of ecological transitions: methods for spatial patterns, PloS One, 9, e92097, doi:10.1371/journal.pone.0092097, 2014.

Lago, M. E., Miralles-Wilhelm, F., Mahmoudi, M., and Engel, V.: Numerical modeling of the effects of water flow, sediment transport and vegetation growth on the spatiotemporal patterning of the ridge and slough landscape of the Everglades wetland, Adv. Water Resour., 33, 1268–1278, 2010.

Larsen, L. G. and Harvey, J. W.: How vegetation and sediment transport feedbacks drive landscape change in the Everglades and wetlands worldwide, Am. Nat., 176, E66–E79, 2010.

Larsen, L. G. and Harvey, J. W.: Modeling of hydroecological feedbacks predicts distinct classes of landscape pattern, process, and restoration potential in shallow aquatic ecosystems, Geomorphology, 126.3, 279–296, 2011.

Larsen, L. G., Harvey, J. W., and Crimaldi, J. P.: A delicate balance: ecohydrological feedbacks governing landscape morphology in a lotic peatland, Ecol. Monogr., 77, 591–614, 2007.

Larsen, L. G., Aumen, N., Bernhardt, C., Engel, V., Givnish, T., Hagerthey, S., Harvey, J., Leonard, L., McCormick, P., McVoy, C., Noe, G., Nungesser, M., Rutchey, K., Sklar, F., Troxler, T., Volin, J., and Willard, D.: Recent and historic drivers of landscape change in the Everglades ridge, slough, and tree island mosaic, Crit. Rev. Environ. Sci. Technol., 41, 344–381, 2011.

Larsen, L. G., Harvey, J. W., and Maglio, M. M.: Mechanisms of nutrient retention and its relation to flow connectivity in river–floodplain corridors, Freshwater Sci., 34, 187–205, 2015.

Levin, S. A.: The Problem of Pattern and Scale in Ecology: The Robert H. MacArthur Award Lecture, Ecology, 73, 1943–1967, doi:10.2307/1941447, 1992.

Li, H. and Wu, J.: Use and misuse of landscape indices, Landscape Ecol., 19, 389–399, 2004.

Light, S. S. and Dineen, J. W.: Water Control in the Everglades: a Historical Perspective, Everglades: the Ecosystem and its Restoration, St. Lucie Press, Delray Beach, Florida, 47–84, 1994.

Ludwig, J. A., Tongway, D. J., and Marsden, S. G.: Stripes, strands, or stipples: modelling the influence of three landscape banding patterns on resource capture and productivity in semi-arid woodlands, Australia, Catena, 37, 257–273, 1999.

Mandelbrot, B. B.: The Fractal Geometry of Nature, Freeman, New York, 1983.

Manor, A. and Shnerb, N. M.: Facilitation, competition, and vegetation patchiness: from scale free distribution to patterns, J. Theor. Biol., 253, 838–842, 2008a.

Manor, A. and Shnerb, N. M.: Origin of Pareto-like spatial distributions in ecosystems, Phys. Rev. Lett., 101, 268104, doi:10.1103/PhysRevLett.101.268104, 2008b.

McVoy, C., Park Said, W., Obeysekera, J., VanArman, J., and Dreschel, T.: Landscapes and Hydrology of the Predrainage Everglades, University Press of Florida, Gainesville, FL, 2011.

Mugglestone, M. A. and Renshaw, E.: Detection of geological lineations on aerial photographs using two-dimensional spectral analysis, Comput. Geosci., 24, 771–784, 1998.

Newman, M. E.: Power laws, Pareto distributions and Zipf's law, Contemp. Phys., 46, 323–351, 2005.

Noe, G. B., Childers, D. L., and Jones, R. D.: Phosphorus biogeochemistry and the impact of phosphorus enrichment: why is the Everglades so unique?, Ecosystems, 4, 603–624, 2001.

Nungesser, M. K.: Reading the landscape: temporal and spatial changes in a patterned peatland, Wetland. Ecol. Manage., 19, 475–493, 2011.

Pascual, M. and Guichard, F.: Criticality and disturbance in spatial ecological systems, Trends Ecol. Evol., 20, 88–95, 2005.

Pascual, M., Roy, M., Guichard, F., and Flierl, G.: Cluster size distributions: signatures of self-organization in spatial ecologies, Philos. T. Roy. Soc. B, 357, 657–666, 2002.

Pickett, S. T. and Cadenasso, M. L.: Landscape ecology: spatial heterogeneity in ecological systems, Science, 269, 331–334, 1995.

Pisarenko, V. F. and Sornette, D.: Characterization of the frequency of extreme earthquake events by the generalized Pareto distribution, Pure Appl. Geophys., 160, 2343–2364, 2003.

RECOVER: 2014 System Status Report, Restoration Coordination and Verification Program, c/o US Army Corps of Engineers, Jacksonville, FL, and South Florida Water Management District, West Palm Beach, FL, 2014.

Remmel, T. K. and Csillag, F.: When are two landscape pattern indices significantly different?, J. Geogr. Syst., 5, 331–351, 2003.

Rietkerk, M. and Van de Koppel, J.: Regular pattern formation in real ecosystems, Trends Ecol. Evol., 23, 169–175, 2008.

Ross, M. S., Mitchell-Bruker, S., Sah, J. P., Stothoff, S., Ruiz, P. L., Reed, D. L., Jayachandran, K., and Coultas, C. L.: Interaction of hydrology and nutrient limitation in the Ridge and Slough landscape of the southern Everglades, Hydrobiologia, 569, 37–59, 2006.

Rutchey, K., Vilchek, L., and Love, M.: Development of a vegetation map for Water Conservation Area 3, Technical Publication ERA Number 421, South Florida Water Management District, West Palm Beach, FL, USA, 2005.

Scanlon, T. M., Caylor, K. K., Levin, S. A., and Rodriguez-Iturbe, I.: Positive feedbacks promote power-law clustering of Kalahari vegetation, Nature, 449, 209–212, 2007.

SCT – Science Coordination Team: The Role of Flow in the Everglades Ridge and Slough Landscape, South Florida Ecosystem Restoration Working Group, West Palm Beach, FL, 2003.

Stauffer, D. and Aharony, A.: Introduction to percolation theory, Taylor and Francis, London, 1991.

Sullivan, P. L., Price, R. M., Miralles-Wilhelm, F., Ross, M. S., Scinto, L. J., Dreschel, T. W., Sklar, F. H., and Cline, E.: The role of recharge and evapotranspiration as hydraulic drivers of ion concentrations in shallow groundwater on Everglades tree islands, Florida (USA), Hydrol. Process., 28, 293–304, 2014.

Todd, M. J., Muneepeerakul, R., Pumo, D., Azaele, S., Miralles-Wilhelm, F., Rinaldo, A., and Rodriguez-Iturbe, I.: Hydrological drivers of wetland vegetation community distribution within Everglades National Park, Florida, Adv. Water Resour., 33, 1279–1289, 2010.

Turner, M. G.: Landscape Ecology in Theory and Practice: Pattern and Process, Springer-Verlag, New York, 2001.

Turner, M. G.: Landscape ecology: what is the state of the science?, Annu. Rev. Ecol. Evol. S., 36, 319–344, 2005.

von Hardenberg, J., Kletter, A. Y., Yizhaq, H., Nathan, J., and Meron, E.: Periodic vs. scale-free patterns in dryland vegetation, Philos. Roy. Soc. B, 277, 1771–1776, 2010.

Watts, D. L., Cohen, M. J., Heffernan, J. B., and Osborne, T. Z.: Hydrologic modification and the loss of self-organized patterning in the ridge-slough mosaic of the Everglades, Ecosystems, 13, 813–827, 2010.

Weerman, E. J., Van Belzen, J., Rietkerk, M., Temmerman, S., Kéfi, S., Herman, P. M. J., and de Koppel, J. V.: Changes in diatom patch-size distribution and degradation in a spatially self-organized intertidal mudflat ecosystem, Ecology, 93, 608–618, 2012.

Wetzel, P. R., van der Valk, A. G., Newman, S., Gawlik, D. E., Troxler Gann, T., Coronado-Molina, C. A., Childers, D. L., and Sklar, F. H.: Maintaining tree islands in the Florida Everglades: nutrient redistribution is the key, Front. Ecol. Environ., 3, 370–376, 2005.

Wetzel, P. R., van der Valk, A. G., Newman, S., Coronado, C. A., Troxler-Gann, T. G., Childers, D. L., Orem, W. H., and Sklar, F. H.: Heterogeneity of phosphorus distribution in a patterned landscape, the Florida Everglades, Plant Ecol., 200, 83–90, 2009.

Wu, Y., Wang, N., and Rutchey, K.: An analysis of spatial complexity of ridge and slough patterns in the Everglades ecosystem, Ecol. Complex., 3, 183–192, 2006.

Yuan, J., Cohen, M. J., Kaplan, D. A., Acharya, S., Larsen, L. G., and Nungesser, M. K.: Linking metrics of landscape pattern to hydrological process in a lotic wetland, Landscape Ecol., 30, 1893–1912, doi:10.1007/s10980-015-0219-z, 2015

Zweig, C. L. and Kitchens, W. M.: Effects of landscape gradients on wetland vegetation communities: information for large-scale restoration, Wetlands, 28, doi:10.1672/08-96.1, 2008.

Assimilation of SMOS brightness temperatures or soil moisture retrievals into a land surface model

Gabriëlle J. M. De Lannoy[1] **and Rolf H. Reichle**[2]

[1]KU Leuven, Department of Earth and Environmental Sciences, Heverlee, Belgium
[2]NASA Goddard Space Flight Center, Global Modeling and Assimilation Office, Greenbelt, Maryland, USA

Correspondence to: Gabriëlle J. M. De Lannoy (gabrielle.delannoy@kuleuven.be)

Abstract. Three different data products from the Soil Moisture Ocean Salinity (SMOS) mission are assimilated separately into the Goddard Earth Observing System Model, version 5 (GEOS-5) to improve estimates of surface and root-zone soil moisture. The first product consists of multi-angle, dual-polarization brightness temperature (Tb) observations at the bottom of the atmosphere extracted from Level 1 data. The second product is a derived SMOS Tb product that mimics the data at a 40° incidence angle from the Soil Moisture Active Passive (SMAP) mission. The third product is the operational SMOS Level 2 surface soil moisture (SM) retrieval product. The assimilation system uses a spatially distributed ensemble Kalman filter (EnKF) with seasonally varying climatological bias mitigation for Tb assimilation, whereas a time-invariant cumulative density function matching is used for SM retrieval assimilation. All assimilation experiments improve the soil moisture estimates compared to model-only simulations in terms of unbiased root-mean-square differences and anomaly correlations during the period from 1 July 2010 to 1 May 2015 and for 187 sites across the US. Especially in areas where the satellite data are most sensitive to surface soil moisture, large skill improvements (e.g., an increase in the anomaly correlation by 0.1) are found in the surface soil moisture. The domain-average surface and root-zone skill metrics are similar among the various assimilation experiments, but large differences in skill are found locally. The observation-minus-forecast residuals and analysis increments reveal large differences in how the observations add value in the Tb and SM retrieval assimilation systems. The distinct patterns of these diagnostics in the two systems reflect observation and model errors patterns that are not well captured in the assigned EnKF error parameters. Consequently, a localized optimization of the EnKF error parameters is needed to further improve Tb or SM retrieval assimilation.

1 Introduction

Microwave satellite missions are collecting large amounts of data for soil moisture monitoring. It is not yet clear, however, how this wealth of data can be used in the most efficient way to obtain global estimates of soil moisture that can improve, e.g., weather prediction, flood and drought modeling, agricultural yield monitoring, or landslide predictions. Many such applications require knowledge of soil moisture in a deeper layer, where water is extracted by plant roots or stored to buffer drainage and runoff, not the approximately 5 cm surface layer to which the current L-band (~ 1.4 GHz) microwave missions are sensitive. Moreover, L-band satellite observations have a fairly coarse spatial resolution (about 40 km) and are available only at particular overpass times, typically once every 2–3 days for a given location. The challenge is thus to derive soil profile moisture information at all times and locations through data assimilation, that is, through the merger of satellite observations with information from a dynamical land surface model.

The Soil Moisture Ocean Salinity (SMOS; Kerr et al., 2010) mission and the Soil Moisture Active Passive (SMAP; Entekhabi et al., 2014) mission are the two L-band observatories currently orbiting in space with the specific aim of measuring global soil moisture. These missions supply Level 1

(L1) brightness temperature (Tb) data, Level 2 (L2) surface soil moisture (SM) retrievals, and derived Level 3 (L3) products. The SMAP mission also provides an operational Level 4 surface and root-zone soil moisture product (L4_SM; Entekhabi et al., 2014; Reichle et al., 2016) that is based on the assimilation of L1 SMAP Tb data into Goddard Earth Observing System Model, version 5 (GEOS-5) land surface simulations. Alternatively, a soil moisture assimilation system could ingest L2 SM retrievals instead of L1 Tb observations.

In this paper, we compare Tb and SM retrieval assimilation using a historical (5-year) record of SMOS observations over North America in an assimilation system similar to that of the SMAP L4_SM system. The main differences between the SMAP L4_SM system and the experiments in this paper pertain to the differences in assimilated data, to the difference in spatial resolution of the resulting soil moisture products (36 km in the current paper; see below; 9 km for the L4_SM product), and to differences in meteorological forcing input (re-analysis meteorology in the current paper; operational forecast meteorology corrected with gauge-based precipitation in the L4_SM product).

It is more difficult to assimilate Tb observations than SM retrievals because brightness temperatures are only indirectly connected with the land surface variables of interest and the Tb data come in multiple polarizations. SMOS Tb observations are even more complex because of their multi-angular nature. Some of the SMOS L1 Tb data complexity is reduced in the L3 SMOS Tb product and further addressed in Munoz-Sabater et al. (2014) and De Lannoy et al. (2015), who prepared the L1 SMOS Tb data for assimilation into (quasi-)operational systems.

Successful examples of SMOS Tb assimilation using a variety of simplifying assumptions are illustrated in Lievens et al. (2015); De Lannoy and Reichle (2016); Kornelsen et al. (2016). These studies use a radiative transfer model (RTM) to dynamically invert Tb information into corrections to modeled soil moisture estimates. In this paper, we advance the spatially distributed multi-angle and dual-polarization Tb assimilation of De Lannoy and Reichle (2016) in the GEOS-5 land surface model with a new version of Tb observations and an improved spatial support and forward simulation of the Tb observation predictions. Moreover, to mimic SMAP Tb assimilation we also assimilate dual-polarization single-angle 40° SMOS Tb observations after fitting the multi-angle Tb data (De Lannoy et al., 2015).

A key disadvantage of a system that assimilates SM retrievals is that the SM retrievals may be produced with inconsistent ancillary data, such as for example soil temperature simulated by another model than that used in the assimilation system. The current SMOS SM retrievals by themselves have been found to be skillful (Al-Yaari et al., 2014; Fascetti et al., 2016), and research is ongoing to further improve them (Rodriguez-Fernandez et al., 2015; Ye et al., 2015; Zhao et al., 2015; van der Schalie et al., 2016; Wigneron et al.,

2016). The use of these SMOS SM retrievals has been manifold, e.g., to derive enhanced estimates of precipitation (Wanders et al., 2015; Koster et al., 2016), to derive offline root-zone soil moisture estimates (Ford et al., 2014), or to offline downscale the data to higher-resolution soil moisture estimates (Piles et al., 2014). Other studies have assimilated SMOS SM retrievals online into land surface models to possibly downscale the retrievals and consistently improve soil moisture and other land surface variables (Ridler et al., 2014; Zhao et al., 2014; Lievens et al., 2015), leading to, e.g., improved estimates of floods (Alvarez-Garreton et al., 2015) and crop growth (Chakrabart et al., 2014). In this paper, we use a spatially distributed assimilation system to integrate SMOS SM retrievals into the GEOS-5 land surface model with the aim of inferring improved surface and root-zone soil moisture estimates. Our study mainly differs from the above SMOS SM retrieval studies in the continental and multi-year scale of the experiments, in the advanced quality screening and spatial support of the SM retrieval observations, and in the comparison between Tb and SM retrieval assimilation (also discussed in Lievens et al., 2015).

To assess the potential of Tb and SM retrieval assimilation, 5 years of SMOS Tb data or SM data are assimilated into the GEOS-5 land surface model using a careful data quality control and data preprocessing. The observations are associated with a realistic antenna pattern, containing 50 % of the signal power in a circular area with 20 km radius. Special attention is paid to large-scale patterns of random and persistent forecast and observation errors in the different assimilation systems, and to the impact of the different assimilation schemes on the skill of surface and root-zone soil moisture estimates. Section 2 describes the SMOS observations, the various modeling components, and the in situ validation data. Section 3 highlights the technical differences between the various assimilation schemes, and Sect. 4 presents the results.

2 Data and model

2.1 SMOS Tb observations

The Microwave Imaging Radiometer with Aperture Synthesis (MIRAS) onboard SMOS provides multi-angle Tb data, with a nominal (3 dB) spatial resolution of 43 km and a global coverage approximately every 3 days (at either 06:00 or 18:00 local time, i.e., ascending or descending half-orbits, separately). The most recent version (v620) of the SCLF1C Tb data is used. Observations are retained for further processing only (a) in the alias-free zone, (b) when the data are not contaminated by point source radio frequency interference (RFI) or tails thereof, (c) when the values fall within the range 100–320 K, and (d) when valid data are available for both horizontal (H) and vertical (V) polarization. The flag for snapshot RFI is not activated, because it is currently

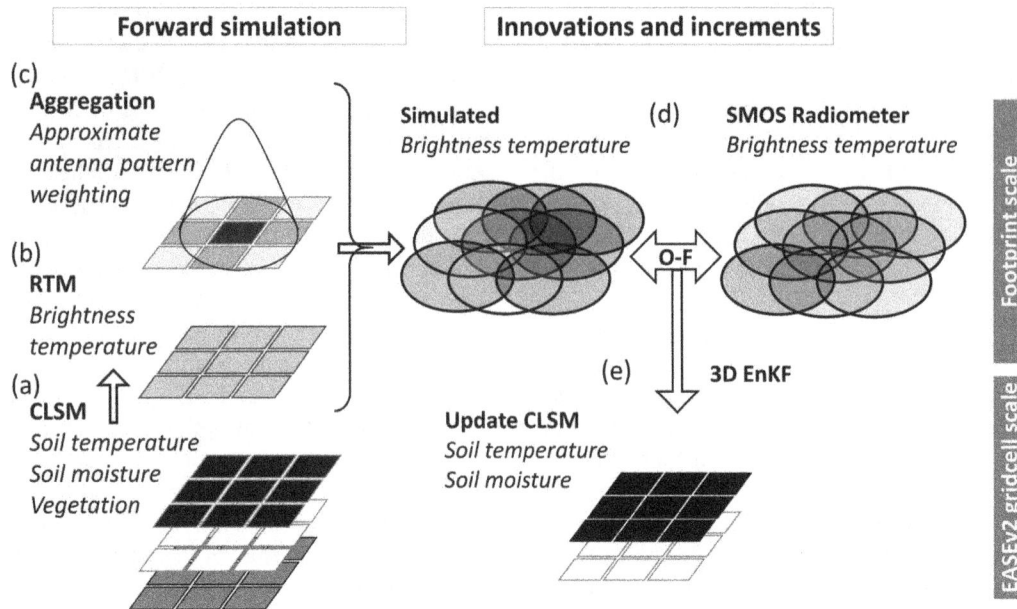

Figure 1. Flowchart of Tb assimilation. The forward simulation consists of **(a)** land surface model simulations and **(b)** Tb simulations on the 36 km EASEv2 grid. The Tb simulations are subsequently **(c)** aggregated using weights based on an approximate antenna pattern. The resulting footprint-scale brightness temperature observation predictions are compared to **(d)** SMOS observations to calculate innovations (O–F) at the footprint scale. **(e)** The three-dimensional EnKF maps the footprint-scale innovations to the 36 km EASEv2 grid based on the modeled error correlations between the footprint-scale Tb and the 36 km soil moisture and soil temperature state variables (per Eqs. 1 and 2).

too sensitive (R. Oliva and Y. Kerr, personal communication, 2016). After the initial screening, we correct the L1 Tb values for geometric and Faraday rotation and for atmospheric and reflected extraterrestrial radiation (De Lannoy et al., 2015) using Modern-Era Retrospective Analysis for Research and Applications (MERRA) version 2 (MERRA2; Bosilovich et al., 2015) background fields. The resulting Tb values at the bottom of the atmosphere are then binned into 41 evenly spaced angular bins with the center angle ranging from 20 through 60°. Next, the data are regridded from the 15 km discrete global grid (DGG) on which they are posted to the 36 km cylindrical Equal-Area Scalable Earth (EASEv2) grid (Brodzik et al., 2014), and the data are screened for excessive sub-36 km heterogeneity (spatial standard deviation > 7 K), which is indicative of open water bodies or RFI. Tb values for a given 36 km EASEv2 grid cell are computed only if at least two valid DGG observations are available.

From these preprocessed Tb data, two datasets are derived for assimilation: (i) a seven-angle Tb dataset, with incidence angles $\theta = [30, 35, 40, 45, 50, 55, 60°]$ (De Lannoy et al., 2013), and (ii) a fitted Tb dataset (De Lannoy et al., 2015) from which only the Tb at a 40° incidence angle is used to mimic the single-angle nature of SMAP Tb observations. We refer to these datasets as Tb_7ang and Tb_fit, respectively. Tb_fit data are only retained when the fitting error is less than 5 K and a minimum of 15 data points contribute to the entire fitted angular signature, with at least 5 data points above and

below the 40° incidence angle and at least 10 data points in the incidence angle interval between 30 and 50°.

2.2 SMOS SM retrieval observations

The SMOS SM retrievals are extracted from the SMUDP2 product v552. Because this product version ends in early May 2015, we limit our study period to 1 July 2010–1 May 2015. (The reprocessed v620 version of the SM retrievals was not yet available at the time we conducted the experiments.) The SMOS retrieval algorithm simultaneously retrieves soil moisture and vegetation opacity, by fitting multi-angle Tb observations at both H- and V-polarization with simulations of the L-band Microwave Emission of the Biosphere Model (L-MEB, Wigneron et al., 2007). Based on the quality information provided within the SMOS products, the SM data are retained only if (a) all retrieved variables fall within a realistic range (0–0.6 $m^3 m^{-3}$ for soil moisture), (b) the SM uncertainty estimated by the SMOS retrieval algorithm is less than 0.1 $m^3 m^{-3}$, (c) the RFI probability for both H- and V-polarization is less than 0.3, and (d) SM retrieval flags are not raised for high topographic complexity, high urban fraction, high open water fraction, sea ice, coastal areas, and high total electron content. Further screening for frozen temperature and snow is based on GEOS-5 model output (Sect. 2.3). After the regridding from the 15 km DGG grid to the 36 km cylindrical EASEv2 grid, the data are screened for excessive sub-36 km heterogeneity (spatial standard deviation $> 0.2\,\mathrm{m}^3\,\mathrm{m}^{-3}$). SM values for a given 36 km EASEv2 grid cell are computed only if at least two valid DGG observations are available.

2.3 Soil moisture and brightness temperature modeling

The land data assimilation system used here employs the GEOS-5 catchment land surface model (CLSM; Koster et al., 2000), along with an L-band tau-omega radiative transfer model (RTM; De Lannoy et al., 2013, 2014b). The CLSM simulations use GEOS-5 parameters (Mahanama et al., 2015; De Lannoy et al., 2014a) similar to those used in the SMAP L4_SM product, and are forced with $1/2° \times 2/3°$ GEOS-5 forcing data from MERRA (Rienecker et al., 2011) bilinearly interpolated to the model grid. The study domain covers most of North America, with the northwestern corner at (125° W, 55° N) and the southeastern corner at (60° W, 24° N).

The computational elements are the 36 km EASEv2 grid cells. The land model computation time step is 7.5 min, and output is saved at 3 h intervals. At each grid cell, the surface soil moisture content (sfmc, 0–5 cm) and root-zone soil moisture content (rzmc, 0–100 cm) are diagnosed based on three prognostic variables: catchment deficit (catdef), root-zone excess (rzexc), and surface excess (srfexc). Similarly, the surface (skin) temperature is diagnosed from the prognostic land surface temperatures across the saturated (tc1), unsaturated (tc2), and wilting (tc4) sub-grid areas. Finally, the soil temperature (tp1 for the topmost layer) is diagnosed from the prognostic ground heat content (ght1 for the top layer). An overview of the model variables is given in Reichle et al. (2015); Koster et al. (2000) and Ducharne et al. (2000).

The L-band tau-omega RTM converts the 36 km CLSM soil moisture and temperature simulations into 36 km L-band Tb estimates when the soil is not frozen or covered with snow, when precipitation is less than $10\,\mathrm{mm\,day}^{-1}$, and where the open water fraction is less than 5 %. For each 36 km grid cell, key parameters of the RTM are estimated by minimizing Eq. (B.1) in De Lannoy et al. (2014b), using a 5-year history of SMOS v620 Tb data, and computing observation predictions (see below) at the footprint scale. Specifically, all 36 km grid cells within one footprint area are initially assigned the same set of RTM parameters, while the dynamic background information is spatially variable. For each 36 km grid cell, the calibration estimates a spatially homogeneous set of RTM parameters for the entire associated footprint area, and the resulting values are assigned to the central (and typically dominant) 36 km grid cell only. For the forward calculation of the Tb observation predictions during the data assimilation, all 36 km pixels have a unique set of RTM parameters. The RTM is calibrated using all 5 years of available Tb data and aims at minimizing climatological biases. The data assimilation is performed over the same 5 years and aims at addressing random (or short-term) errors. The methodology is very similar to that in De Lannoy and

Reichle (2016), but with the difference that, here, the RTM does not simulate atmospheric contributions (because the Tb observations are now a priori corrected for atmospheric contributions) and the observation predictions are now spatially aggregated using a realistic (but approximate) antenna pattern.

For the computation of differences between SMOS observations and footprint-scale model simulations in the RTM calibration and for the computation of the "observation-minus-forecast" (O–F) residuals in the assimilation system (Sect. 3.1, Fig. 1), the modeled 36 km soil moisture or Tb simulations are aggregated to the footprint scale by spatial convolution with weights given by an approximation of the SMOS antenna pattern. We also refer to these spatially aggregated model estimates as "observation predictions". The SMOS antenna pattern is approximated by a two-dimensional Gaussian function containing 50 % of the signal within a circle with a radius of 20 km. The simulations outside a radius of 40 km are discarded in the computation of the footprint-scale estimates.

The number of 36 km EASEv2 grid cells included in one footprint area varies with latitude. The circular footprint shape is preserved everywhere on the globe. In contrast, the shape of the EASEv2 grid cells projected on the globe varies with the latitude, with an aspect ratio of 1 at 30° (north–south) latitude, larger than 1 towards the poles and less than 1 towards the Equator. Therefore, at higher latitudes multiple EASEv2 grid cells with the same latitude and various longitudes belong to one circular footprint, whereas towards the Equator, several EASEv2 grid cells with the same longitude and various latitudes contribute to the footprint. Overall, the difference between single 36 km simulations and footprint-scale values is small, but the number of valid Tb observation predictions at the footprint scale is reduced, because of the increased likelihood of finding a 36 km grid cell with a non-negligible water fraction, snow amount, or precipitation within the footprint area.

2.4 In situ soil moisture data and metrics

The assimilation results are evaluated using independent in situ measurements of surface and root-zone soil moisture from two sparse networks across the US: the US Natural Resources Conservation Service Soil Climate Analysis Network (SCAN; Schaefer et al., 2007) and the US Climate Reference Network (USCRN; Diamond et al., 2013; Bell et al., 2013). Surface soil moisture measurements are taken at approximately 5 cm depth. Root-zone soil moisture measurements are a weighted average of measurements at 5, 10, 20, and 50 cm depth, with respective weights of 0.1, 0.1, 0.27, and 0.53. Given the difference in spatial support between these point measurements and the 36 km gridded model and assimilation results, the skill is quantified in terms of anomaly time series correlation (anomR) and unbiased root-mean-square difference (RMSD$_{ub}$; Entekhabi et al., 2010), using all 3 h forecast and analysis time steps in the period 1 July 2010–1 May 2015, excluding times when the soil is frozen (top layer soil temperature < 274.15 K) or snow covered (snow water equivalent > 0 kg m^{-2}). The anomaly correlation is based on anomaly time series obtained by subtracting a multi-year smoothed climatology from both the simulations and in situ observations. Note that the assimilation and open-loop simulations have, by design, the same climatological variability; the assimilation only corrects for random errors. Metrics at a single site are only calculated if at least 200 data points are available. Skill metrics across an entire network are calculated by clustering the sites within SCAN and USCRN to avoid densely sampled areas dominating the validation metrics and to ensure realistic confidence intervals (De Lannoy and Reichle, 2016). The number of clusters is estimated a priori after prescribing an average cluster radius of 3°, which approximately reflects the autocorrelation length of large-scale topographic and meteorological phenomena, or of large-scale soil moisture patterns (Vinnikov et al., 1996). The actual size of the clusters that results from the clustering algorithm varies strongly in space.

3 Data assimilation

3.1 Distributed ensemble Kalman filter

For both Tb and SM retrieval assimilation, a spatially distributed (or three-dimensional, 3-D) ensemble Kalman filter (EnKF; Reichle and Koster, 2003; De Lannoy and Reichle, 2016) is used. This system simultaneously assimilates multiple spatially distributed observation sets, using horizontal and vertical error covariance structures, to update the simulations at each 36 km model grid cell. The details of the Tb assimilation system are explained in De Lannoy and Reichle (2016) and differ only in that the observations are here associated with a spatially variable antenna pattern reaching out to a radius of 40 km.

During the model integration, a data assimilation step is activated every 3 h. All the SMOS observations y_i collected within 1.5 h of the analysis time i are assimilated simultaneously to update the forecasted state $\hat{x}_{k,i}^{j-}$ at location k as follows:

$$\hat{x}_{k,i}^{j+} = \hat{x}_{k,i}^{j-} + \mathbf{K}_{k,i}[y_i^j - \hat{y}_i^{j-}], \tag{1}$$

with j denoting the ensemble member, $\mathbf{K}_{k,i}$ the Kalman gain, y_i^j the perturbed observations, $\hat{y}_i^{j-} = h_i(\hat{x}_i^{j-})$ the observation predictions, and $h_i(.)$ the observation operator mapping the simulated land surface variables to observed quantities. Bias in the observation-minus-forecast residuals is addressed prior to the analysis (Sect. 3.2). The ensemble is created by perturbing the model forcing, the model forecasts, and the observations (Sect. 3.3). The Kalman gain is calculated as

$$\mathbf{K}_{k,i} = \mathrm{Cov}(\hat{x}_{k,i}^-, \hat{y}_i^-)\left[\mathrm{Cov}(\hat{y}_i^-, \hat{y}_i^-) + \mathbf{R}_i\right]^{-1}, \tag{2}$$

where $\mathrm{Cov}(\hat{x}_{k,i}^-, \hat{y}_i^-)$ is the (sample) error covariance (across the ensemble) between the forecasted land surface state and the forecasted Tb or SM. Similarly, $\mathrm{Cov}(\hat{y}_i^-, \hat{y}_i^-)$ is the (sample) error covariance of the Tb or SM forecasts, and $\mathbf{R}_i$ is the Tb or SM observation error covariance. The Kalman gain is identical for all ensemble members.

In the case of SM retrieval assimilation, the observation operator $h_i(.)$ performs the spatial aggregation of soil moisture simulations from the 36 km grid cells to the satellite footprint; in the case of Tb data assimilation, the observation operator includes both the RTM and the spatial aggregation of gridded Tb simulations to the footprint (Sect. 2.3). For the Tb_7ang assimilation, one observation set at location κ contains Tb observations at a maximum of seven angles and both H- and V-polarization, i.e., up to 14 individual observations $y_{\lambda,\kappa,i} \in \boldsymbol{y}_{\kappa,i}$. The subscript λ refers to the polarization and incidence angle of the individual Tb observations. In the middle part of the swath, all 14 observations are typically available, whereas slightly fewer observations are available in the outer portions of the swath, where the observations with lower incidence angles are missing.

For the Tb_fit assimilation, one observation set usually contains two observations, i.e., both H- and V-polarization Tb at a 40° incidence angle. For the SM retrieval assimilation, each observation set contains only one observation. In all cases, the observation vector $\boldsymbol{y}_i^j$ collects multiple perturbed observation sets that are spatially distributed within an influence radius of 1.25° around the model grid cell k, and each observation vector $\boldsymbol{y}_i^j$ has a forecasted counterpart $\hat{\boldsymbol{y}}_i^{j-}$. After removal of the persistent errors (Sect. 3.2) from the O–F residuals (or innovations), the increments $\mathbf{K}_{k,i}[\boldsymbol{y}_i^j - \hat{\boldsymbol{y}}_i^{j-}]$ are calculated and applied to the state variables. Figure 1 illustrates the forward simulation from 36 km gridded land surface simulations to footprint-scale observation predictions of Tb and the downscaling of the footprint-scale Tb innovations to 36 km gridded land surface increments.

The subset of prognostic variables updated in Eq. (1) differs depending on the assimilation experiment. The state vector for Tb assimilation ($x = $ [catdef, srfexc, rzexc, tc1, tc2, tc4, ght1]^T) includes prognostic variables related to soil moisture and soil temperature (Sect. 2.3), because Tb observations are by definition sensitive to surface soil moisture and temperature. In contrast, the state vector for SM retrieval assimilation ($x = $ [catdef, srfexc, rzexc]^T) contains only model prognostic variables related to soil moisture, because the SM retrievals do not carry direct information about the soil temperature. The selected updates will be propagated to all other variables within the land surface modeling system through energy and water exchange between various soil layers and land–vegetation–atmosphere compartments. For the discussion of the soil moisture increments we will focus on the total profile water increments (Δwtot$=\Delta$srfexc$+\Delta$rzexc$-\Delta$catdef) in units of kg m^{-2} (that is, mm of water equivalent). This

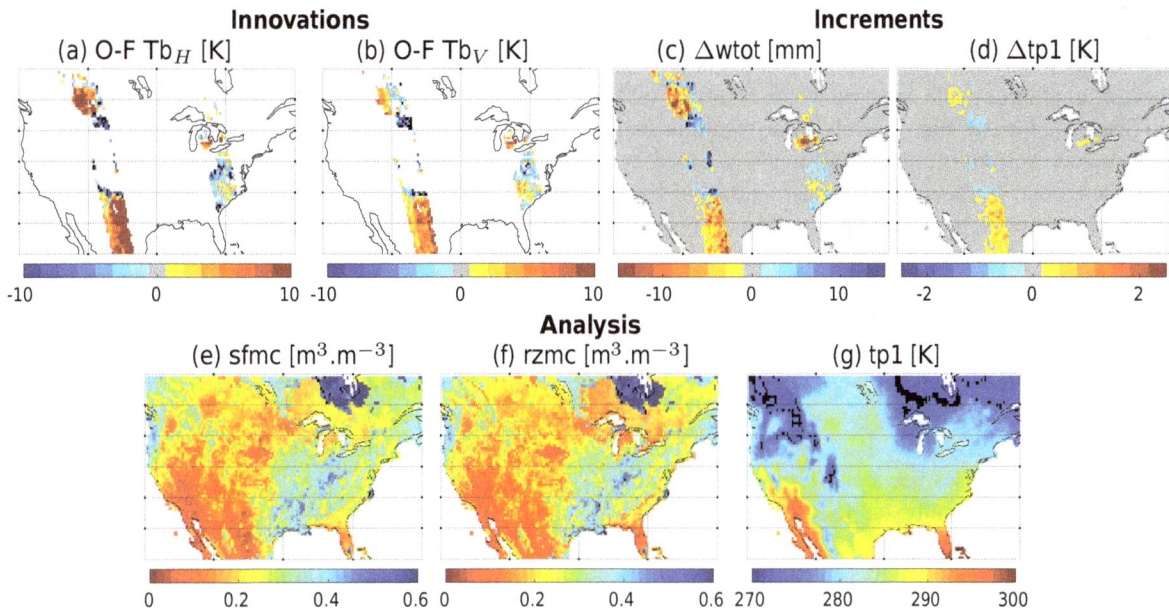

Figure 2. Soil moisture and temperature analysis on 30 April 2015 at 12:00 UTC for the Tb_fit assimilation system. (**a, b**) Tb innovations (O–F) at a 40° incidence angle for H- and V-polarization respectively; (**c, d**) increments in total profile water (Δwtot) and first soil layer temperature (Δtp1), respectively; (**e, f, g**) assimilation analyses of surface soil moisture (sfmc), root-zone soil moisture (rzmc), and soil temperature (tp1), respectively.

quantity is easily understandable and thus simplifies the discussion.

Figures 2 and 3 illustrate the concept for Tb assimilation and SM retrieval assimilation, respectively. Figure 2a–b show swaths of footprint-scale bias-corrected Tb_fit innovations (mapped onto the 36 km EASEv2 grid), for H- and V-polarization at a 40° incidence angle from the single-angle Tb assimilation system. The Tb innovations are then transformed into soil moisture and temperature increments using Eq. (1). Where Tb innovations are warm, the soil water is reduced and the temperature is increased. Figure 2c shows the total profile water increments Δwtot and Fig. 2d shows increments to the first soil layer temperature Δtp1. Increments to the surface temperature prognostic variables (Sect. 2.3; Δtc1, Δtc2, Δtc4) are similar (not shown). Finally, the increments are added to the forecasted fields to create spatially complete analysis maps of surface and root-zone soil moisture, as well as surface temperature and soil temperature (Fig. 2e–g).

Similarly, Fig. 3a shows the SM innovations from the SM retrieval assimilation at the same time as in Fig. 2. Areas with positive (wet) SM innovations in the SM retrieval assimilation roughly correspond to negative (cold) Tb innovations in the Tb assimilation system (Fig. 2a–b). Note that the color bars for Tb and SM throughout the paper are chosen according to the rule of thumb that a 2–3 K change in Tb corresponds to a 0.01 m^3 m^{-3} change in soil moisture, but keep in mind that the relationship between Tb and SM is nonlinear and varies with time, location, and incidence angle. Next, the SM innovations are converted to soil moisture increments

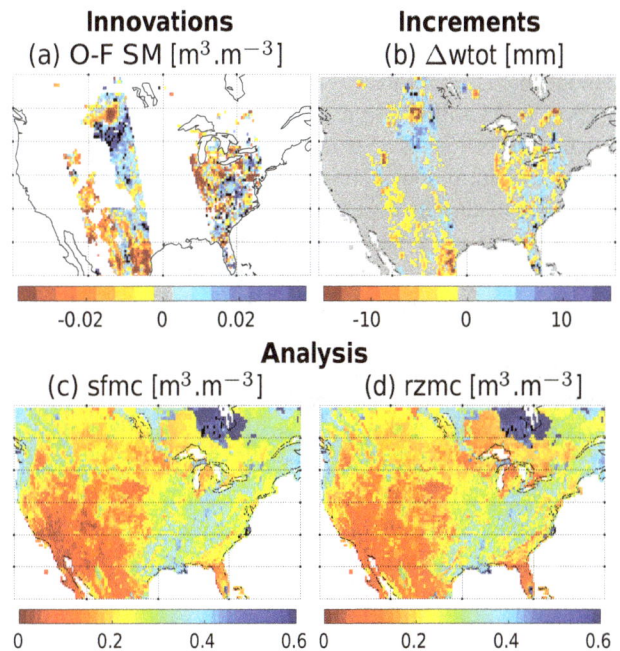

Figure 3. Soil moisture analysis on 30 April 2015 at 12:00 UTC for the SM retrieval assimilation system. (**a**) SM innovations (O–F); (**b**) increments in total profile water (Δwtot); (**c, d**) assimilation analyses of surface soil moisture (sfmc) and root-zone soil moisture (rzmc).

(Δwtot; Fig. 3b); no increment to surface or soil temperature is calculated. Figures 2c and 3b show that the Tb and SM retrieval assimilation systems produce wtot increments with somewhat different large-scale patterns, which is further discussed in Sect. 4.2. Finally, Fig. 3c–d show the resulting surface and root-zone soil moisture analysis fields obtained by adding the increments to the model forecast fields. For both the Tb and SM retrieval assimilation systems, the analysis increments blend smoothly into the forecast fields; that is, the analysis maps do not reveal sharp spatial edges that would reveal the geometry of the assimilated satellite swaths. Further details about this figure are discussed in Sect. 4.1.

3.2 Tb and SM innovation bias

To limit the long-term biases between Tb observations and simulations, the RTM was calibrated (Sect. 2.3). The 5-year average absolute bias between SMOS Tb and forecasted Tb is about 2 K across the domain. In general, slightly warm model biases are found in the boreal zones and cold model biases over the central part of the US (not shown), but larger seasonal Tb biases remain, primarily due to systematic errors in the modeled temperature and vegetation. The seasonally varying climatological Tb bias is removed prior to data assimilation for each angle, polarization, and overpass time separately, as described in De Lannoy and Reichle (2016). The Tb innovation biases are calculated over the period 1 July 2010–1 May 2015 for each individual 36 km grid cell without spatial sampling.

The CLSM soil moisture was not calibrated for lack of global observations that would support such an effort and because modeled soil moisture does not necessarily represent soil moisture as observed in the field anyway (Koster et al., 2009). Unlike biases in Tb innovations, the biases in the SM innovations are more stationary and do not depend on seasonal temperature variations. Therefore, the SM innovation biases are not corrected seasonally, but instead cumulative distribution function (CDF) matching between the observations and simulations is performed (Reichle and Koster, 2004) to reconcile the differences in long-term mean, variance, and higher moments, as in earlier retrieval assimilation studies (Liu et al., 2011; Draper et al., 2012). The observed and simulated SM CDFs are computed for the entire study period, i.e., for 1 July 2010–1 May 2015, at each 36 km grid cell individually.

3.3 Random forecast and observation error

The imposed ensemble forecast perturbations for Tb and SM retrieval assimilation are identical to those of De Lannoy and Reichle (2016) and not repeated here. The total observation error standard deviation for SMOS Tb_7ang is set to 6 K, which yields near-optimal assimilation diagnostics on average across the globe. However, the diagnostics are not necessarily near-optimal in individual regions (De Lannoy

and Reichle, 2016). The input observation error standard deviation for SM retrievals is $0.04\,\mathrm{m^3\,m^{-3}}$, in line with the soil moisture accuracy requirement for the recent SMOS and SMAP missions. The SM retrieval error standard deviation is rescaled following the CDF matching of the SM observations and results in an effective mean error standard deviation of $0.02\,\mathrm{m^3\,m^{-3}}$, with larger values in the wetter eastern part, which exhibits a higher temporal variability in soil moisture simulations, and lower values in the drier, western part of the study domain (not shown). In all cases, the spatial observation error correlation length is $0.25°$. In the case of multi-angle Tb_7ang assimilation, interangular error correlations are imposed as in De Lannoy and Reichle (2016).

Observation errors in Tb data or SM retrievals are a combination of instrument error and representation error (Cohn, 1997; van Leeuwen, 2015). The 6 K Tb error consists of a radiometric error of about 4 K for individual incidence angles (instrument error) plus 4.5 K representation inaccuracies (in our system, i.e., based on the near-optimal 6 K observation error) due to errors in the RTM, the spatial aggregation, or other discrepancies between Tb observations and forecasts ($6 = \sqrt{4^2 + 4.5^2}$). For Tb_fit observations, the instrument error may be slightly reduced compared to that for Tb_7ang after the angular smoothing, but the representation error remains similar. SM observations contain retrieval errors due to errors in the RTM and in the input L1 Tb observations, as well as representation error due to, e.g., the inherently different nature of simulated and observed soil moisture (Koster et al., 2009). In either case, the representation error depends on the soil moisture and temperature dynamics and should ideally be modeled as a function of time and location, but we chose a constant input observation error standard deviation in this paper for simplicity. For SM retrieval assimilation, some spatial error variability is introduced after rescaling in line with the CDF matching.

3.4 Tb or SM retrieval assimilation

In our experiments, we do not expect the SMOS Tb and SM retrieval assimilation systems to yield the same results. During the SMOS L2 SM retrieval optimization, the Tb data are used to estimate surface soil moisture and vegetation opacity, given soil temperature background fields provided by the European Center for Medium-Range Weather Forecasts (ECMWF) and look-up parameter information that differs significantly from the NASA GEOS-5 land data assimilation system. In contrast, our SMOS Tb assimilation scheme estimates soil moisture and temperature, given vegetation information. Furthermore, the data screening is necessarily different for Tb data and SM retrievals, and the approach for bias correction is intentionally different. The soil moisture information extracted during the L2 retrieval process or Tb assimilation is thus by design expected to be different. Finally, differences in the Tb and SM retrieval assimilation results could also be due to differences in how close each of the systems is to an optimal calibration of its model and observation error parameters.

4   Results

4.1   Observation and forecast diagnostics

4.1.1   Number of assimilated observations

Let us revisit Figs. 2a–b and 3a to further highlight some differences between the various assimilated SMOS observations. First, the swath width for Tb innovations is much narrower than that of the SM innovations because the assimilated Tb observations are strictly limited to the alias-free zone within the full swath, while the assimilated SM retrievals are retained in the extended alias-free zone. Furthermore, the swath width of the Tb_fit innovations is narrower than that of the multi-angle assimilation (not shown) because the fitting requires sufficient data at a range of incidence angles and lower angle data are not available at the outer edges of the swaths. Note that SMAP provides useable Tb measurements over a much wider swath (not shown).

The different swath widths result in different numbers of observation sets assimilated in each of the three experiments. Figure 4a–c show the average number of assimilated observation sets (defined in Sect. 3.1) over the study period 1 July 2010–1 May 2015. The number of observation sets is smallest (one every 4 days) for Tb_fit and largest for SM retrievals (one every 2 days), because the swath width is narrowest for Tb_fit and widest for SM retrievals. The northern areas and the western mountain ranges have the fewest observations, because data are not used when the soil is frozen or snow covered. Tb observations are not assimilated in many small areas scattered around the study domain, where more than 5 % of open water is found in the footprint, based on the underlying GEOS-5 land mask. For the SM retrievals, the screening for an excessive (> 5 %) water fraction is only based on the product science flags, not on GEOS-5 information. Data gaps in the SM retrievals are found in the western mountain ranges and in the vegetated southeastern part of the US. The data coverage is also different for Tb and SM retrieval assimilation because the availability of the climatological information needed for the innovation bias correction (Sect. 3.2) is different for the Tb and SM retrieval observations.

4.1.2   Actual observation and forecast errors

The long-term mean observation-minus-forecast differences (O–F, or innovations) are unbiased by design (Sect. 3.2). The Hovmüller plots for two data assimilation cases in Fig. 5 reveal that the temporal pattern in area-averaged biases is fairly random for the Tb_7ang assimilation case (very similar for Tb_fit assimilation, not shown), whereas it shows a slight seasonal pattern in the SM retrieval assimilation case. This small difference is not surprising, given that the Tb innova-

tion bias is seasonally corrected, whereas the SM innovation bias is not.

The time series standard deviation of the innovations, that is, the root-mean-square difference (RMSD) between SMOS observations and simulations, represents the total observation and forecast error that is present in the assimilation system (Desroziers et al., 2005). The spatial patterns of this diagnostic are very different for Tb and SM retrieval assimilation. Figure 4d–e show values of about 7.4 K for Tb_7ang and Tb_fit, with larger values (exceeding 10 K) in the central plains and along the Mississippi, where agricultural practices, such as altering crop rotation and irrigation, are observed by SMOS, whereas interannual variations in vegetation are not simulated by the model or provided as input to the model. Along the eastern coast and in the southeast, the temporal standard deviation in the innovations is low (2–3 K): forests show a limited interannual variability, and under dense vegetation Tb is only marginally sensitive to soil moisture and depends primarily on vegetation characteristics and (physical) temperature.

The standard deviation in the SM innovations in the SM retrieval assimilation (Fig. 4f) is 0.03 $m^3 m^{-3}$, showing larger values in the wetter vegetated east and smaller values in the drier west, with the exception of the western coast. Surprisingly, even though altering crop rotation and irrigation are not simulated, the values over the central agricultural area are not higher than elsewhere in the domain. This good agreement between SMOS SM retrievals and our simulations is partly due to the bounded nature of SM (unlike Tb) and the CDF matching between both.

Our current system has a Tb sensitivity to soil moisture of about 1.3 K/0.01 $m^3 m^{-3}$ across the domain, averaged over all incidence angles and polarizations. A standard deviation in SM innovations of 0.03 $m^3 m^{-3}$ would thus roughly correspond to a standard deviation in Tb innovations of about 4 K, but instead we find 7.4 K across the study domain in the Tb assimilation systems. The Tb observations thus either have a comparably higher observation (including representation) error or they contain more information than the SM retrievals. At this point, we anticipate that the larger Tb innovations in the central plains may indicate that the Tb observations contain more unfiltered information about soil moisture (e.g., irrigation) and that the Tb observation error is higher due to shortcomings, e.g., in the vegetation modeling (representation error).

4.1.3   Actual vs. simulated observation and forecast errors

In a near-optimal filtering system, that is, a system that correctly simulates the actual model and observation errors, the standard deviation of the normalized innovations $[y_{\kappa,i} - \hat{y}_{\kappa,i}^-]_\lambda / \sqrt{ [\mathbf{R}_{\kappa,i} + \mathrm{Cov}(\hat{y}_{\kappa,i}^-, \hat{y}_{\kappa,i}^-)]_{\lambda\lambda} }$ is close to unity (Reichle et al., 2002). Figure 4g–i show that, averaged across the do-

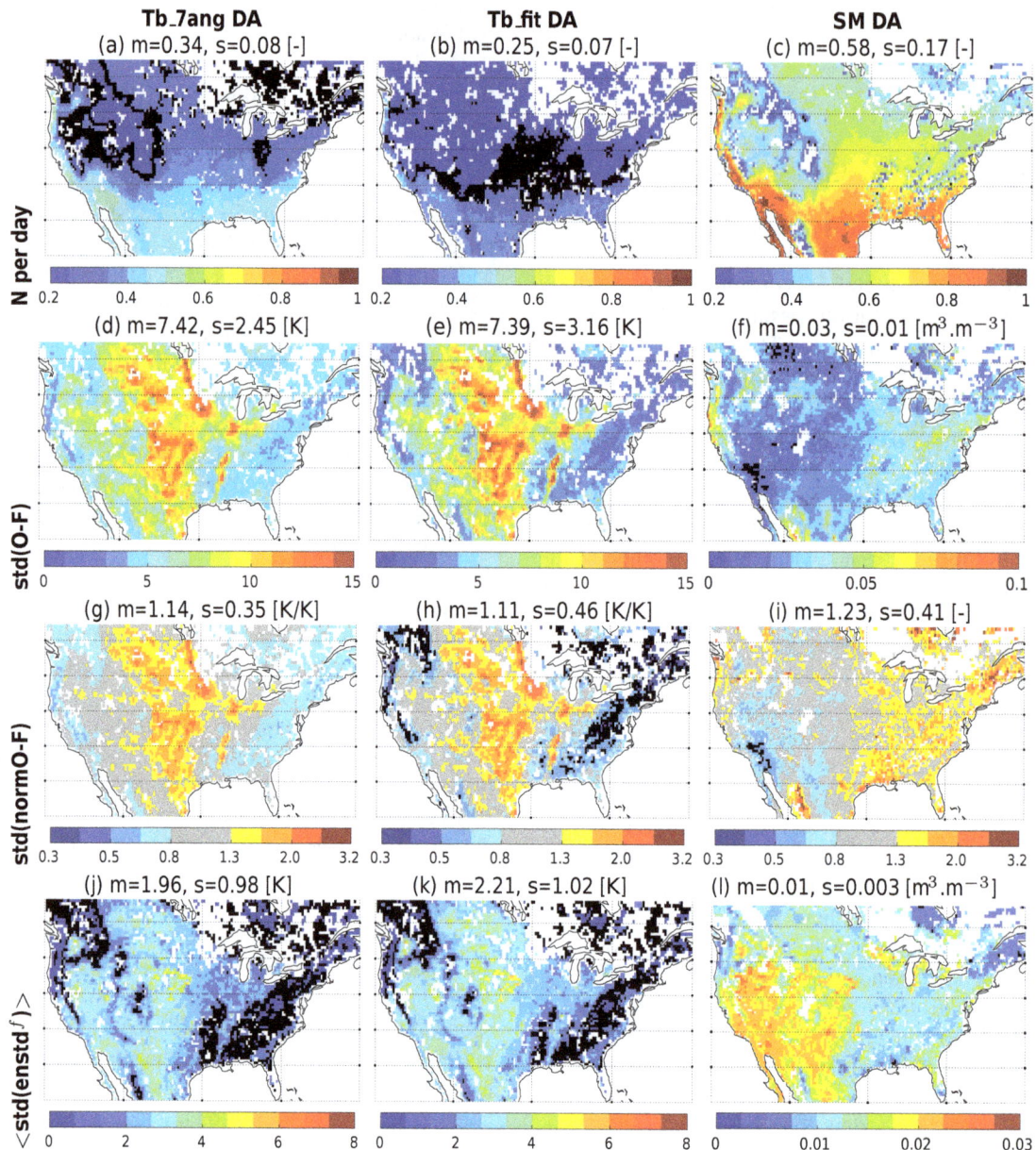

Figure 4. Observation-space assimilation diagnostics for the period from 1 July 2010 to 1 May 2015. Number of assimilated observation sets for **(a)** Tb_7ang assimilation, **(b)** Tb_fit assimilation, and **(c)** SM retrieval assimilation. Standard deviation of the **(d)** Tb innovations from Tb_7ang assimilation, **(e)** Tb innovations from Tb_fit assimilation, and **(f)** SM innovations from SM retrieval assimilation. (**g, h, i**) Same as (**d, e, f**), but for normalized innovations (normO–F). Ensemble standard deviation of the (j) Tb forecast error for Tb_7ang assimilation, **(k)** Tb forecast error for Tb_fit assimilation, and **(l)** surface soil moisture forecast error for SM retrieval assimilation. The titles show the spatial mean (m) and standard deviation (s) across each map.

main (and across all angles and polarizations for Tb assimilation), this metric is 1.14, 1.11, and 1.23 (–) for Tb_7ang, Tb_fit, and SM retrieval assimilation, respectively. The figure thus suggests that, on average, the simulated errors in the assimilation system only slightly underestimate the actual errors. But the figures also show that the metric varies strongly across the domain and exhibits very different spatial patterns for Tb and SM retrieval assimilation. For Tb_7ang

and Tb_fit assimilation, values are much larger than 1 in the central area and much smaller than 1 in the eastern forested area. This indicates that the assigned observation and forecast errors are severely underestimated in the central area and overestimated in the eastern forested area. Over forests, it can be assumed that the assigned representation error (part of the observation error) should be smaller. The Tb forecast error is already very small (see below), because the Tb uncertainty is

(a) O-F Tb_7ang [K] (H,V,Asc,Desc,7 angles)

(b) O-F SM [m³.m⁻³] (Asc,Desc)

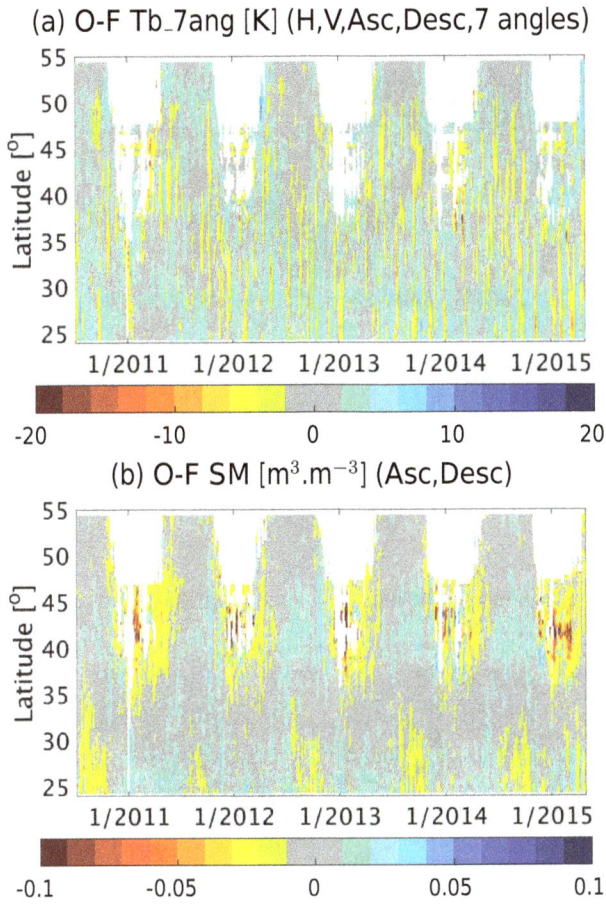

Figure 5. Hovmüller plots showing the temporal evolution of longitudinally averaged innovations (O–F) for the period from 1 July 2010 to 1 May 2015. **(a)** Tb_7ang innovations, averaged over H- and V-polarization, ascending and descending swaths, and over seven incidence angles. **(b)** SM innovations, averaged over ascending and descending swaths.

only marginally sensitive to soil moisture uncertainties under dense vegetation. For SM retrieval assimilation, the pattern is reversed, with the largest values in the eastern half of the domain, suggesting that here the simulated errors underestimate the actual errors. Values less than 1 are found in most of the western half of the domain, where the SM retrieval assimilation seems to overestimate the actual errors.

To further interpret the actual and simulated error magnitudes, Fig. 4j–k show the ensemble spread in the Tb forecasts (that is, the simulated forecast error standard deviation) $\sqrt{[\mathrm{Cov}(\hat{\mathbf{y}}_{\kappa,i}^-, \hat{\mathbf{y}}_{\kappa,i}^-)]_{\lambda\lambda}}$. Averaged across all angles and polarizations λ, the values are around 2 K when averaged across the entire domain. Larger values (3 K) are found in the central and dry western part, and smaller values (1 K) in the wetter eastern part. This pattern is similar for the SM ensemble spread in the SM retrieval assimilation system (Fig. 4l). In dry climates, the root-zone soil moisture often drops to the

wilting point, remains stagnant and no longer replenishes the surface. This results in increased sensitivity of the surface soil moisture to perturbations in meteorological conditions, and thus in higher uncertainty estimates for surface soil moisture in dry climates.

Given that the Tb observation error $\sqrt{[\mathbf{R}_{\kappa,i}]_{\lambda\lambda}}$ is set to 6 K for each individual angle, polarization, and overpass time in the Tb assimilation, the approximate total assigned observation and forecast error is 6.1 K ($\sqrt{6^2 + 2^2}$) across the study domain, 6.7 K ($\sqrt{6^2 + 3^2}$) in the central area, and 6 K ($\sqrt{6^2 + 1^2}$) in the eastern Appalachian area. Because the assigned observation error is uniformly set to 6 K, the spatial variability in the total simulated errors is thus too small compared to the actual errors (Fig. 4d–e), which ranges from more than 10 K in the central area to around 2–3 K in the eastern Appalachian area.

The SM observation error (after rescaling) is 0.02 m³ m⁻³ on average across the domain, with higher values in the eastern part and lower values in the western part, with the exception of Mexico, California, and western Oregon, where higher observation errors are found (Sect. 3.3). This general pattern is reversed in the SM forecast errors. Combined, the spatial variability in the SM observation and forecast errors does not capture the spatial variability in the actual errors (Fig. 4f), which leads to an overestimation of the errors in the west and an underestimation in the east.

4.2 Analysis increments

4.2.1 Spatio-temporal patterns

The Kalman filter translates footprint-scale innovations into 36 km increments. Because of the spatially distributed (3-D) filtering (Sect. 3.1), the number of increments in Fig. 6a–c is about 1.4 times the number of assimilated observation sets (Fig. 4a–c). Many areas with missing observations (or observation predictions) are filled through interpolation and extrapolation. With SM retrieval assimilation, there is almost one increment per day.

Figure 6d–f show the temporal standard deviations in the increments for the total soil profile water (Δwtot=Δsrfexc+Δrzexc−Δcatdef). The area average ($\pm$standard deviation) values are 6.9 ± 3.7 mm for Tb_7ang assimilation, 5.9 ± 3.5 mm for Tb_fit assimilation, and 4.2 ± 1.9 for SM retrieval assimilation. After scaling for the (variable) profile depth, the area-average values in volumetric soil moisture units are $3.4 \pm 1.7 \times 10^{-3}$ for Tb_7ang assimilation, $2.9 \pm 1.7 \times 10^{-3}$ for Tb_fit assimilation, and $2.3 \pm 1.9 \times 10^{-3}$ m³ m⁻³ for SM retrieval assimilation.

The individual components of the wtot increments are shown in Fig. 6g–i for the surface excess increments, Fig. 6j–l for the root-zone excess increments, and Fig. 6m–o for the catchment deficit increments. The patterns in wtot increments are dominated by catdef increments, and they generally reflect the patterns in the respective innovations' stan-

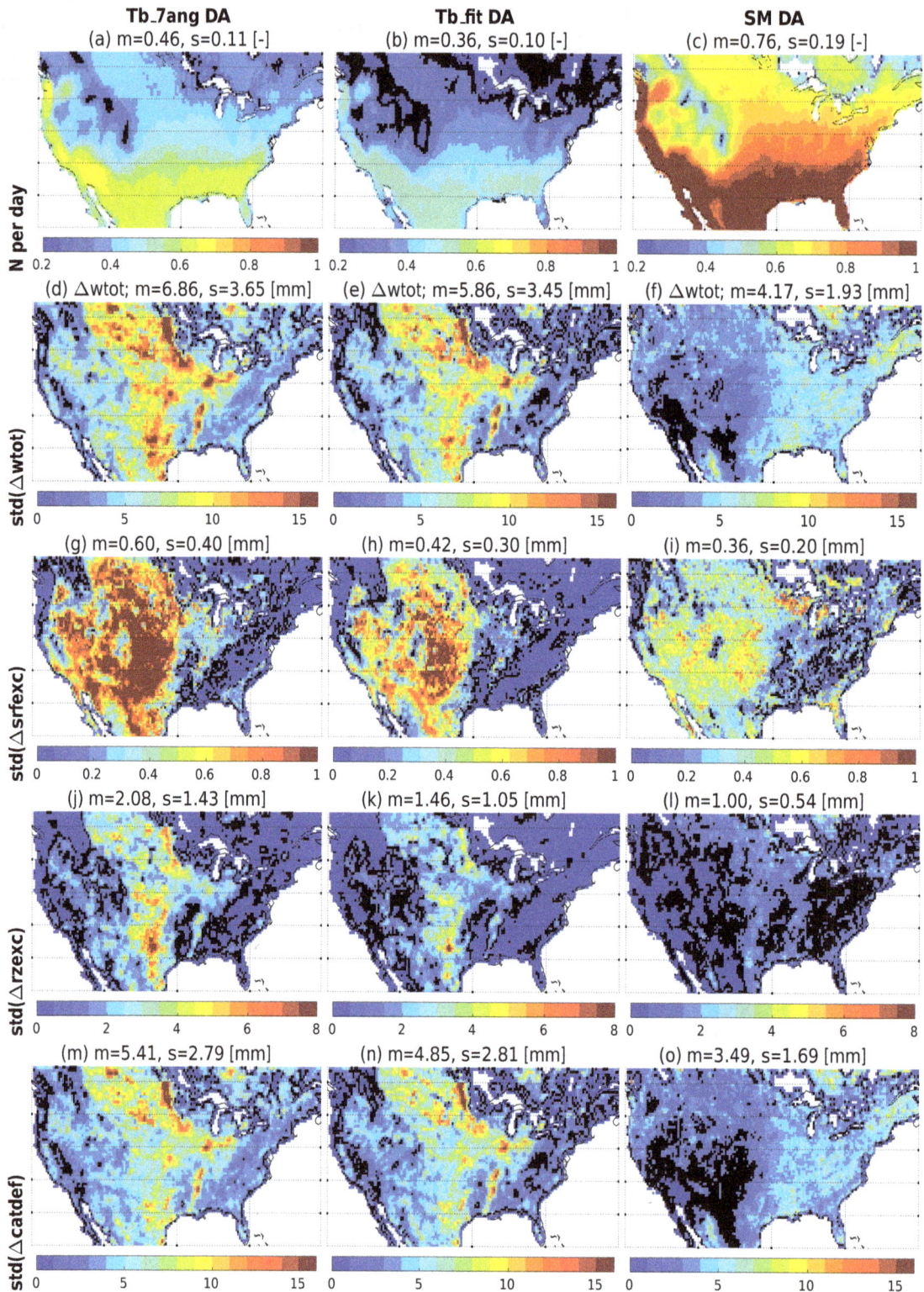

Figure 6. Statistics of the increments, calculated for the period from 1 July 2010 to 1 May 2015. Number of increments per day for **(a)** Tb_7ang assimilation, **(b)** Tb_fit assimilation, and **(c)** SM assimilation. Temporal standard deviation of total profile water (wtot) increments for **(d)** Tb_7ang assimilation, **(e)** Tb_fit assimilation, and **(f)** SM assimilation. **(g, h, i)** Same as **(d, e, f)** but for srfexc increments. **(j, k, l)** Same as **(d, e, f)** but for rzexc increments. **(m, n, o)** Same as **(d, e, f)** but for catdef increments. The titles show the spatial mean (m) and standard deviation (s) across each map.

dard deviations (Fig. 4d–f), which are very different for Tb and SM retrieval assimilation. The catdef increments pertain to the entire profile depth (which typically ranges between 2 and 3 m) and they presumably have a relatively small impact on the upper 5 cm soil layer (surface soil moisture): the domain-averaged magnitude of 5.4, 4.9, and 3.5 mm for catdef increments due to Tb_7ang, Tb_fit or SM retrieval assimilation, respectively (Fig. 6m–o), would linearly scale to about 0.1 mm for a 5 cm soil layer. This is a rough approximation: in reality the part of catdef that contributes to the 5 cm soil moisture cannot be calculated without computing the entire balanced profile. However, the approximate 0.1 mm is considerably less than the 0.6, 0.4, and 0.4 mm for the corresponding srfexc increments (Fig. 6g–i), which are directly applied to the upper 5 cm soil layer. The increments in rzexc (Fig. 6j–l) are relatively the smallest, because this variable is not perturbed by design.

Both Tb and SM retrieval assimilation show similar spatial patterns in the standard deviations of srfexc increments (Fig. 6g–i): the largest increments are found in the dry west and the smallest in the wetter east. The patterns in srfexc increments agree with the patterns in the ensemble forecast uncertainty for this variable (not shown, but implied by the Tb and soil moisture uncertainty in Fig. 4j–l). The srfexc values are small with small uncertainties, and the increments are thus similarly bounded in both Tb and SM retrieval assimilation, yielding comparable spatial increment patterns.

Finally, Fig. 7 compares spatially and temporally collocated wtot, srfexc, and rzexc increments obtained with Tb_7ang assimilation, Tb_fit assimilation, and SM retrieval assimilation; i.e., the figure shows all pairs of increments available from two assimilation cases. The scatter plots show that the increments are usually small and unbiased. The correlation between the wtot increments (Fig. 7a) obtained by Tb_7ang and Tb_fit assimilation is 0.7, and aligns with the expectation that either Tb assimilation experiment roughly corrects for the same events. In contrast, the correlation between the increments obtained by Tb_7ang and SM retrieval assimilation is only 0.3 (Fig. 7b). The figure is similar when comparing the Tb_fit and SM retrieval assimilation (not shown). For srfexc and rzexc (Fig. 7c–f), the increments are again similar for Tb_7ang and Tb_fit assimilation, but different for Tb and SM retrieval assimilation. For all soil moisture prognostic variables, Tb assimilation leads to larger increments than SM retrieval assimilation. The different assimilation systems thus introduce distinct corrections to the modeled soil moisture trajectories.

4.2.2 Discussion

In a nutshell, Eq. (1) states that the increments are given by the product of the Kalman gain and the innovations. To explain the differences in increment patterns between Tb and SM retrieval assimilation, we must therefore consider each system's innovations and Kalman gains. The relatively larger

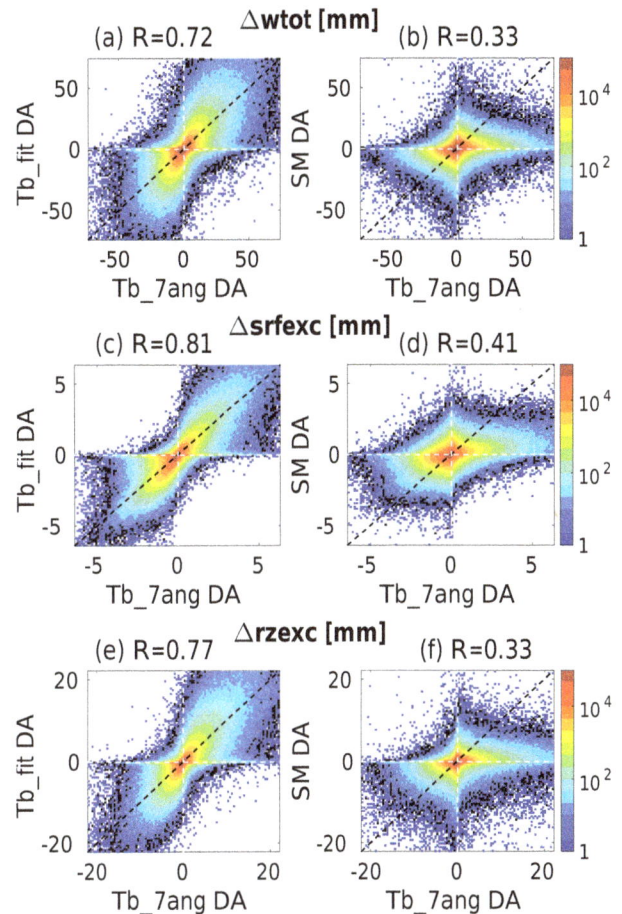

Figure 7. Spatially and temporally collocated analysis increments from (**a, c, e**) Tb_fit assimilation and (**b, d, f**) SM retrieval assimilation vs. the same from Tb_7ang assimilation for (**a, b**) profile-integrated wtot increments, (**c, d**) srfexc increments, and (**e–f**) rzexc increments. Increments are from the period 1 July 2010 to 1 May 2015. The plot range is limited to the maximum value of 10 times the standard deviation in either experiment, and divided into 100 even sample bins. Colors indicate the number of sample points within each 1.5, 0.13, or 0.44 mm bin for Δwtot, Δsrfexc, and Δrzexc, respectively. R is the spatio-temporal Pearson correlation coefficient between the individual increments from two assimilation experiments.

magnitude of the Tb innovations compared to the SM innovations (Sect. 4.1.2) contributes to the fact that the Tb assimilation results in larger soil moisture increments. This is the case even though the SM retrieval assimilation (unlike Tb assimilation) applies increments only to moisture variables and does not adjust modeled temperatures.

Furthermore, the Kalman gain matrices $\mathbf{K}_{k,i}$ (Eq. 2) for Tb and SM retrieval assimilation are different because the two systems employ different observation operators $h_i(.)$ and different observation error covariances $\mathbf{R}_i$. First, we note that the nonlinear inversion of Tb innovations to soil moisture increments, driven by the RTM in the observation operator, is *not* responsible for the larger wtot increments in the central grass and crop areas, because these areas exhibit low values for the microwave roughness parameter ($h < 0.2$, not shown) and a high sensitivity of Tb to soil moisture (as confirmed by the high forecast Tb errors in Fig. 4j–k). That is, in these areas commensurately large Tb innovations (O–F) values result in only small updates to soil moisture.

Second, the choice of a spatially uniform observation error covariance in the Tb assimilation experiment creates an imprint of the innovation pattern in the increment pattern. Higher increments are found in the agricultural areas with large Tb innovation standard deviations (Fig. 4d–e), because irrigation is not modeled and vegetation is not accurately parameterized. Since the filter is not set up to correct the latter, occasional excessive increments to soil moisture and temperature may be introduced. Such shortcomings could be mitigated by a more sophisticated assignment of Tb observation (representation) errors.

For SM retrieval assimilation, the pattern of the SM innovation standard deviation (RMSD) is similarly visible in the increments, with smaller values in the west and higher values in the east. Here again, the true spatio-temporal nature of the observation errors is not captured in the assigned observation error covariance and therefore propagated into the increments. Note also that the $0.03 \, \text{m}^3 \, \text{m}^{-3}$ SM innovation standard deviation (top 5 cm, Fig. 4f) is translated into a standard deviation of profile moisture increments of $0.002 \, \text{m}^3 \, \text{m}^{-3}$ (Fig. 6f rescaled by profile depth), but these increments are not equally distributed; i.e., larger increments are found for surface soil moisture and smaller increments for the deeper profile.

4.3   In situ validation

The above discussion highlights similarities and stark contrasts in how the Tb and SM retrieval assimilation systems operate. In this section, we look at the effect of these differences on the skill of the assimilation estimates vs. in situ observations. Figure 8 shows the RMSD_{ub} (Sect. 2.4) for the model-only open-loop (OL) simulation, and the change in RMSD_{ub} (Sect. 2.4) between the OL simulation and either the Tb_7ang or SM retrieval data assimilation (DA) experiment ($\Delta\text{RMSD}_{\text{ub}} = \text{RMSD}_{\text{ub}}(\text{DA}) - \text{RMSD}_{\text{ub}}(\text{OL})$) at individual SCAN and USCRN sites, for the period 1 July 2010–1 May 2015. The gray background shading indicates areas with modest topographic complexity and vegetation cover and where the satellite observations are most sensitive to surface soil moisture (details in De Lannoy and Reichle, 2016). The OL simulation has an average RMSD_{ub} value of $0.054 \, \text{m}^3 \, \text{m}^{-3}$ for surface soil moisture and $0.039 \, \text{m}^3 \, \text{m}^{-3}$ for root-zone soil moisture. Looking more closely, the RMSD_{ub} values are generally higher in the central and wetter eastern regions. In dry areas, the RMSD_{ub} is limited, because the time series show a limited variability for lack of much precipitation. On average, both assimilation experiments introduce

improvements at about 80 % of the sites for surface soil moisture, with spatially averaged $\Delta\text{RMSD}_{\text{ub}}$ values of -0.004 and $-0.003 \, \text{m}^3 \, \text{m}^{-3}$ for Tb_7ang and SM retrieval assimilation, respectively. (Spatial average metrics are computed using a cluster-based algorithm, Sect. 2.4.) The improvements are also propagated to the root-zone soil moisture (65 % of sites improved) with smaller average $\Delta\text{RMSD}_{\text{ub}}$ values of -0.002 and $-0.001 \, \text{m}^3 \, \text{m}^{-3}$, respectively.

The domain-average $\Delta\text{RMSD}_{\text{ub}}$ values caused by assimilation are only barely statistically significant for surface soil moisture in "favorable" areas, i.e., where the satellite observations are most sensitive to soil moisture (indicated with green background shading in Fig. 8). The differences between Tb_7ang, Tb_fit, or SM retrieval assimilation are not significant. The assimilation contributes an average relative improvement in surface soil moisture of 7 % of the OL RMSD_{ub} in favorable locations and 4 % in non-favorable areas. Both Tb and SM retrieval assimilation show improvements in the central and eastern parts of the US, but perform poorly in the western dry mountain areas, where the RMSD_{ub} for the OL was small and the assimilation may have introduced some additional noise. The Tb_7ang assimilation shows the largest improvements in the central US, whereas the SM retrieval assimilation shows the largest improvements in the southeastern part, for both surface and root-zone soil moisture. It is possible that the Tb assimilation has a larger impact in the central US than the SM retrieval assimilation, because irrigation events may be filtered in the SM retrievals (and perhaps partly assigned to vegetation opacity retrievals).

The bar plots in Fig. 9 summarize the average anomR values for the open-loop and data assimilation experiments, after stratifying all SCAN and USCRN sites into "favorable" and "non-favorable" categories (gray vs. white background in Fig. 8). The figures show that the open-loop anomR values for surface soil moisture are similar for both the favorable and non-favorable areas (0.51 and 0.50, respectively). However, data assimilation has a larger impact in favorable areas, where all assimilation schemes introduce significant improvements (anomR = 0.63, 0.61, and 0.59 for Tb_7ang, Tb_fit, and SM retrieval assimilation). In non-favorable areas, the improvements are smaller but still significant (anomR = 0.57, 0.56, and 0.54, for Tb_7ang, Tb_fit, and SM retrieval assimilation).

In the root zone, data assimilation also improves the skill over the open-loop simulations, but without statistical significance. The open-loop simulations yield anomR values of 0.56 and 0.50 in favorable and non-favorable areas, respectively. In favorable areas, the assimilation increases the anomR to 0.64, 0.64, and 0.62, for Tb_7ang, Tb_fit, and SM retrieval assimilation. In non-favorable areas, the skill improvement is limited and the anomR values are 0.54, 0.54, and 0.52, for Tb_7ang, Tb_fit, and SM retrieval assimilation. In any case, with assimilation, all anomR values exceed 0.5, meaning that the skill becomes better than a climatological forecast (Brier skill score larger than 0).

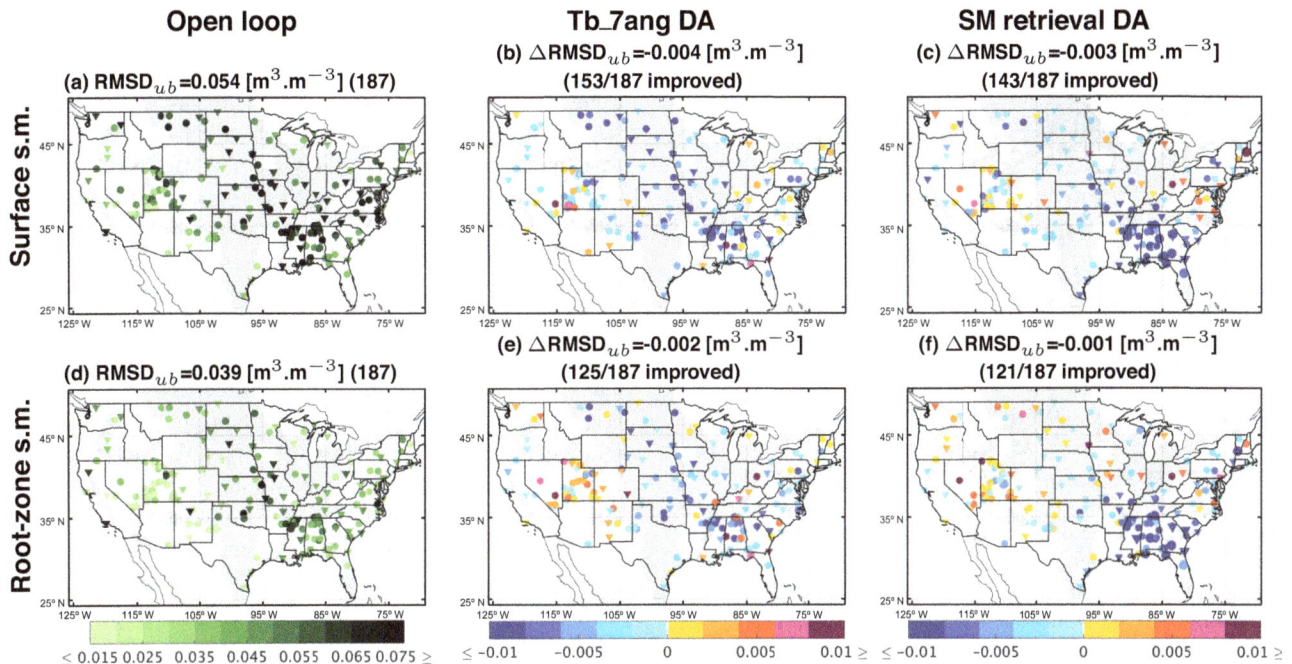

Figure 8. Unbiased RMSD (RMSD$_{ub}$) for the model-only open-loop (OL) simulation, and change in unbiased RMSD (ΔRMSD$_{ub}$) due to data assimilation at (circles) SCAN and (triangles) USCRN sites for (**a, b, c**) surface and (**d, e, f**) root-zone soil moisture. The skill of (**a, d**) the open-loop simulation is the reference value for the changes in skill due to (**b, e**) Tb_7ang and (**c, f**) SM retrieval assimilation. Statistically significant changes are marked by larger symbols (e.g., the southeastern US for SM retrieval assimilation). Metrics are calculated across 3 h time steps during the period from 1 July 2010 to 1 May 2015. The titles indicate the spatial mean (Δ)RMSD$_{ub}$ across all sites with clustering (31 clusters). The gray background shading marks areas with limited vegetation and topographic complexity based on model parameters.

Overall, the skill metrics are comparable for the Tb_7ang and Tb_fit assimilation (Fig. 9). The results from SM retrieval assimilation are slightly worse than those from Tb assimilation, which may indicate that Tb observations indeed still contain more information (Sect. 4.2) than the SM retrievals, which are implicitly filtered during the retrieval process. However, the differences between the domain-averaged skill values of the various assimilation schemes are minimal. Furthermore, when running the assimilation scheme with different spatially constant Tb observation error parameters, the skill metrics only changed marginally. This shows that our skill metrics are relatively insensitive to uniform changes in the data assimilation parameters. One reason for this is that the skill metrics are presented as (clustered) spatial averages, which compensate for large local differences. It is expected that the skill of our data assimilation systems can only be further improved by using a more localized (in space and time) approach to optimizing the assimilated observations (e.g., L2 SM retrievals) and the forecast and observation error parameters in the EnKF.

Finally, unlike Liu et al. (2011), the skill improvements in this study are smaller when we correct the re-analysis precipitation input with gauge-based precipitation data (Reichle and Liu, 2014). This and other recent improvements in the GEOS-5 modeling system make it increasingly chal-

lenging to obtain significant skill improvements from the assimilation of microwave observations over areas for which high-quality forcing data are available, such as the domain studied here. The benefits of the microwave-based soil moisture assimilation system are expected to be greater in areas with poorer ancillary inputs to the modeling system. This aspect will be further investigated through the validation of the global SMAP L4_SM data product.

5 Conclusions

The SMOS and SMAP satellite missions currently provide a wealth of L-band data to monitor large-scale soil moisture. A key question is how to make the best use of these data in current land surface data assimilation systems. The L1 Tb data from these missions are often complex, because of their multi-polarization and possibly multi-angle nature and their indirect connection with soil moisture. In theory, the best approach is to directly assimilate Tb observations using a consistent data assimilation system, but a correct global characterization of the Tb forecast and observation errors remains difficult. The L2 SM retrievals are easily handled products, but their assimilation is impacted by errors introduced by inconsistent ancillary information in the SM retrieval algorithm and the assimilation system. With further improvements in

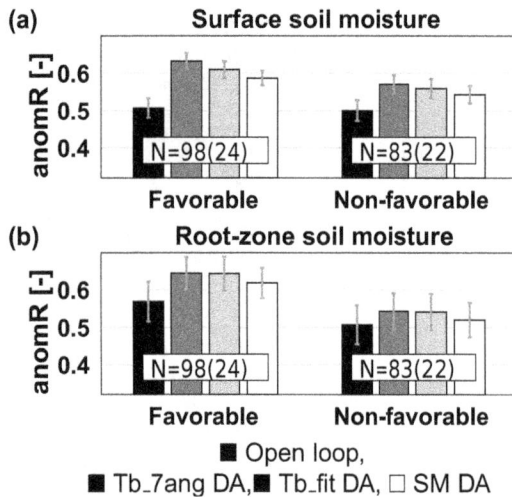

Figure 9. Performance of open-loop and data assimilation experiments in terms of anomaly correlations (anomR) calculated across 3 h analyses and forecast time steps from 1 July 2010 to 1 May 2015 for **(a)** surface and **(b)** root-zone soil moisture. The bars show skill metrics averaged over sites in either favorable or non-favorable areas, where favorable areas refer to the areas indicated by the gray background shading in Fig. 8. The variable N is the total number of SCAN and USCRN sites considered for each category, with the number of clusters in parentheses. The error bars reflect cluster-averaged 95 % confidence intervals.

the assimilated retrievals and careful selection of the ancillary data, SM retrieval assimilation may become a coequal alternative.

Three different data products from the SMOS mission are assimilated separately into the GEOS-5 land surface model to improve estimates of surface and root-zone soil moisture and to study the workings of each assimilation system. The first product consists of L1-based data of multi-angle, dual-polarization Tb observations at the bottom of the atmosphere. The second product is a derived 40° Tb product that mimics SMAP data. The third product is the operational L2 SM dataset. Special care is taken during quality control and processing of the satellite observations prior to assimilation and within the assimilation system. The Tb assimilation uses a distributed EnKF with a temporally variable Tb bias mitigation, a system that is also used for the SMAP L4_SM product (Reichle et al., 2016). The SM retrieval assimilation uses a similar system, but with CDF matching instead to eliminate the more stationary SM innovation biases. The study covers most of North America for the period of 1 July 2010–1 May 2015.

The Tb and SM innovations show very different spatial patterns and the number of assimilated observations differs because of different needs for data screening and bias mitigation. Based on the average sensitivity of Tb to soil moisture, the magnitude of the Tb innovations is comparably larger than that of the SM innovations, which may either intro-

duce more information or more error into the Tb assimilation system. The Tb and SM retrieval assimilation schemes also yield surprisingly different spatio-temporal increment patterns, leading to very different adjustments to the modeled soil moisture trajectories. Despite these stark differences, the various assimilation schemes yield soil moisture estimates with similar average skill metrics, computed from a set of 187 SCAN and USCRN sites across the US. Compared to in situ observations, both Tb and SM retrieval assimilations yield anomaly correlations around or larger than 0.6 for both the surface and root-zone soil moisture in "favorable" areas, where the satellite data are expected to better represent the soil moisture conditions, i.e., in areas with limited topographic complexity and limited vegetation. The anomaly correlation with data assimilation is between 0.5 and 0.6 in non-favorable areas. The data assimilation introduces significant improvements over the model-only simulations for surface soil moisture everywhere, but the improvements are much larger in favorable areas. For the root zone, improvements are also found, but without statistical significance. While no significant differences in domain-averaged skills can be found between the various assimilation systems, there are large local differences in performance between the Tb and SM retrieval assimilation which may be due to differences in information content and screening of the observations, and differences in how close each of the systems is to an optimal calibration of its model and observation error parameters. Therefore, we expect that soil moisture data assimilation systems can be further improved only if the systems manage to better simulate the spatial and temporal variations of the actual errors in the model and the observations. Furthermore, the SM retrieval assimilation results will benefit from any future improvement in the SM retrievals.

In line with our findings for the SMOS data assimilation, we anticipate that future versions of the Tb assimilation system for the SMAP L4_SM product may benefit from an improved characterization of spatial model and observation error structures, and from a better representation of some modeling components, such as, e.g., vegetation. In addition, given that SMOS and SMAP both provide L-band Tb observations, future assimilation systems should consider a joint assimilation of SMOS and SMAP Tb data. In such a system, it is important to consider the different instrument, Tb processing, and Tb error characteristics of the two L-band missions (De Lannoy et al., 2015).

Acknowledgements. The NASA Soil Moisture Active Passive (SMAP) mission supported this study. The NASA Center for Climate Simulation (NCCS) at the Goddard Space Flight Center provided computational resources through the NASA High-End Computing (HEC) program. The authors thank the editors and reviewers for their input.

Edited by: B. Su

References

Al-Yaari, A., Wigneron, J.-P., Ducharne, A., Kerr, Y., Wagner, W., Lannoy, G. D., Reichle, R., Bitar, A. A., Dorigo, W., Richaume, P., and Mialon, A.: Global-scale comparison of passive (SMOS) and active (ASCAT) satellite based microwave soil moisture retrievals with soil moisture simulations (MERRA-Land), Remote Sens. Environ., 152, 614–626, 2014.

Alvarez-Garreton, C., Ryu, D., Western, A. W., Su, C.-H., Crow, W. T., Robertson, D. E., and Leahy, C.: Improving operational flood ensemble prediction by the assimilation of satellite soil moisture: comparison between lumped and semi-distributed schemes, Hydrol. Earth Syst. Sci., 19, 1659–1676, doi:10.5194/hess-19-1659-2015, 2015.

Bell, J., Palecki, M., Baker, C., Collins, W., Lawrimore, J., Leeper, R., Hall, M., Kochendorfer, J., Meyer, T., Wilson, T., and Diamond, H.: US climate reference network soil moisture and temperature observations, J. Hydrometeorol., 14, 977–988, 2013.

Bosilovich, M. G., Akella, S., Coy, L., Cullather, R., Draper, C., Gelaro, R., Kovach, R., Liu, Q., Molod, A., Norris, P., Wargan, K., Chao, W., Reichle, R., Takacs, L., Vikhliaev, Y., Bloom, S., Collow, A., Firth, S., Labow, G., Partyka, G., Pawson, S., Reale, O., Schubert, S. D., and Suarez, M.: MERRA-2: Initial Evaluation of the Climate, Tech. rep., National Aeronautics and Space Administration, Goddard Space Flight Center, Greenbelt, Maryland, USA, 2015.

Brodzik, M. J., Billingsley, B., Haran, T., Raup, B., and Savoie, M.: Correction: Incremental but Significant Improvements for Earth-Gridded Data Sets, ISPRS Int. J. Geo.-Inf., 3, 1154–1156, 2014.

Chakrabart, S., Bongiovanni, T., Judge, J., Zotarelli, L., and Bayer, C.: Assimilation of SMOS Soil Moisture for Quantifying Drought Impacts on Crop Yield in Agricultural Regions, IEEE J. Sel. Top. Appl., 7, 3867–3879, 2014.

Cohn, S.: An Introduction to Estimation Theory, J. Meteorol. Soc. Jpn., 75, 257–288, 1997.

De Lannoy, G. and Reichle, R.: Global Assimilation of Multi-Angle and Multi-Polarization SMOS Brightness Temperature Observations into the GEOS-5 Catchment Land Surface Model for Soil Moisture Estimation, J. Hydrometeorol., 17, 669–691, 2016.

De Lannoy, G., Reichle, R., and Pauwels, V.: Global Calibration of the GEOS-5 L-band Microwave Radiative Transfer Model over Non-Frozen Land Using SMOS Observations, J. Hydrometeorol., 14, 765–785, doi:10.1175/JHM-D-12-092.1, 2013.

De Lannoy, G., Koster, R., Reichle, R., Mahanama, S., and Liu, Q.: An Updated Treatment of Soil Texture and Associated Hydraulic Properties in a Global Land Modeling System, J. Adv. Model. Earth Syst., 6, 23, doi:10.1002/2014MS000330, 2014a.

De Lannoy, G., Reichle, R., and Vrugt, J.: Uncertainty Quantification of GEOS-5 L-Band Radiative Transfer Model Parameters using Bayesian Inference and SMOS Observations, Remote Sens. Environ., 148, 146–157, doi:10.1016/j.rse.2014.03.030, 2014b.

De Lannoy, G., Reichle, R., Peng, J., Kerr, Y., Castro, R., Kim, E., and Liu, Q.: Converting Between SMOS and SMAP Level-1 Brightness Temperature Observations Over Nonfrozen Land, IEEE Geosci. Remote Sens. Lett., 12, 1908–1912, 2015.

Desroziers, G., Berra, L., Chapnik, B., and Poli, P.: Diagnosis of observation, background and analysis-error statistics in observation space, Q. J. Roy. Meteorol. Soc., 131, 3385–3396, 2005.

Diamond, H., Karl, T., Palecki, M., Bell, J., Leeper, R., Easterling, D., Lawrimore, J., Meyers, T., Helfert, M., Goodge, G., and Thorne, P.: US climate reference network after one decade of operations: status and assessment, BAMS, 94, 485–498, 2013.

Draper, C. S., Reichle, R. H., Lannoy, G. J. M. D., and Liu, Q.: Assimilation of passive and active microwave soil moisture retrievals, Geophys. Res. Lett., 39, L04401, doi:10.1029/2011GL050655, 2012.

Ducharne, A., Koster, R., Suarez, M., Stieglitz, M., and Kumar, P.: A catchment-based approach to modeling land surface processes in a GCM, Part 2, Parameter estimation and model demonstration, J. Geophys. Res., 105, 24823–24838, 2000.

Entekhabi, D., Reichle, R. H., Koster, R. D., and Crow, W. T.: Performance Metrics for Soil Moisture Retrievals and Application Requirements, J. Hydrometeorol., 11, 832–840, 2010.

Entekhabi, D., Yueh, S., O'Neill, P., and Kellogg, K.: SMAP Handbook, JPL Pasadena, CA, USA, 400–1567, 2014.

Fascetti, F., Pierdicca, N., Crapolicchio, L. P. R., and Munoz-Sabater, J.: A comparison of ASCAT and SMOS soil moisture retrievals over Europe and Northern Africa from 2010 to 2013, Int. J. Appl. Earth Obs., 45, 135–142, 2016.

Ford, T. W., Harris, E., and Quiring, S. M.: Estimating root zone soil moisture using near-surface observations from SMOS, Hydrol. Earth Syst. Sci., 18, 139–154, doi:10.5194/hess-18-139-2014, 2014.

Kerr, Y., Waldteufel, P., Wigneron, J.-P., Delwart, S., Cabot, F., Boutin, J., Escorihuela, M.-J., Font, J., Reul, N., Gruhier, C., Juglea, S., Drinkwater, M., Hahne, A., Martin-Neira, M., and Mecklenburg, S.: The SMOS Mission: New Tool for Monitoring Key Elements of the Global Water Cycle, P. IEEE, 98, 666–687, 2010.

Kornelsen, K. C., Davison, B., and Coulibaly, P.: Application of SMOS Soil Moisture and Brightness Temperature at High Resolution With a Bias Correction Operator, IEEE JSTAR, 9, 1590–1605, 2016.

Koster, R., Guo, Z., Yang, R., Dirmeyer, P., Mitchell, K., and Puma, M.: On the Nature of Soil Moisture in Land Surface Models, J. Climate, 22, 4322–4335, 2009.

Koster, R., Brocca, L., Crow, W., Burgin, M., and De Lannoy, G.: Precipitation Estimation Using L-Band and C-Band Soil Moisture Retrievals, Water Resour. Res., 52, 7213–7225, doi:10.1002/2016WR019024, 2016.

Koster, R. D., Suarez, M. J., Ducharne, A., Stieglitz, M., and Kumar, P.: A catchment-based approach to modeling land surface processes in a general circulation model 1. Model structure, J. Geophys. Res., 1050, 24809–24822, 2000.

Lievens, H., Tomer, S., Bitar, A. A., De Lannoy, G., Drusch, M., Dumedah, G., Hendricks-Franssens, H.-J., Kerr, Y., Pan, M., Roundy, J., Vereecken, H., Walker, J., Wood, E., Verhoest, N., and Pauwels, V.: SMOS soil moisture assimilation for improved stream flow simulation in the Murray Darling Basin, Australia, Remote Sens. Environ., 168, 146–162, 2015.

Liu, Q., Reichle, R. H., Bindlish, R., Cosh, M. H., Crow, W. T., de Jeu, R., Huffman, G., De Lannoy, G. J. M., and Jackson, T.: The contributions of precipitation and soil moisture observations to the skill of soil moisture estimates in a land data assimilation system, J. Hydrometeorol., 12, 750–765, 2011.

Mahanama, S. P., Koster, R., Walker, G., Tackacs, L., Reichle, R., Lannoy, G. D., Liu, Q., Zhao, B., and Suarez, M.: Land Boundary

Conditions for the Goddard Earth Observing System Model Version 5 (GEOS-5) Climate Modeling System – Recent Updates and Data File Descriptions, Tech. rep., National Aeronautics and Space Administration, Goddard Space Flight Center, Greenbelt, Maryland, USA, 2015.

Munoz-Sabater, J., de Rosnay, P., Jiminnez, C., Isaksen, L., and Albergel, C.: SMOS Brightness Temperature Angular Noise: Characterization, Filtering, and Validation, IEEE T. Geosci. Remote Sens., 52, 5827–5839, 2014.

Piles, M., Sanchez, N., Vall-llossera, M., Camps, A., Martinez-Fernandez, J., Martinez, J., and Gonzaliez-Gambau, V.: A Downscaling Approach for SMOS Land Observations: Evaluation of High-Resolution Soil Moisture Maps Over the Iberian Peninsula, IEEE JSTAR, 7, 3845–3857, 2014.

Reichle, R. H. and Koster, R.: Assessing the impact of horizontal error correlations in background fields on soil moisture estimation, J. Hydrometeorol., 4, 1229–1242, 2003.

Reichle, R. H. and Koster, R.: Bias reduction in short records of satellite soil moisture, Geophys. Res. Lett., 31, L19501, doi:10.1029/2004GL020938, 2004.

Reichle, R. H. and Liu, Q.: Observation-Corrected Precipitation Estimates in GEOS-5, Tech. rep., National Aeronautics and Space Administration, Goddard Space Flight Center, Greenbelt, Maryland, USA, 2014.

Reichle, R. H., Walker, J. P., Houser, P. R., and Koster, R. D.: Extended vs. Ensemble Kalman filtering for land data assimilation, J. Hydrometeorol., 3, 728–740, 2002.

Reichle, R. H., Lucchesi, R. A., Ardizzone, J., Kim, G.-K., Smith, E., and Weiss, B.: Soil Moisture Active Passive (SMAP) Mission Level 4 Surface and Root Zone Soil Moisture (L4_SM) Product Specification Document, Tech. rep., NASA Goddard Space Flight Center, GMAO Office Note No. 10 (Version 1.4), 2015.

Reichle, R. H., De Lannoy, G. J. M., Liu, Q., Ardizzone, J., Chen, F., Colliander, A., Conaty, A., Crow, W., Jackson, T., Kimball, J., Koster, R., and Smith, E. B.: Soil Moisture Active Passive Mission L4_SM Data Product Assessment (Version 2 Validated Release), NASA GMAO Office Note, No. 12 (Version 1.0), 2016.

Ridler, M., Madsen, H., Sitsen, S., Bircher, S., and Fensholt, R.: Assimilation of SMOS-derived soil moisture in a fully integrated hydrological and soil-vegetation-atmosphere transfer model in Western Denmark, Water Resour. Res., 50, 8962–8981, 2014.

Rienecker, M. M., Suarez, M. J., Gelaro, R., Todling, R., Bacmeister, J., Liu, E., Bosilovich, M. G., Schubert, S. D., Takacs, L., Kim, G.-K., Bloom, S., Chen, J., Collins, D., Conaty, A., da Silva, A., Gu, W., Joiner, J., Koster, R. D., Lucchesi, R., Molod, A., Owens, T., Pawson, S., Pegion, P., Redder, C. R., Reichle, R., Robertson, F. R., Ruddick, A. G., Sienkiewicz, M., and Woollen, J.: MERRA – NASA's Modern-Era Retrospective Analysis for Research and Applications, J. Climate, 24, 3624–3648, doi:10.1175/JCLI-D-11-00015.1, 2011.

Rodriguez-Fernandez, N., Aires, F., Richaume, P., Kerr, Y., Prigent, C., Kolassa, J., Cabot, F., Jimenez, C., Mahmoodi, A., and Drush, M.: Soil moisture retrieval using neural networks: Application to SMOS, IEEE T. Geosci. Remote Sens., 53, 5991–6007, 2015.

Schaefer, G. L., Cosh, M. H., and Jackson, T. J.: The USDA Natural Resources Conservation Service Soil Climate Analysis Network (SCAN), J. Atmos. Ocean. Technol., 24, 2073–2077, 2007.

van der Schalie, R., Kerr, Y., Wigneron, J., Rodriguez-Fernandez, N., Al-Yaari, A., and Jeu, R.: Global SMOS Soil Moisture Retrievals from The Land Parameter Retrieval Model, Int. J. Appl. Earth Observ. Geoinf., 45, 125–134, 2016.

van Leeuwen, P. J.: Representation errors and retrievals in linear and nonlinear data assimilation, Q. J. Roy. Meteorol. Soc., 141, 1612–1623, 2015.

Vinnikov, K., Robock, A., Speranskaya, N., and Schlosser, A.: Scales of temporal and spatial cariability of midlatitude soil moisture, J. Geophys. Res., 101, 7163–7174, 1996.

Wanders, N., Pan, M., and Wood, E.: Correction of real-time satellite precipitation with multi-sensor satellite observations of land surface variables, Remote Sens. Environ., 160, 206–221, 2015.

Wigneron, J., Kerr, Y., Waldteufel, P., Saleh, K., Escorihuela, M.-J., Richaume, P., Ferrazzoli, P., de Rosnay, P., Gurney, R., Calvet, J., Grant, J., Guglielmetti, M., Hornbuckle, B., Mätzler, C., Pellarin, T., and Schwank, M.: L-band Microwave Emission of the Biosphere (L-MEB) Model: Description and calibration against experimental data sets over crop fields, Remote Sens. Environ., 107, 639–655, 2007.

Wigneron, J.-P., Jackon, T. J., O'Neill, P., De Lannoy, G. J. M., Walker, J. P., Ferrazzoli, P., Mironov, V., Bircher, S., Grant, J. P., Kurum, M., Schwank, M., Das, N., Royer, A., Al-Yaari, A., Al Bitar, A., Fernandez-Moran, R., Lawrence, H., Mialon, A., Parrens, M., Richaume, P., Delwart, S., and Kerr, Y.: Modelling the passive microwave signature from land surfaces: a review of recent results and application to the SMOS & SMAP soil moisture retrieval algorithms, Remote Sens. Environ., in review, 2016.

Ye, N., Walker, J., Guerschman, J., Ryu, D., and Gurney, R.: Standing water effect on soil moisture retrieval from L-band passive microwave observations, Remote Sens. Environ., 169, 232–242, 2015.

Zhao, L., Yang, K., Qin, J., Chen, Y., Tanga, W., Lud, H., and Yang, Z.-L.: The scale-dependence of SMOS soil moisture accuracy and its improvement through land data assimilation in the central Tibetan Plateau, Remote Sens. Environ., 152, 345–355, 2014.

Zhao, T., Shi, J., Bindlish, R., Jackson, T. J., Kerr, Y. H., Cosh, M. H., Cui, Q., Li, Y., Xiong, C., and Che, T.: Refinement of SMOS Multiangular Brightness Temperature Toward Soil Moisture Retrieval and Its Analysis Over Reference Targets, JSTAR, 8, 589–603, 2015.

3

Toward seamless hydrologic predictions across spatial scales

Luis Samaniego[1], **Rohini Kumar**[1], **Stephan Thober**[1], **Oldrich Rakovec**[1], **Matthias Zink**[1], **Niko Wanders**[2,6],
Stephanie Eisner[3,a], **Hannes Müller Schmied**[4,5], **Edwin H. Sutanudjaja**[6], **Kirsten Warrach-Sagi**[7], and **Sabine Attinger**[1]

[1]Department of Computational Hydrosystems, UFZ-Helmholtz Centre for Environmental Research, Leipzig, Germany
[2]Department of Civil and Environmental Engineering, Princeton University, Princeton, NJ 08544, USA
[3]Center for Environmental Systems Research, University of Kassel, Kassel, Germany
[4]Institute of Physical Geography, Goethe-University Frankfurt, Frankfurt, Germany
[5]Senckenberg Biodiversity and Climate Research Centre (BiK-F), Frankfurt, Germany
[6]Universiteit Utrecht, Department of Physical Geography, Utrecht, the Netherlands
[7]Institute of Physics and Meteorology, University of Hohenheim, Stuttgart, Germany
[a]now at: Division for Forestry and Forest Resources, Norwegian Institute of Bioeconomy Research, Ås, Norway

Correspondence to: Luis Samaniego (luis.samaniego@ufz.de)

Abstract. Land surface and hydrologic models (LSMs/HMs) are used at diverse spatial resolutions ranging from catchment-scale (1–10 km) to global-scale (over 50 km) applications. Applying the same model structure at different spatial scales requires that the model estimates similar fluxes independent of the chosen resolution, i.e., fulfills a flux-matching condition across scales. An analysis of state-of-the-art LSMs and HMs reveals that most do not have consistent hydrologic parameter fields. Multiple experiments with the mHM, Noah-MP, PCR-GLOBWB, and WaterGAP models demonstrate the pitfalls of deficient parameterization practices currently used in most operational models, which are insufficient to satisfy the flux-matching condition. These examples demonstrate that J. Dooge's 1982 statement on the unsolved problem of parameterization in these models remains true. Based on a review of existing parameter regionalization techniques, we postulate that the multiscale parameter regionalization (MPR) technique offers a practical and robust method that provides consistent (seamless) parameter and flux fields across scales. Herein, we develop a general model protocol to describe how MPR can be applied to a particular model and present an example application using the PCR-GLOBWB model. Finally, we discuss potential advantages and limitations of MPR in obtaining the seamless prediction of hydrological fluxes and states across spatial scales.

1 Introduction

"If it disagrees with experiment, it's wrong".
Richard P. Feynman

Land surface and hydrologic models (LSMs/HMs) are currently used at diverse spatial resolutions ranging from 1 to 10 km in catchment-scale impact analysis and forecasting (Christensen and Lettenmaier, 2007; Addor et al., 2014) to over 50 km in global-scale climate change simulations to estimate land surface boundary conditions of key state variables (Haddeland et al., 2011; Bierkens, 2015; Wanders and Wada, 2015). The fundamental conditions behind the applicability of the same LSM/HM model structure at different spatial scales requires that the model parameterizations are scale invariant and that the model estimates similar fluxes across a range of spatial resolutions. In other words, it must fulfill the flux-matching condition across scales so that the mass conservation principle can be ensured (Wood, 1997).

A parameterization is a simplified and idealized representation of subgrid physical phenomenon that is either "too small, too brief, too complex, or too poorly understood" to be explicitly represented by a model at a given resolution (Edwards, 2010). Parameterizations require variables called predictors, effective parameters and constants also called transfer, global, or super parameters (Pokhrel and Gupta, 2010). Super parameters are often parameters in empirical relationships that have been found with measurements in the field or in the laboratory, e.g., regression parameters in pedotransfer functions (Cosby et al., 1984). They are often tuned to represent observed variables and often have no physical meaning. These parameters constitute simplified surrogates to compensate for the missing subgrid processes that are not accounted for within a modeling system (Brynjarsdottir and O'Hagan, 2014).

Effective parameters of LSMs/HMs are usually obtained by ad hoc procedures (e.g., automatic calibration) at a given spatial resolution for a given modeling domain. As a consequence of this standard practice, parameter fields of LSMs/HMs often exhibit artificial spatial "discontinuities" such as calibration imprints circumscribing river basin boundaries, and consequently they are not seamless (Merz and Blöschl, 2004; Li et al., 2012). Inconsistent patterns of effective parameter fields for land surface geophysical properties across spatial scales constitute a clear indication that their parameterizations are not scale invariant. There are several reasons explaining this parameterization deficiency. With the advent of electronic computers, the performance of general circulation models (GCMs), numerical weather prediction (NWP) models (Pielke Sr, 2013), land surface models (Liang et al., 1994; Sellers et al., 1997; Niu et al., 2011), and hydrologic models (Batjes, 1996; Lindstrom et al., 1997; van Beek et al., 2011; Samaniego et al., 2010b) has been increased mainly by improving model conceptualization (i.e., the number of process descriptions) and/or spatial resolution since the storage capacity and computational power allowed for it (Le Treut et al., 2007; Wood et al., 2011; Bierkens et al., 2014). As a result, parameterizations in LSMs have also increased in their complexity during the past decades (Sellers et al., 1997; Fisher et al., 2014). The procedures to estimate effective parameters required for the parameterizations, however, remained unchanged. For example, LSMs evolved from simple aerodynamic bulk transfer schemes with uniform description of surface parameters during the 1970s to detailed LSMs having a consistent description of the exchange of energy and matter between the atmosphere, the vegetation, and the land surface (Sellers et al., 1997). State-of-the-art LSMs, such as the Community Land Model version 4 (Bonan et al., 2011) and Noah-MP (Niu et al., 2011), however, still use quite simple pedotransfer functions based on work of Clapp and Hornberger (1978) and Cosby et al. (1984) to estimate fundamental soil properties such as porosity (Oleson et al., 2013).

Further reasons that have prevented the improvement of parameterization techniques are

- the lack of procedures and theories for linking physical properties (e.g., soil porosity) that can be measured at the field scale with "effective" parameter values that represent the aggregate behavior of the land characteristics at the scale of a grid cell required in LSMs or HMs,

- poor understanding of the scaling of parameters (Dooge, 1982) and its influence on the hydrological response of the system (Wood, 1997; Wood et al., 1988),

- limited inclusion of subgrid heterogeneity in hydrological parameterizations and multiscale modeling of hydrologically relevant variables as suggested by Famiglietti and Wood (1995, 1994); Liang et al. (1996),

- lack of significant progress on the applicability of seminal upscaling theories (Miller and Miller, 1956; Dagan, 1989; Gelhar, 1993; Neuman, 2010; Kitanidis and Vomvoris, 2010) developed for subsurface hydrologic problems into LSMs/HMs, and

- lack of transparency in most of the existing LSM/HM source codes with respect to the meaning, origin, and uncertainty associated with the hard-coded numerical values (i.e., parameters) either in the code or in the look-up tables (Mendoza et al., 2015; Cuntz et al., 2016).

Consequently, it is possible to assert that model parameterization is an old, ubiquitous, and recurring problem in land surface and hydrologic modeling. Considering this lack of coherent development during the past decades, we can still concur with Dooge (1982, p. 269) and say that the "parameterization of hydrologic processes to the grid scale of general circulation models is a problem that has not been approached, let alone solved."

There are potential methods available in the literature that may lead toward coherent parameterizations and prediction of water and energy fluxes in LSMs/HMs. For example, (1) sidestepping the scaling problem of key model parameters by assuming scale-independent distribution functions with regionalized distribution parameters (Intsiful and Kunstmann, 2008), (2) finding strong links between model parameters to mapped geophysical attributes via regularization procedures (Pokhrel and Gupta, 2010), and (3) finding strong links between of observed functional responses of hydrological systems and geophysical characteristics (Yadav et al., 2007). These methods, however, alone may not satisfy the flux-matching criteria.

In contrast to these existing methods, we argue that the multiscale parameter parameterization (MPR) technique (Samaniego et al., 2010b) offers a framework to link the field scale (observations) with the catchment scale (Dooge, 1982). MPR also accounts for the effect of the spatial variability and non-linearity of geophysical characteristics in the parameterization of hydrologic processes that operate at a range of spatial resolutions (Dooge, 1982; Wood et al., 1988). Depending on the conditions imposed on the parameter estimation technique, MPR can lead to parameterizations that satisfy the flux-matching criteria and hence contributes to obtaining seamless parameter and water flux fields. Because MPR relies on empirical transfer functions and upscaling operators to link geophysical properties with model parameters, it provides a very effective procedure to transfer "global parameters" to scales and locations other than those used in calibration (Samaniego et al., 2010a, b; Kumar et al., 2013b). This dependency on several transferable coefficients also contributes to minimizing a serious drawback of spatially explicit models called "overparameterization" (Beven, 1995).

In this study, we analyze to which extent existing LSM/HM parameterizations are limited to obtain seamless predictions of water fluxes and states across multiple spatial resolutions. Through several modeling experiments addressing Wood (1990)'s query (i.e., "What modeling experiments need to be performed to resolve the scale question …"), we demonstrate that a large portion of the predictive uncertainty in existing LSMs/HMs originates from the deficient estimation of effective parameters, which leads to a lack of scale invariance and thus to their poor transferability across scales and locations. These experiments also aim to help the modeler to reveal poor-performing parameterizations, i.e., those that exhibit non-seamless fields. Finally, based on our past experiences and aiming to address the challenges stated above, we develop a protocol that systematizes the application of the MPR technique for any LSM/HM and demonstrate its effectiveness by implementing it into the PCR-GLOBWB model.

2 Current parameterization techniques

2.1 The state of the art

The most common parameterization techniques found in the literature are (1) look-up tables (LUTs), (2) manual or automatic calibration, (3) hydrologic response units (HRUs), (4) representative elementary watersheds (REWs), (5) a priori regularization functions, (6) simultaneous regionalization/regularization functions, and (7) dissimilarity-based metrics to transfer model parameters.

The simplest technique to assign a parameter value to a modeling unit (e.g., grid cell, HRU, or subcatchment) is based on a LUT. In this case, a categorical index associated with a modeling unit links it with information taken from an external reference file (i.e., the LUT) which maps this index with parameter values that are usually taken from the literature. This technique is commonly used in most of the (operational) LSMs such as CABLE, CHTESSEL, CLM, JULES, and Noah-MP (Kowalczyk et al., 2006; Viterbo and Beljaars, 1995; ECMWF, 2016; Oleson et al., 2013; Best et al., 2011; Niu, 2011). A disadvantage of this method is the difficulty to perform sensitivity analysis (Cuntz et al., 2016). Moreover, the number of classes defined in LUT is often limited to a few (e.g., 13 soil classes in Noah-MP) resulting in non-seamless parameter fields that are not continuous.

Manual or automatic calibration is a commonly used technique to parameterize spatially lumped hydrologic models (e.g., Crawford and Linsley, 1966; Burnash et al., 1973;

Lindstrom et al., 1997; Edijatno et al., 1999; Fenicia et al., 2011; Martina et al., 2011; Andréassian et al., 2014; Singh et al., 2014) and semi-distributed hydrologic models (e.g., Leavesley et al., 1983; Kavetski et al., 2003; Lindström et al., 2010; Hundecha and Bárdossy, 2004; Merz and Blöschl, 2004; Hundecha et al., 2016). The aim is to minimize the disagreement between model simulations and observations. In the majority of the cases, the target variable is streamflow. The main drawback of this parameterization technique is that the parameter fields, which are obtained by colocating lumped model parameters from sub-basins, are doubtful because they exhibit sharp discontinuities along individually calibrated sub-basin boundaries despite having spatial continuity in basin physical attributes like soil, vegetation, and geological properties that govern spatial dynamics of hydrological processes (Merz and Blöschl, 2004; Li et al., 2012; Blöschl et al., 2013). In addition, the "patchwork quilt" parameter fields shown in these references exhibit significant sensitivity to the calibration conditions as demonstrated by Merz and Blöschl (2004). Thus, models that are parameterized with this technique may exhibit (1) poor predictability of state variables and fluxes at locations and periods not considered in calibration and (2) sharp discontinuities along subbasin boundaries in state, flux, and parameter fields (e.g., Merz and Blöschl, 2004; Lindström et al., 2010). Parameter fields derived from basin-wise "calibrated" lumped models lack spatial seamlessness and thus are "inadequate representations of real-world systems" (Savenije and Hrachowitz, 2017). Moreover, excessive reliance on parameter calibration leads to deficient performance at interior points of the basin or at other locations at which the model was not calibrated (Pokhrel and Gupta, 2010; Lerat et al., 2012; Brynjarsdottir and O'Hagan, 2014).

There have been many attempts to improve the parameterization of lumped and semi-distributed models by further discretizing the sub-basins into a given number of regions that exhibit nearly similar hydrologic behavior, i.e., the so-called HRU concept initially proposed by Leavesley et al. (1983) and further developed by others (e.g., Flügel, 1995; Beldring et al., 2003; Blöschl et al., 2008; Viviroli et al., 2009; Zehe et al., 2014). Unfortunately, results obtained in these parameterization attempts have not been very successful in realistically representing the spatial variability of model parameters, states, and fluxes because of the lack of regionalized parameters and the unabridged reliance on parameter calibration to improve model performance (Kumar et al., 2010). Commonly, the effective parameters estimated for the HRUs are found by automatic calibration. Efforts have been made to enforce continuity on parameter fields (Gotzinger and Bárdossy, 2007; Singh et al., 2012) but with somewhat limited success during the transferability of parameters across scales and locations. In addition, models parameterized using HRUs do not lead to mass conservation of water fluxes (i.e., flux-matching) when applied to scales other than those used for calibration (Kumar et al., 2010, 2013b). Recent attempts have been made to improve the HRU concept to increase the seamless representation of parameters, states, and fluxes (Chaney et al., 2016a). However, this concept has not been tested for scalability and seamlessness of the estimated fields at coarse resolutions. Lately, a thermodynamic reinterpretation of the HRU concept was proposed by Zehe et al. (2014), but to date, the implementation of this approach has not found its way into meso-scale to macro-scale LSMs/HMs.

The representative elementary watershed approach (Reggiani et al., 1998) is an interesting theoretical concept, which scales mass and momentum balance equations. Unfortunately, to the best of our knowledge, it has not been used to estimate effective parameters at meso- and regional scales.

A priori regularization functions (e.g., pedotransfer functions) were introduced by Koren et al. (2013) to ensure the "inappropriate randomness in the spatial patterns of model parameters", i.e., the lack of seamlessness. Unfortunately, in this case, the parameters (or coefficients) of regularization functions were not subject to parameter estimation or to the verification of their ability to predict fluxes and states across various scales. The use of empirical point-scale-based relationships to link geophysical characteristics with LSM/HM parameters and the assumption that their coefficients are universally applicable with certainty (e.g., the coefficients in the Clapp and Hornberger (1978) pedotransfer functions) are the major reasons for the proliferation of hidden parameters in LSM/HM code (Mendoza et al., 2015; Cuntz et al., 2016). It is of pivotal importance to understand that these point-scale relationships should not be applied beyond the scale at which they were derived.

Many types of regionalization (or regularization) approaches have been tested for semi-distributed and distributed models. According to Samaniego et al. (2010b), these approaches can be broadly classified into post-regionalization and simultaneous regionalization approaches, depending on if the regionalization function parameters (or global parameters) are estimated after (Abdulla and Lettenmaier, 1997; Seibert, 1999; Wagener and Wheater, 2006; Livneh and Lettenmaier, 2013) or during the model calibration (Fernandez et al., 2000; Hundecha and Bárdossy, 2004; Gotzinger and Bárdossy, 2007; Pokhrel and Gupta, 2010). None of these procedures consider the subgrid variability of the model parameters or geophysical characteristics. Livneh and Lettenmaier (2013) noted that most of these regionalization procedures exhibit limited transferability because of the use of discrete soil texture classes as predictors, and very likely discontinuous parameter fields.

Recently, a dissimilarity-based regionalization technique was used by Beck et al. (2016) to generate an ensemble of global parameters of the Hydrologiska Byråns Vattenbalansavdelning (HBV) model at a 0.5° resolution for global-scale hydrological modeling. A shortcoming of this approach is the use of ad hoc nearest-neighbor interpolation of parameter fields to fill gaps where no donor basins are available

in (geographically) surrounding regions. Following a similar concept of that of Beck et al. (2016), the parameterization method proposed by Bock et al. (2016) for the contiguous United States (CONUS) will likely lead to discontinuous parameter fields for reasons similar to those mentioned above.

Many attempts have been made in the land surface modeling community to address Dooge's challenges, especially with respect to the transferability of model parameters across locations and scales, and to obtain seamless parameter fields. One of the earliest prominent experiments was conducted in the Project for Intercomparison of Land-surface Parameterizations (PILPS) (Wood et al., 1998). In this project, calibrated LSM parameters were transferred from small catchments to their nearest computational grid cells. The results indicated that LSMs exhibited poor transferability across space, leading to significant differences in the partitioning of water and energy fluxes. For instance, Troy et al. (2008) used calibrated variable infiltration capacity (VIC) model parameters from small basins to generate parameter fields for continental-scale land surface modeling by "linearly interpolating to fill in those grid cell not calibrated" on a sparse grid. As noted by Samaniego et al. (2010b), this type of regionalization is inadequate because of the nonlinearity of soil and geological formations. The spatial patterns of model parameters that would be obtained by ad hoc extrapolations based on calibrated parameters from small basins or grid cells would most likely lead to unrealistic parameter fields with spatial discontinuities circumscribing river basins, as shown in recent studies by Wood and Mizukami (2014) and Mizukami et al. (2017) for the VIC model parameters.

Recent community-driven efforts, such as the Protocol for the Analysis of Land Surface Models (PALS) and the Land Surface Model Benchmarking Evaluation Project (PLUMBER) (Haughton et al., 2016), indicate that the hurdles noted in PILPS have not been overcome. Thus, it is required to gain understanding on whether the inferior predictability of many LSMs evaluated with empirical benchmarks in the PLUMBER project (e.g., CABLE, CHTESSEL, JULES, Noah) may be the result of deficient parameterizations, among other factors.

2.2 Parameterization of soil porosity and available water capacity in selected LSMs/HMs

The above-mentioned challenges that we face in estimating key physical parameters in LSMs/HMs have been intensively discussed in many studies (Gupta et al., 2014; Bierkens et al., 2014; Bierkens, 2015; Clark et al., 2016, 2017; Mizukami et al., 2017; Peters-Lidard et al., 2017). To further visualize the problems and to understand the deficiencies of current parameterization techniques, we selected a representative sample of LSMs/HMs used for research and/or operational purposes, namely CABLE, CLM, JULES, LISFLOOD, Noah-MP, mHM, PCR-GLOBWB, WaterGAP2 (30 arcmin), WaterGAP3 (5 arcmin), CHTESSEL, and HBV. These models vary in process complexity and spatial resolution.

We selected soil porosity as an example to visualize existing shortcomings because it is one of the most common parameters in many LSMs/HMs. This parameter controls the dynamic of several state variables and fluxes such as soil moisture, latent heat, and soil temperature, and its sensitivity has been demonstrated in various studies (Goehler et al., 2013; Cuntz et al., 2015; Mendoza et al., 2015; Cuntz et al., 2016). A representation of the porosity of the top 2 m soil column in these models over the Pan-European domain (Pan-EU) is shown in Fig. 1. The Pan-EU domain was selected for depiction, but we note that the problem is general and persistent across other domains (Mizukami et al., 2017). For cases in which a HM does not use this parameter, the "available water capacity" (WaterGAP) or the "field capacity" (HBV) were selected as a surrogate due to their similarity with porosity. Both surrogate fields are normalized (in space) to ease their comparison with the porosity fields. Soil porosity is expressed in $m^3 m^{-3}$ to ease the comparison among different models.

The following lessons can be learned from Figure 1:

- There is a large variability in the parameterization of this key physical parameter because none of the analyzed models have comparable spatial patterns or comparable estimates at a given location. It should be noted that the definition of the selected parameter is rather simple: it represents the ratio of the volume of voids to the total volume in the soil column. One can now wonder how large the uncertainty of other parameters would be (e.g., hydraulic conductivity) whose relationship with soil properties is very nonlinear.

- The degree of seamlessness strongly depends on the level of aggregation and the upscaling of underlying soil texture fields. For example, the proxy of porosity for WaterGAP is substantially different in spatial pattern and magnitude for 30 arcmin and 5 arcmin simulations. On the contrary, the spatial pattern and magnitude for porosity used in mHM remain almost unchanged for application at 30 and 5 arcmin resolution.

- A parameter field becomes highly discontinuous and patchy when, for a given model, the parameter is calibrated in a limited domain (or basins) and then extrapolated to other regions (e.g., as shown in the panel corresponding to the HBV).

- These experimental results confirm the postulation of Dooge (1982) that the parameterization of the existing state-of-the-art LSMs/HMs at large and continental scales is still an unsolved problem.

The analysis of current parameterization techniques allow us to put forward the following questions:

- Why are there such large differences between models in estimating a parameter that has a physical meaning?

- What are the consequences of poor parameterizations on the spatiotemporal dynamics of state variables and fluxes?

- What are the consequences of model calibration on parameter fields?

- Are current model parameterizations scale invariant?

- Do the fluxes estimated with these models at various scales satisfy the fundamental mass conservation criterion (hereafter denoted as the flux-matching test)?

3 Seamless parameterization framework

3.1 The flux-matching postulation

The key postulation aiming at obtaining scalable (global) parameters that are transferable across locations and scales was proposed by Samaniego et al. (2010b) and further tested in Kumar et al. (2013a, b) and Rakovec et al. (2016b). We hypothesize that flux matching across scales leads to quasi-scale-invariant global parameters $\hat{\gamma}$; thus,

$$\sum_i \sum_t \left| W_i(\hat{\gamma}, t) a_i - \sum_{k \in i} w_k(\hat{\gamma}, t) a_k \right| \to 0, \quad \forall i \in \Omega. \quad (1)$$

Here, k denotes the subgrid elements constituting a given modeling cell i with area a_k. i denotes a modeling grid cell i with area a_i. W_i and w_k denote fluxes at two modeling scales ℓ_1 and ℓ'_1, respectively, with $\left(\frac{\ell_1}{\ell'_1} \right)^2 = \frac{a_i}{a_k}$. Ω denotes the modeling domain, e.g., a river basin, and t a point in time. It should be noted that the topology of the cells at either level is not specified. Normally, rectangular grid cells are used for convenience, but this is not a necessary condition. This strong flux-matching condition can be used as a penalty function or as an additional test to discriminate parameter sets obtained with conventional parameter estimation approaches.

3.2 The MPR approach

MPR, proposed by Samaniego et al. (2010b), aims to estimate model parameters that are seamless across scales, satisfy the flux-matching conditions (see Sect. 3.1), and enable the transferability of global or transfer-function parameters across scales and locations (Samaniego et al., 2010a, b; Kumar et al., 2013a; Wöhling et al., 2013; Livneh et al., 2015; Rakovec et al., 2016a). The development of MPR is ongoing. Regionalization functions used in MPR for the mHM model (www.ufz.de/mhm) by Samaniego et al. (2010a) were further improved by Kumar et al. (2013b). More recently, a model-agnostic implementation of MPR has been proposed

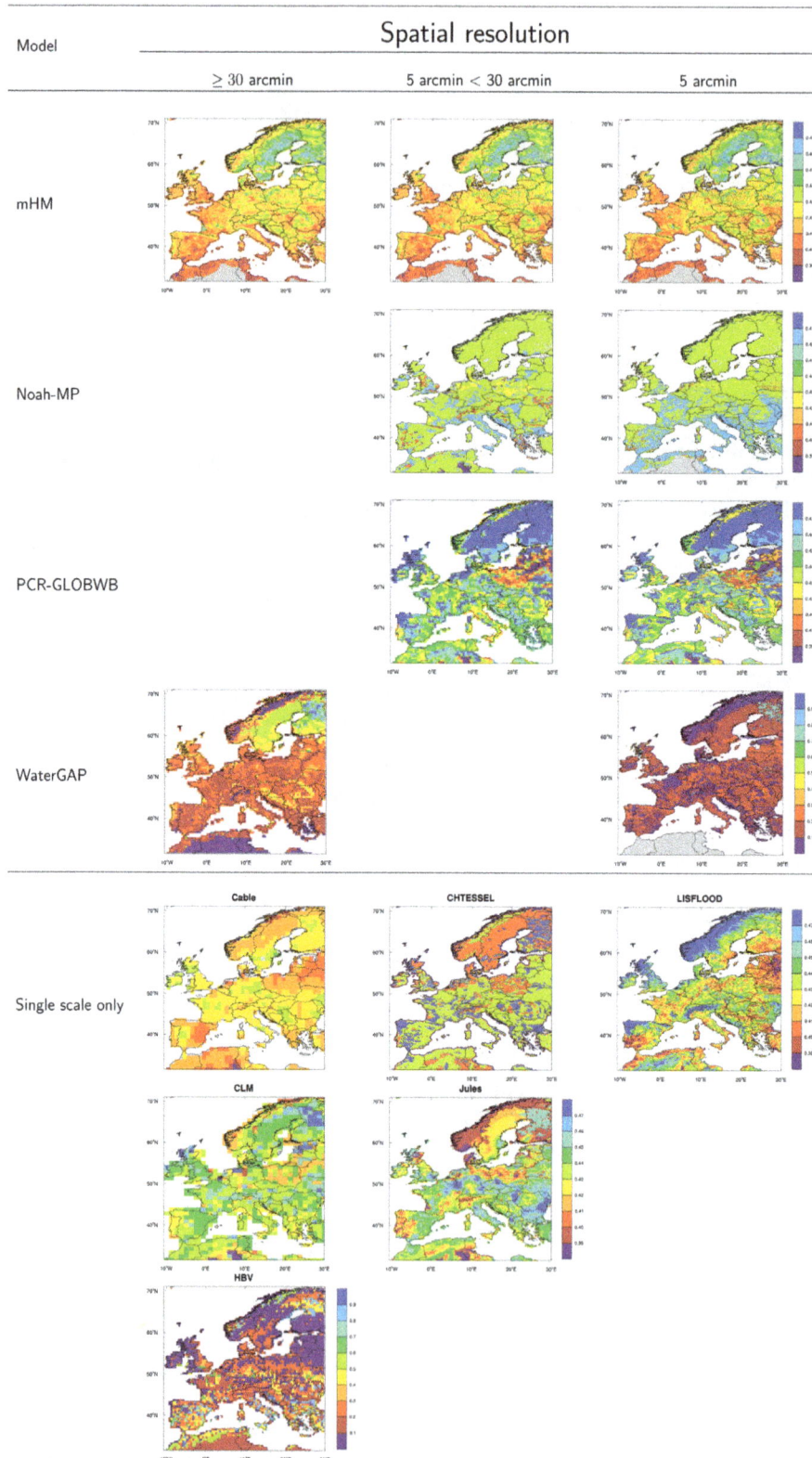

Figure 1. Porosity fields (top 2 m) of typical LSM/HM over Pan-EU at various resolutions: CABLE (1°), CLM (1°), CHTESSEL (0.11°), JULES (35 km), LISFLOOD (EFAS, 5 km), mHM (EDgE-C3S, 5 km), Noah-MP (CORDEX-EU, 0.11°), and PCR-GLOBWB (EDgE-C3S, 5 km). Normalized available water capacity of WaterGAP2 (HyperHydro, 30 arcmin), [3, 536] mm, WaterGAP3 (HyperHydro, 5 arcmin), [1, 960] mm, and HBV [50, 698] mm. In brackets, the normalization values, denoted as [min, max], are provided only for HBV and WaterGAP.

by Mizukami et al. (2017) and tested in the VIC model in over 500+ basins in the CONUS. The study of Mizukami et al. (2017), in contrast to the present study, does not include flux-matching tests nor the evaluation of model skill across different spatial scales.

The scaling problem in MPR is addressed by using process-specific representative elementary areas (REAs) that determine the minimum computational grid size ℓ_1 at which the continuum assumptions can be used without explicit knowledge of the actual patterns of the topography, soil, or rainfall fields (Wood et al., 1988). The REA of a specific process, such as streamflow, can be determined by conducting a careful sensitivity analysis as shown by Samaniego et al. (2010b). To estimate an "effective" model parameter (e.g., total soil porosity) at the selected modeling scale, it is first necessary to estimate its variability at a much finer scale $\ell_0 \ll \ell_1$ such that the effects of its spatial heterogeneity can be adequately represented. In other words, the parameter at the fine scale ℓ_0 represents the minimum support at which the proposed equations are still valid. Barrios and Francés (2011) indicated that a suitable estimate of ℓ_0 for a given parameter could be near its correlation length. The subgrid variability of a parameter β_0 depends, in turn, on the spatial heterogeneity of geophysical and biophysical characteristics ($\mathbf{u}_0$), such as terrain elevation, slope and aspect, soil texture, geological formation, and land cover, which are now available at hyper-resolution for the entire globe. The mathematical relationships that link model parameters with these characteristics at the finer resolution are called pedotransfer, regionalization, or regularization functions f (Clapp and Hornberger, 1978; Cosby et al., 1984; Wösten et al., 2001). The constants required in these functions are usually denoted as global parameters $\hat{\boldsymbol{\gamma}}$; thus, $\boldsymbol{\beta}_0 = f\left(\mathbf{u}_0, \hat{\boldsymbol{\gamma}}\right)$. Note that the fields $\boldsymbol{\beta}_0$ and $\mathbf{u}_0$ are dependent on space and time, but the vector $\hat{\boldsymbol{\gamma}}$ is not.

Regularization functions are commonly used in mathematics and statistics to solve ill-posed problems (which is the case when the parameters of a distributed LSM/HM are determined by calibration) and/or to prevent overfitting. The direct consequence of the regularization is the substantial decrease in degrees of freedom of the optimization problem because the cardinality of the gridded parameter fields #$\{\boldsymbol{\beta}_0\}$ is orders of magnitude larger than that of the vector of the global parameters #$\{\hat{\boldsymbol{\gamma}}\}$. Hence, MPR is a parsimonious parameterization technique that offers spatially continuous parameter fields and removes spatial discontinuities in water fluxes and states, as observed by Gotzinger and Bárdossy (2007) and discussed by Mizukami et al. (2017). From the Bayesian point of view, the regularization functions impose a prior distribution on the model parameters. Consequently, greater care should be taken in their selection.

The second step of the MPR approach consists of upscaling the subgrid distribution of a regionalized parameter to the modeling scale. In other words, $\boldsymbol{\beta}_1 = \langle \boldsymbol{\beta}_0 \rangle$. Here, the symbol $\langle \cdot \rangle$ represents an averaging or scaling operator that is parameter specific, and thus $\boldsymbol{\beta}_1$ denotes the upscaled effective pa-

rameter field. It is important to note that this scaling operator is not necessarily the arithmetic mean.

A schematic representation of the MPR procedure can be seen in Fig. 2. In short, the motto of MPR is "estimate first, then average", whereas other existing regionalization methods follow the opposite approach of "average first, then estimate." Because the processes in LSMs/HMs are highly nonlinear, this sequence of operations does not commute. The consequences can be dramatic (to be shown in the results section). The latter, which is the standard approach, does not preserve fluxes/states across scales, whereas MPR does to a considerable extent. The key question here is in finding the right scaling rule for the model parameters such that the fluxes/states are preserved across a range of spatial scales.

Model parameters at the ℓ_1 scale (i.e., 1 to 100 km) are called "effective" parameters because they cannot be measured by physical means at this resolution and can only be inferred by heuristic relationships $f(\cdot)$. Thus, it is essential that the inequality $\ell_0 \ll \ell_1$ is fulfilled so that the law of large numbers leads to stable estimates of the effective parameter $\boldsymbol{\beta}_1$ having low uncertainty. Since every LSM/HM (e.g., those mentioned in Sect. 2) contains "effective" model parameters, depending on heuristic relationships (that are hidden in the source code in many cases; Mendoza et al., 2015; Cuntz et al., 2016), it is logical that existing LSMs/HMs are subject to parameter uncertainty. These models can be treated as stochastic models, even though their governing equations are deterministic in nature and based on physical principles such as the conservation of mass and energy (Clark et al., 2015; Nearing et al., 2016). Effective parameters should not be the pure result of a blind calibration algorithm. MPR varies from other regionalization approaches in that the introduced relationships may lead to seamless parameter fields and model simulations fulfilling the flux-matching condition.

Currently, MPR is the only method that consistently and simultaneously addresses the scale, nonlinearity, and overparameterization issues if global parameters are estimated simultaneously at multiple locations (i.e., basins). The MPR approach also addresses the principle of scale-dependent subgrid parameterization (i.e., "net fluxes must satisfy the conservation of mass" proposed by Beven, 1995) but does not adhere to Beven's other principles, such as that subgrid parameterizations may be data and scale dependent (principles 3 and 4 in Beven, 1995), because exhaustive tests reported in the above-mentioned references carried out over hundreds of river basins do not appear to support them. We find MPR to be a robust technique that has the ability to provide "effective parameters" and is capable of addressing the scaling problem; in this sense, it diverges from the Beven's view (Beven, 1995, p. 507) that these "effective parameters" are an "inadequate approach to the scale problem". Furthermore, MPR differs on the regionalization and aggregation scheme (i.e., patch model areal weighting) proposed by Beven (1995, p. 520).

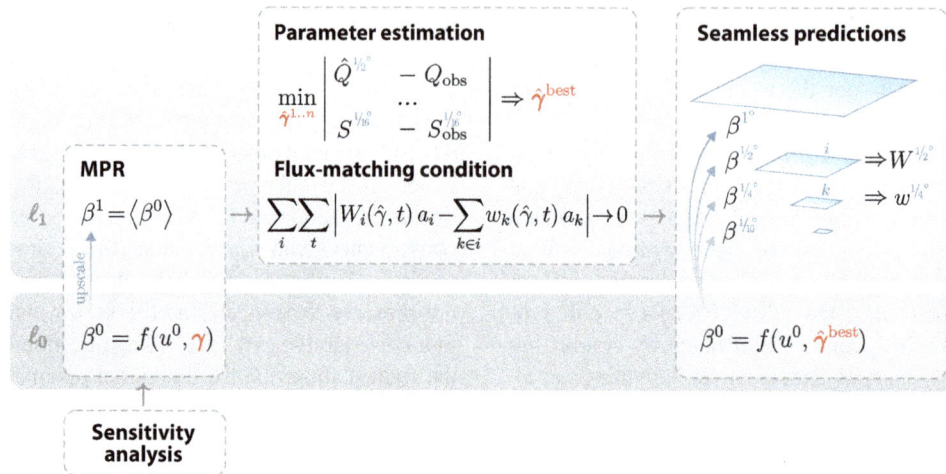

Figure 2. Schematic representation of the proposed seamless prediction framework based on Rakovec et al. (2016b). It includes a preliminary sensitivity analysis, MPR estimation, global-parameter estimation, a flux-matching test, and multiscale seamless prediction. W_i and w_k are the fluxes at the i and k cells of the $1/2°$ and $1/4°$ resolutions, respectively (as an example). Q_{obs} and S_{obs} are the observed time series of streamflow and soil moisture, respectively. The operator $|\cdot|$ is a compromise dissimilarity metric composed of many independent observations at various scales.

The selection of regionalization functions and scaling operators is fundamental to ensuring the transferability of global parameters across scales and to guarantee the seamlessness of parameter fields across scales, e.g., from ℓ_1 to $2\ell_1$ and so on. Samaniego et al. (2010b) proposed that the key to determining them is the flux-matching condition mentioned above. A seamless parameter field $\boldsymbol{\beta}_1$ can be interpreted as the corollary of the flux-matching condition. Moreover, MPR employs geophysical properties at ℓ_0 that allow for a representative sample at the hyper-resolution promoted by Wood et al. (2011) and Bierkens et al. (2014).

3.3 Protocol for implementing the MPR approach

The development of LSMs/HMs and their parameterizations should be guided by a strict hypothesis-driven framework (Nearing et al., 2016) that aims at finding parsimonious and robust parameter sets that fulfill the flux-matching condition and a number of efficiency metrics that are not used during the parameter estimation phase. A multivariate, multiscale evaluation assessing the reliability of model simulations should follow the scheme presented in Rakovec et al. (2016a). Based on our previous experiences, we synthesize a formalized scheme (i.e., protocol) for systematically implementing the MPR technique in other LSMs/HMs with the aim to obtain a robust and seamless parameterization. A graphical depiction of the estimation procedure at multiple scales is shown in Fig. 2.

1. Retrofit the source code of an LSM/HM so that all model parameters are exposed to analysis algorithms. Parameters are the values of a model that can be considered random variables, i.e., those that are subject to

various outcomes and can be fully defined by a probability density function. Parameters should not be confused with numerical or physical constants.

2. Determine a set of the most sensitive model parameters through a sensitivity analysis (SA). For computationally expensive LSMs such as CLM or Noah-MP, computationally frugal methods such as the elementary effects method (Morris, 1991), its enhanced version such as that proposed by Cuntz et al. (2015), or the distributed evaluation of local sensitivity analysis (DELSA; Rakovec et al., 2014; Mendoza et al., 2015) are of particular interest because use of the popular standard Sobol' method (Sobol', 2001) can be computationally expensive although still possible (Cuntz et al., 2016).

3. Regionalize sensitive model parameters that exhibit marked spatial variabilities. The selection of the regionalization function $f(\cdot)$ can be guided by existing literature or by step-wise methods (e.g., Samaniego and Bárdossy, 2005). This regularization step should be conducted at the highest available spatial resolution for all predictor fields. This resolution is denoted as level ℓ_0. The output of the regularization is the parameter field $\boldsymbol{\beta}_0$.

4. Estimate effective parameter fields $\boldsymbol{\beta}_1$ using upscaling operators based on the underlying subgrid variability $\boldsymbol{\beta}_0$. The scale ℓ_1 is determined by synthetic experiments aimed at finding the optimal REA for processes related to the parameter in question (Samaniego et al., 2010b; Kumar et al., 2013b).

5. Estimate the global parameters $\hat{\gamma}$ using standard optimization algorithms (simulated annealing, shuffled complex evolution (SCE), dynamically dimensioned search (DDS)) by minimizing a compromise metric that includes observations at multiple scales and locations (Duckstein and Opricovic, 1980; Rakovec et al., 2016a). The compromise metric could also include hydrologic signatures to extract as much information from a time series as possible (Nijzink et al., 2016).

6. Perform multi-basin, multiscale, multivariate cross-validation tests to evaluate the robustness of the regionalization functions, scaling operators, and global parameters (Rakovec et al., 2016a).

7. Evaluate the parameter seamlessness and the preservation of the statistical moments of fluxes and states across scales (seamless prediction step in Fig. 2).

8. If the cross-validation tests provide satisfactory results (e.g., Kling–Gupta efficiency (KGE) of the compromise solution > 0.6), then evaluate the flux-matching condition given by Eq. (1). If the total error is too large to be tolerated, repeat steps 3 to 8.

It should be noted that any of the steps above can be tested within a sequential hypothesis-testing framework (Clark et al., 2016). A substantial difference from a standard model optimization exercise is that the transfer function $f(\cdot)$ (step 3) and the upscaling operator (step 4) can also be modified in the modeling protocol.

Failure to satisfy the imposed condition, such as the flux-matching test, after exhaustively testing the options in steps 3 to 6 may indicate deficits in process understanding and/or poor data. Consequently, the evaluation step should also provide guidance on detecting and separating the errors stemming from process conceptualization (modeling) and input data.

3.4 Seamless parameter fields across multiple scales using MPR

In Sect. 3.2, it was postulated that the MPR technique aims at estimating seamless parameter fields across scales which minimize the occurrence of artificial discontinuities and ease the transferability of model parameters across scales and locations. The latter has been tested and reported in many studies in Europe, USA, and other basins worldwide (Samaniego et al., 2011; Kumar et al., 2013a, b; Rakovec et al., 2016a, b). In this study, we provide evidence in favor of the former postulation.

To achieve this goal, the mHM model is parameterized using MPR (Samaniego et al., 2010b) with hyper-resolution fields of geophysical characteristics at $\ell_0 = 500$ m resolution as input. Among them, the land cover data were obtained from the Corine datasets (http://land.copernicus.eu/

pan-european/corine-land-cover), and the soil texture information was derived from SoilGrids (soilgrids.org). These very detailed and homogenized soil texture fields provide the fractions of clay and sand, mineral bulk density, and fraction of organic matter for six soil horizons up to 2 m deep. A hyper-resolution digital elevation model (DEM) over Europe (approximately 30 m) from the GMES RDA project (EU-DEM; www.eea.europa.eu/data-and-maps/data/eu-dem) was used to derive terrain characteristics such as slope, aspect, and flow direction. The underlying hydrogeological characteristics are based on the International Hydrogeological Map of Europe (IHME; www.bgr.bund.de/ihme1500), available at a 1 : 1 500 000 scale. Details on the pedotransfer function used for these simulations can be found in Livneh et al. (2015). mHM global parameters were obtained by closing the water balance over selected river basins in Europe (Rakovec et al., 2016a).

Based on these settings, which constitute the basis for the EDgE project (edge.climate.copernicus.eu), we estimated porosity fields at three modeling resolutions of $\ell_1 = 5$, 10, and 25 km, based on the same ℓ_0 support information. Following the MPR procedure depicted in Fig. 2, the parameter fields for the mHM model at these three resolutions can be estimated. Results are shown in Fig. 3.

The results illustrate that the MPR approach can preserve the spatial pattern of the porosity fields (see Fig. 3a, b, and c) and the first and second moments of its probability density function shown in Fig. 3e–g. Two-sample Kolmogorov–Smirnov tests indicate that there is insufficient evidence to reject the null hypothesis that any of the three possible pairs of empirical distributions were drawn from the same unknown distribution. This highlights that the MPR approach leads to consistent parameter fields across scales. In this case, the mean porosity is estimated to be $0.42\,\mathrm{m}^3\,\mathrm{m}^{-3}$ independent of the scale.

3.5 Limitations of the MPR approach

The MPR approach, as any method, has some limitations. One of the crucial aspects of MPR is the selection of transfer functions and upscaling operators. Existing theories could be the first guess, but in the event that nothing is available, the protocol proposed in Sect. 3.3 could be used to guide the search of robust transfer functions. Testing the model parameterization for flux-matching conditions across a range of basin and spatial scales may help to identify adequate upscaling operators. This procedure, although tedious, is the only solution for the moment.

In the event that some state variables change over time (e.g., land cover/use), or during parameter estimation, the MPR algorithm has to be linked to the model because every time a global parameter ($\hat{\gamma}$) is re-estimated, all related model parameters (β_1) have to be updated as illustrated in Fig. 2. The computational cost of performing MPR is there-

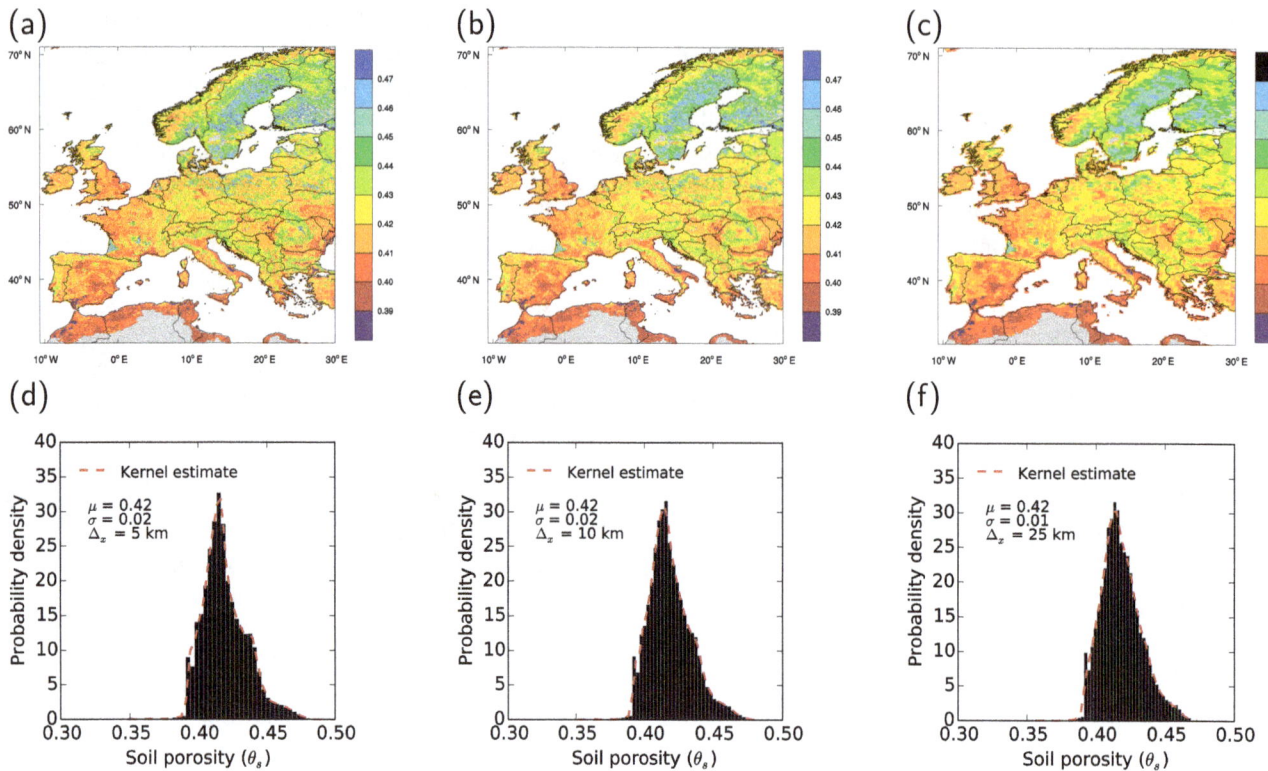

Figure 3. Seamless soil porosity (top 2 m) fields obtained using MPR at three spatial resolutions ℓ_1: **(a)** 5 km, **(b)** 10 km, and **(c)** 25 km, respectively. Lower panels **(d)**–**(f)** show the empirical distribution function of porosity at the respective resolution and method.

fore larger than other parameterization method discussed before.

Another limitation of the applicability of the MPR technique until recently was its availability only as an intrinsic module of the mHM model (www.ufz.de/mhm). This implies that tailored algorithms (i.e., source code) to perform the regionalization and upscaling of parameters for a target LSM/HM have to be developed from scratch, as it is demonstrated here as a case study for the PCR-GLOBWB model. This activity is of course time-consuming and not pleasing due to its complexity. For this reason, Mizukami et al. (2017) have started a community effort to develop a model-agnostic MPR implementation (MPR-flex), which has been so far evaluated for the VIC model.

The availability of high-resolution biophysical characteristics at the spatial scale ℓ_0 constitutes another limitation of the applicability of MPR. Since the subgrid variability is fundamental to estimating robust effective parameter values at coarser scales, the minimum scale at which a model can be applied (ℓ_1) is strongly determined by the data availability. For example, if the soil data are available for the Pan-EU domain at $\ell_0 = 250$ m, the ℓ_1 should not be lower than 1000 m, so that each modeling cell (ℓ_1) has a representative number of underlying subgrid cells (ℓ_0).

MPR has been mainly developed for a hydrologic model representing the water cycle. However, land surface mod-

els also include the energy and carbon cycles and thus have greater complexity. In particular, they have more detailed representation of vegetation. It is a topic for future research to develop a MPR approach (i.e., transfer functions and up-scaling operators) for plant functional-type-specific parameters such as carboxylation rate and the slope of the Ball–Berry equation for stomatal conductance (Ball et al., 1987), which are required for a successful implementation of MPR in LSMs.

Finally, the computational effort for MPR is also considerably larger in comparison with other methods, because of its requirement to estimate model parameters ($\boldsymbol{\beta}_0$) at the highest resolution at which the biophysical characteristics are available. The computational time, however, could be substantially reduced by using a restart file (i.e., a dataset containing a copy of all parameters, state variables, and fluxes of a model at a given point in time). If this capability is available, the MPR estimation can be greatly reduced for operational simulations because the effective parameter fields and past modeled states do not need to be estimated often.

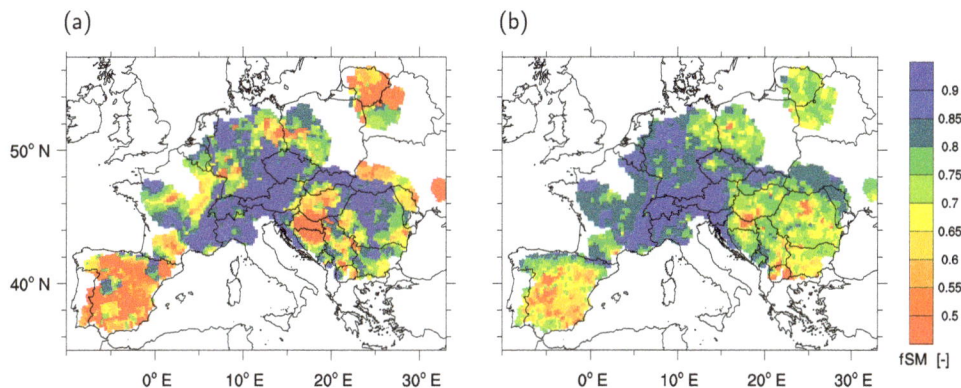

Figure 4. mHM simulations of soil moisture as the fraction from saturation $\frac{\theta}{\theta_s}$ for a day in August 2005 conducted with **(a)** basin-wise parameter estimation and **(b)** seamless parameter estimation. Panel **(b)** shows a seamless soil moisture field.

4 Experiments to reveal non-seamless parameterizations

In this section, we perform four modeling experiments, inspired by Wood (1990)'s recommendation to investigate

- the effects of the overcalibration of global parameters on the spatial patterns of modeled state variables,

- the effects of a parameterization technique on the spatial pattern of effective parameters,

- the effects of a parameterization technique on the dynamics of a state variable, and

- the effects of not satisfying the flux-matching condition on simulated flux across different spatial scales. In these experiments, four models are employed: mHM, Noah-MP, PCR-GLOBWB, and WaterGAP.

4.1 Effects of on-site model calibration

As noted in the introduction, on-site (basin-specific) parameter estimation based on HRU or similar techniques (such as clustering grid cells or sub-basins into regions that exhibit quasi-similar hydrological behavior) leads to non-seamless parameter fields such as those reported in Merz and Blöschl (2004). Here, we go one step further to show the consequences of this common practice on state variables such as soil moisture. Our postulation is that an on-site calibration of global parameters $\hat{\gamma}$ leads to biased state variables even with regularization techniques such as MPR. To falsify this postulation, we performed two model simulations denoted "on-site" and "multisite" calibration schemes. In both cases, we used the mHM setup described in Rakovec et al. (2016b) over the Pan-EU domain at a 0.25° resolution.

In the first simulation, we perform on-site calibrations at 400 river basins in the Pan-European domain. Subsequently, the respective optimized parameter sets are used in each corresponding basin to generate the target variable, in this case,

the daily soil moisture of the top 1 m soil column. Lastly, daily soil moisture fields are assembled using the independent basin simulations for the entire Pan-EU domain. The results of this experiment are shown in Fig. 4a for a day in August 2005. In the second simulation, the global parameters $\hat{\gamma}$ are estimated simultaneously for a set of 13 basins covering various hydroclimatic regimes in the Pan-EU domain. The corresponding soil moisture field for the same point in time is depicted in Fig. 4b.

The first simulation shows clear evidence of strong spatial imprint in the soil moisture fields that is easily identifiable because the shapes of the constituent river basins (Fig. 4a) are apparent. Another interesting feature is a strong wet bias in a basin located in center of the Iberian Peninsula compared to its neighboring regions. Wet soils during this period are very unlikely because the entire region was enduring a prolonged and extreme drought. Moderate dry bias is apparent in basins in southwest Germany, and a strong dry bias was detected in basins in west Croatia, south Lithuania, south Hungary, and north Bosnia and Herzegovina. Conversely, the soil moisture field obtained with the multi-basin parameter estimation does not exhibit these nuisances and thus can be regarded as a spatially seamless field. In this case, parameter estimation with a large sample of geophysical characteristics and many streamflow time series to estimate efficiency measures leads to a well-posed parameter estimation problem.

Based on these results, it can be concluded that parameter sets obtained using the on-site parameter estimation technique do not lead to seamless parameter fields or state variables. Moreover, automatic optimization algorithms, such as SCE or DDS, tend to overlearn from time series with large observational errors, which in turn leads to poor identifiability of parameters (Brynjarsdottir and O'Hagan, 2014) and biased simulations, as demonstrated above. Consequently, parameter estimation should be performed with a representative sample of basins that adequately cover the variability of hydrological regimes and geophysical properties (e.g., soil types) (Kumar et al., 2015). It is worth noting that if the pa-

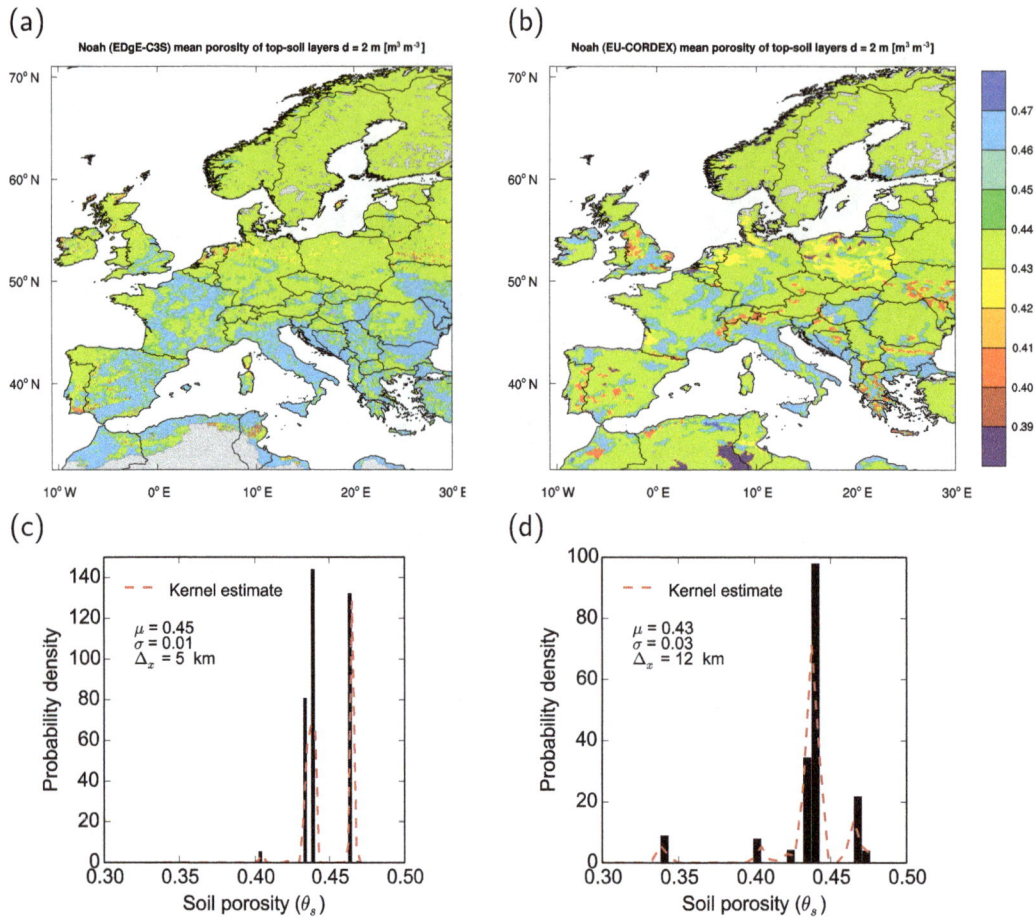

Figure 5. Porosity fields obtained using the majority upscale operator for spatial resolutions of (a) 5 km and (b) 12 km with the Noah-MP model used in the EDgE and EURO-CORDEX projects, respectively. Lower panels (c)–(d) show the empirical distribution function of porosity at the respective resolution and method.

rameters of a model are estimated in a small basin with very few soil types, a single geological formation, or very flat terrain, then it is very likely that some parameters cannot be constrained during calibration. The obtained parameter set is biased to the specific basin in which it has been estimated, and hence it is not skillful for seamless and continental-scale simulations.

4.2 Effects of a parameterization technique on spatial patterns of effective parameters

The effects of the commonly used parameterization techniques to generate the porosity fields of LSMs (such as CHTESSEL and Noah-MP depicted in Fig. 1) are important to investigate. These fields are obtained by combining the majority (or dominant) upscaling operator and a look-up table containing categorical values of model parameters tabulated for a limited set of dominant soil types (e.g., Niu, 2011, p. 20., ECMWF, 2016, p. 137). The majority-based operator is mostly used for estimating grid-specific vegetation classes in LSMs (Li et al., 2013).

The porosity field, based on a majority upscaling for the Noah-MP model used in EURO-CORDEX (www.euro-cordex.net) at an approximately 12 km resolution, is depicted in Fig. 1. Compared with the other model-derived porosity fields, the Noah-MP field appears to be most homogeneously distributed in space. It is very likely that the spatial heterogeneity is underrepresented in this case as the default soil LUT contains only 13 soil classes. It should be noted that a model such as CABLE that uses a porosity field with an approximately 100 km resolution has a larger variability than that of Noah-MP at 12 km.

The following experiment is carried out to evaluate whether the variability of the soil map or the upscaling operator has a larger effect on the derived porosity field. The highest resolution soil map available for Europe is used and applied in the same manner to derive porosity fields as described above. The texture field is provided by the SoilGrids dataset (http://soilgrids.org) at 1000 m resolution (level-0). The upscaled porosity field is generated at 5 km for the EDgE project. The soil characteristics for

Noah-MP are estimated using the same look-up table as in the EURO-CORDEX–Noah-MP case. The comparison of both parameter fields (i.e., EDgE–Noah-MP and EURO-CORDEX–Noah-MP) and the main statistical moments describing the spatial variability of the porosity fields are shown in Fig. 5. The results clearly indicate the inappropriateness of the majority-based upscaling operator for this parameter in both cases. It leads to reduction of the variance of the porosity field and thus can be considered the least sensitive operator. This means that the informational content of the hyper-resolution soil maps, commonly available globally, is almost lost.

Notably, although the overall mean of the porosity estimated using MPR over the Pan-EU domain for mHM (Fig. 3a) is only 6.6 % lower than that calculated using the majority-based approach for Noah-MP (Fig. 5a), the spatial patterns obtained by both models are very different. The evidence of this remarkable dissimilarity can also be visualized by comparing the empirical density functions shown in Figs. 3d and 5c, both corresponding to a field at $\ell_1 = 5$ km and with the same input data. A detailed evaluation conducted by Samaniego et al. (2012) in Germany showed that large porosity values estimated with the majority-based approach could overestimate those obtained with MPR by up to 40 %, whereas in other locations, underestimation up to 15 % from those estimated by MPR can be found.

Other upscaling operators, such as the weighted arithmetic mean, are commonly used in LSMs in combination with the mosaic approach. For example, in CLM (Oleson et al., 2013, see p. 160), the texture class of the subunits of the cell, called tiles, are provided in a look-up table. The upscaled porosity field obtained using this approach is shown in Fig. 1 at a 1° (100 km) resolution. Methods based on the majority and weighted arithmetic mean operators exhibit some similarity and lack spatial variability. In both cases, the spatial mean is approximately 0.43 m^3 m^{-3}.

Hydrologic models that do not use soil porosity tend to use a similar conceptualization and values denoted as the total available water capacity (TAWC; WaterGAP versions 2 and 3) and field capacity (FC; HBV). For these types of conceptual models, normalized values of these parameters are used as surrogates for soil porosity. The consistency of the spatial patterns of TAWC and FC are compared here instead of their actual values. A distinctive difference in the patterns can be observed. For example, WaterGAP3 exhibits lower values than WaterGAP2, whereas the pattern of the normalized FC in HBV is the opposite in many locations (e.g., Spain, Germany, and Scandinavia).

Details of the parameterization schemes used to estimate TAWC and FC are beyond the scope of this study. Interested readers may refer to Müller Schmied et al. (2014) or Beck et al. (2016), respectively. However, the TAWC in WaterGAP is obtained by linking the soil type provided by the FAO soil map with available water capacity values estimated by Batjes (1996). Thus, no scaling rule or form of regularization is used

in this case. The field capacity parameters used in HBV were determined using an ad hoc nearest-neighbor interpolation technique that relies on calibrated parameters from nearby similar donor basins that might exhibit very different geophysical characteristics. The parameter fields obtained for two versions of WaterGAP (30 and 5 arcmin) and HBV are depicted in Fig. 1. It can be concluded that the parameterization technique employed is not scale invariant as revealed by distinct parameter sets from WaterGAP model versions, which are operated at different resolutions. The regionalization proposed by Beck et al. (2016) leads to a patchwork-quilt field that does not resemble to any other field presented. Evident from Fig. 1, the HBV field lacks seamlessness that may result in non-seamless fields of water fluxes and states.

4.3 Effects of a parameterization technique on the dynamics of a state variable

There is a complex interplay between soil moisture (SM) and latent heat (LH) in LSMs/HMs. Improving our understanding of soil–land–atmosphere feedback is fundamental for making reliable predictions of water and energy fluxes. In this context, we carry out a sensitivity experiment to investigate the effects of soil-related parameterizations (e.g., soil porosity) on latent heat and soil moisture. Two contrasting modeling paradigms (Noah-MP and mHM) are employed.

The WRF/Noah-MP system is forced with ERA-Interim at the boundaries of the rotated CORDEX grid (www.meteo.unican.es/wiki/cordexwrf) at a spatial resolution of 0.11° covering Europe from 1989 to 2009. To ease the comparison, the process-based hydrological model mHM (www.ufz.de/mhm) is driven with daily precipitation and temperature fields generated by the WRF/Noah-MP system during the same period. The spatial resolution of mHM is fixed at 5×5 km^2. The main geophysical characteristics in WRF/Noah-MP of land cover and soil texture are represented with a 1×1 km^2 MODIS and a single-horizon, coarse-resolution FAO soil map with 16 soil texture classes, respectively. The porosity field of Noah-MP is estimated by applying a majority-based operator to values for different soil classes, as shown in Fig. 5b.

The settings of the mHM model used in this experiment are described in Sect. 3.4. In contrast to those of Noah-MP, the global parameters of mHM estimated using the MPR technique are obtained by closing the water balance over selected river basins in Europe (Rakovec et al., 2016a). The porosity fields obtained for mHM over the Pan-EU are depicted in Fig. 3.

The phase diagrams of the monthly fraction of soil water saturation fSM = $\frac{\theta}{\theta_s}$ (i.e., plots of monthly fSM(t) vs. fSM$(t + 1)$) are subsequently investigated to understand the effect of differences in porosity estimates of the top 2 m soil column on the soil moisture dynamics (Fig. 6). Two locations in Germany are selected in which Noah-MP systematically over- or underestimated the latent heat fluxes with

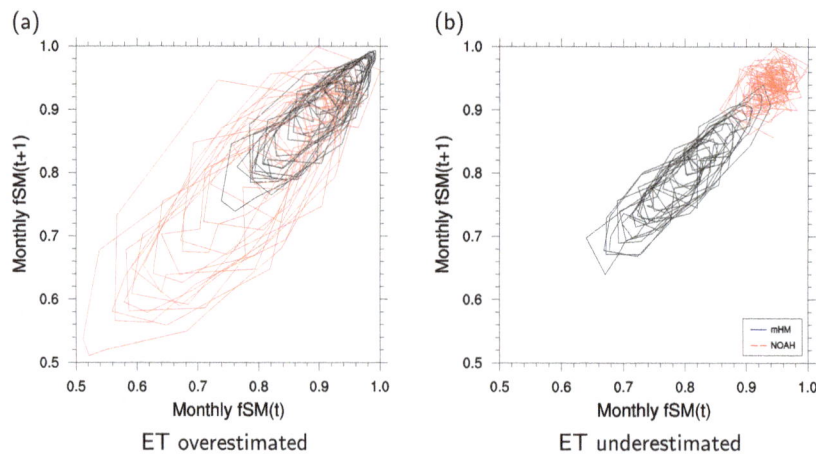

Figure 6. Phase diagrams of monthly soil moisture fraction for two locations in Germany, **(a)** 54° N, 10° E and **(b)** 51° N, 7° E, in which the latent heat estimated by Noah-MP is over- or underestimated with respect to corresponding estimates of mHM. The models have identical forcings.

respect to mHM (the latitude and longitude coordinates of the center of the selected Noah-MP grids are A (54° N, 10° E) and B (51° N, 7° E), respectively). At location A, the majority-based approach underestimates the MPR soil porosity by $-10\,\%$, whereas in location B, it overestimates it by $40\,\%$. This experiment unambiguously shows that, at locations where Noah-MP overestimates latent heat with respect to mHM, the temporal variance (i.e., dynamic) of the monthly SM time series simulated by Noah-MP is almost doubled compared to that of mHM, leading to much lower soil moisture values (Fig. 6a). Conversely, underestimation of latent heat greatly reduces the variance of the soil moisture dynamics (Fig. 6b).

4.4   Effects of not satisfying the flux-matching condition

In Sect. 2, we postulated that ad hoc parameterization schemes do not necessarily fulfill the flux-matching test performed with a flux simulated by a given model at two modeling resolutions ($\ell_1 = 5$ and 30 arcmin). A detailed description of how to perform this test is provided in Samaniego et al. (2010b). The following experiment is conducted with three models (mHM, PCR-GLOBWB, and WaterGAP) in an attempt to falsify the above postulation. All models use the same forcings and geophysical information. The simulations are conducted in the Rhine River upstream of the Lobith gauging station. All three models are driven by daily forcing with a spatial resolution of 5 km, which was kindly provided by the EFAS team at JRC (www.eea.europa.eu). Additional details of the modeling settings of this experiment are provided in Sutanudjaja et al. (2015) and at www.hyperhydro. org/. The KGE and bias values of these three models obtained for both scales at the Lobith station during 2003 are reported in Table 2. The daily streamflow time series during this year is selected for evaluation because it exhibits strong

temporal dynamics, with wet conditions in the beginning of the year followed by a drought during the summer and fall seasons. The performances obtained for the three models are satisfactorily, but the results shown in Table 2 indicate that mHM is the only model that can have higher KGE values regardless of the spatial modeling resolution.

The flux-matching test presented in Sect. 3.1 is performed with simulated evapotranspiration (ET) because it is the largest flux in the water cycle besides precipitation, and is prone to the largest predictive uncertainties (Mueller et al., 2013). To ease the comparison, collocated grids are employed for every model such that every coarser scale grid cell has exactly the same number of underlying cells at finer resolution (5 arcmin). The results of this test are shown in Fig. 7a, b. They reveal that mHM exhibits the best flux-matching between these two scales. This experiment also shows that the MPR technique implemented in mHM leads to ET fields that are of similar magnitude at both scales, indicating a close conservation of mass leading to the lowest relative errors (Fig. 7c) among the three models.

The PCR-GLOBWB and WaterGAP models reveal large inconsistencies in preserving the spatial pattern of annual ET across two modeling scales, although the streamflow performance at the outlet is good (greater than 0.83 in both cases). PCR-GLOBWB at coarse resolution tends to underestimate ET (up to 50 %) compared with those at finer resolution (Fig. 7f). Conversely, the coarser version of WaterGAP tends to overestimate ET (up to 60 %) compared with those at the finer resolution (Fig. 7i). Interestingly, it can be observed that changes in model resolution affect the dynamic of water fluxes in those models that do not use any consistent scaling rules for model parameterization. These results also confirm the postulation that "streamflow-related metrics are a necessary but not sufficient condition to warrant the proper partitioning of incoming precipitation P into various spatially

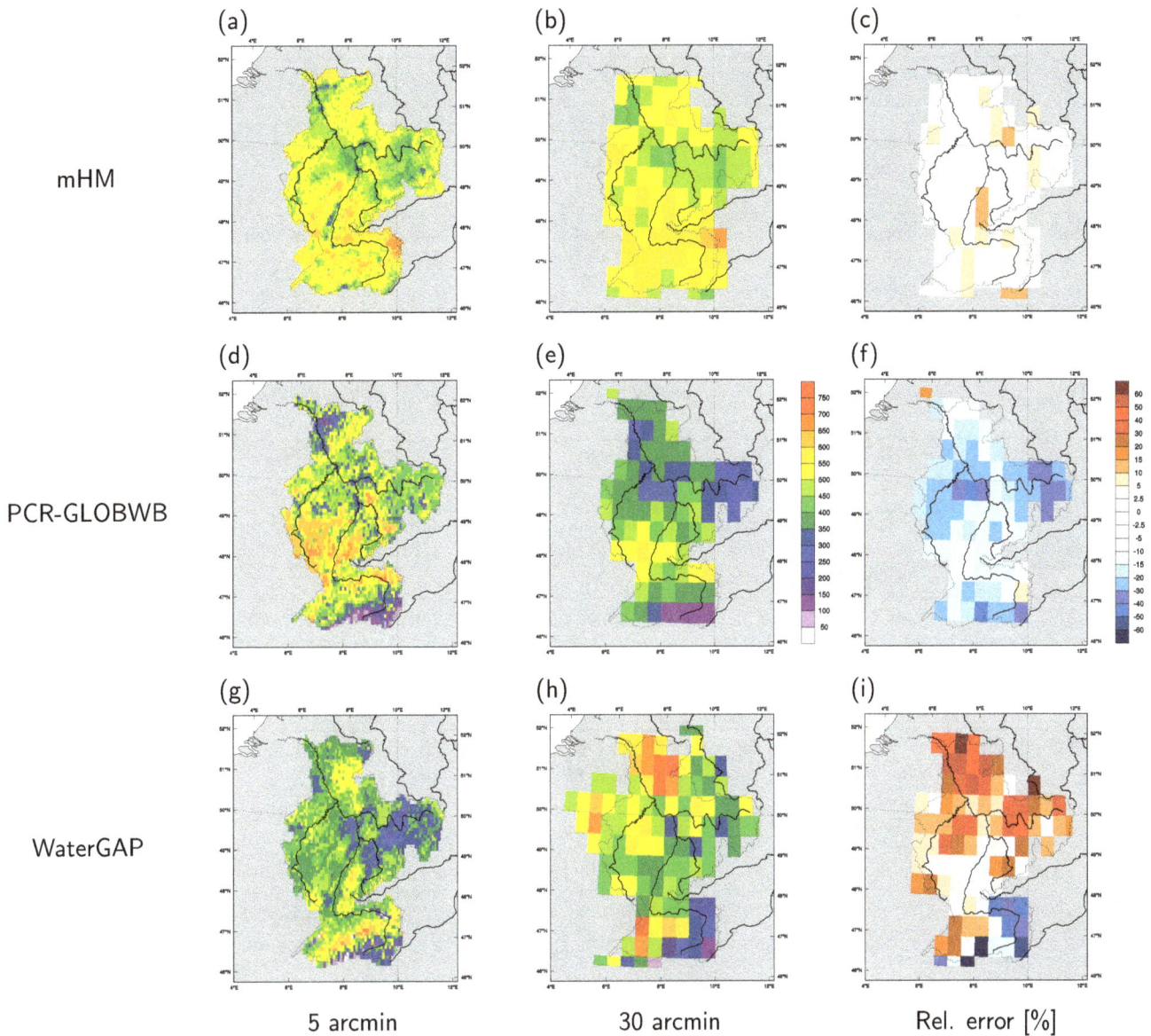

Figure 7. Multiscale simulation of annual ET for the Rhine River in 2003 with mHM, PCR-GLOBWB, and WaterGAP (versions 3 and 2) at spatial resolutions ℓ_1 of 5 and 30 arcmin, respectively. The relative errors in percentage of the coarse field estimates with respect to the finer ones (aggregated to the coarser level) for mHM, PCR-GLOBWB, and WaterGAP are shown in panels **(c)**, **(f)**, and **(i)**, respectively.

distributed water storage components (e.g., SM) and fluxes (e.g., ET)" (Rakovec et al., 2016b). Because all models are forced with the same forcings, share the same geophysical information, and have almost similar hydrological process descriptions, it can be safely concluded that the parameterization method used in the models caused the ET mismatch. To falsify this postulation, the MPR parameterization protocol proposed in Sect. 3.3 is next applied to PCR-GLOBWB.

5 Implementation of the parameterization protocol in PCR-GLOBWB

To evaluate the consistency of land surface fluxes before and after MPR implementation, we analyze the impact of MPR on evaporative fluxes and soil moisture content in PCR-GLOBWB (van Beek et al., 2011; Wada and Bierkens, 2014; Sutanudjaja et al., 2016) over the Rhine River basin during 2003. The model is used to simulate the hydrological states at two different spatial resolutions ($\ell_1 = 5$ and 30 arcmin), and the sensitivity to MPR implementation is evaluated using a field difference method (in line with Eq. 1):

Table 1. Data sources and parameterization methods used by models used in this study.

Model	Parameterization method	References	Source code and projects
CABLE	Pedotransfer functions, look-up table, dominant soil type	Kowalczyk et al. (2006)	http://www.cawcr.gov.au/technical-reports/CTR_057.pdf
CLM	Pedotransfer functions, look-up table, mosaic approach	Oleson et al. (2013)	www.cesm.ucar.edu/models/cesm1.2/clm/
CHTESSEL	Look-up table, dominant soil type	Viterbo and Beljaars (1995); ECMWF (2016)	www.ecmwf.int/search/elibrary
HBV	k-NN interpolation, calibrated parameter	Beck et al. (2016)	www.gloh2o.org/hbv-simreg/
JULES	Look-up table, dominant soil type	Best et al. (2011)	http://jules.jchmr.org
LISFLOOD	Pedotransfer functions, mosaic approach, arithmetic mean	De Roo and Wesseling (2000)	http://publications.jrc.ec.europa.eu/repository/bitstream/JRC78917/lisflood_2013_online.pdf
mHM	MPR	Samaniego et al. (2010b)	http://edge.climate.copernicus.eu www.ufz.de/mhm
Noah-MP	Look-up table, dominant soil type	Niu (2011)	www.jsg.utexas.edu/noah-mp www.meteo.unican.es/wiki/cordexwrf
PCR-GLOBWB	(Original) pedotransfer functions with averaged predictors (New) MPR	van Beek et al. (2011); Wada and Bierkens (2014) Samaniego et al. (2010b)	http://pcraster.geo.uu.nl/projects/applications/pcrglobwb/
WaterGAP (v.2, v.3)	Look-up tables	Müller Schmied et al. (2014); Batjes (1996)	www.uni-kassel.de/einrichtungen/en/cesr/research/projects/active/watergap.html www.uni-frankfurt.de/45218063/WaterGAP

Table 2. Efficiency of mHM, PCR-GLOBWB (Ludovicus et al., 2017), and WaterGAP obtained for the Rhine Basin at the Lobith station during 2003 for spatial resolutions of 5 and 30 arcmin.

Model	5 arcmin		30 arcmin	
	KGE	Bias ($m^3 s^{-1}$)	KGE	Bias ($m^3 s^{-1}$)
mHM	0.96	61.19	0.96	21.74
PCR-GLOBWB	0.93	−20.61	0.86	248.09
WaterGAP (v.3, v.2)	0.83	143.02	0.90	−41.99

$$\Delta = \sqrt{\frac{1}{T}\sum_{t=1}^{T}\left(100\frac{W(t)-w(t)}{w(t)}\right)^2}, \qquad (2)$$

where W and w are the coarse and fine resolution simulations of variable W, respectively, and T is the total time series length.

The original PCR-GLOBWB parameterization does not include consistency in upscaling as enforced by MPR, leading to a larger difference in soil properties. Figure 8 depicts the porosity fields of this model before and after the implementation of MPR. Figure 8a and b clearly show the problems mentioned in Sect. 2, for example, lack of coherence in spatial patterns and the existence of spatial discontinuities of parameter fields at two scales. The porosity fields obtained with the MPR technique shown in Fig. 8c and d, on the contrary, exhibit a typical seamless spatial structure in which the main features of the field can be distinguished across scales.

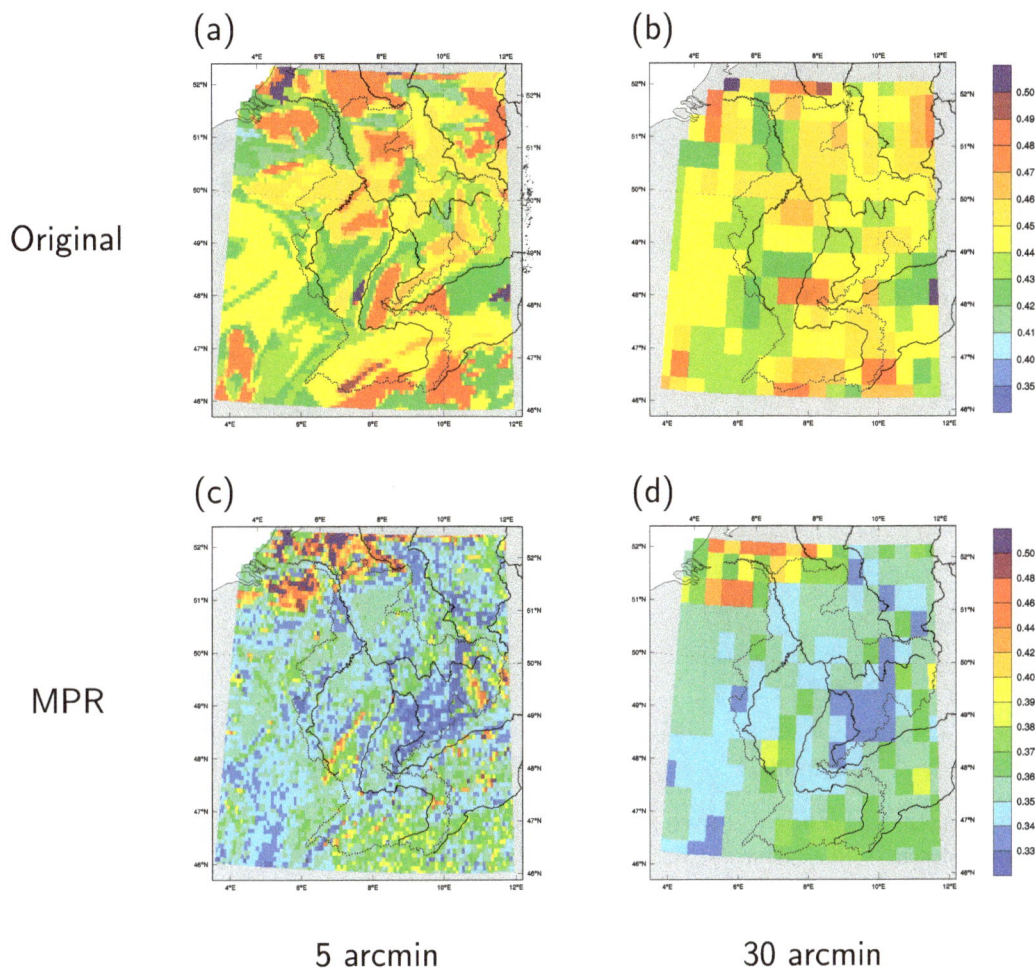

Figure 8. Porosity fields of PCR-GLOBWB before **(a, b)** and after implementing MPR **(c, d)** for two spatial resolutions of 5 and 30 arcmin. Dotted lines denote the Rhine Basin and the continuous line is the main EU river basin network.

It is worth noting that differences seen between Fig. 8a and c are not only due to the improved upscaling procedure, but also due to a modified pedotransfer function. The parameters of the pedotransfer function have also been included in the calibration within the MPR approach.

These differences in soil hydraulic properties influence the derived hydrological properties, leading to changes in saturated conductivity and storage capacity in the unsaturated zone. The considerable differences in ET fluxes are shown in Fig. 9a and b, and are the result of these changes. When MPR is employed, we observe that the difference in actual average Rhine Basin evapotranspiration between the two scales Δ drops from 29 to 9.4 % (Fig. 9d, e). For the total column soil moisture, we find a stronger decrease in Δ from 25 to 6.9 %, clearly indicating the benefits of MPR implementation. The error fields in Fig. 9c and f show a clear benefit of implementing MPR in PCR-GLOBWB. It should be noted, however, that the improvements are not as high as those obtained for mHM as shown in Fig. 7c. This is related to the fact that all effective parameters related to the

evaporation and soil dynamic processes have been scaled with MPR in mHM, whereas in PCR-GLOBWB, only soil porosity has been scaled with this technique. Nevertheless, it is remarkable to see the improvements in flux matching (Fig. 9f) by scaling a single parameter of PCR-GLOBWB using MPR. We also observe a slight increase in the discharge performance (KGE) at Lobith. The original KGEs are 0.86 ($\ell_1 = 5$ arcmin) and 0.93 ($\ell_1 = 30$ arcmin), whereas the KGEs with MPR implementation are 0.91 and 0.93, respectively. Another advantage is that PCR-GLOBWB is calibrated at a coarser resolution, whereas this model is calibrated for each spatial resolution individually in the original setup and with lower consistency in the discharge simulation.

From these evaluations, we conclude that MPR implementation leads to significant improvement in the flux-matching and discharge simulations across scales, allowing for more consistency across scales for hydrological model simulations. Notably, additional parameters in PCR-GLOBWB still need to be regionalized within the MPR framework, which

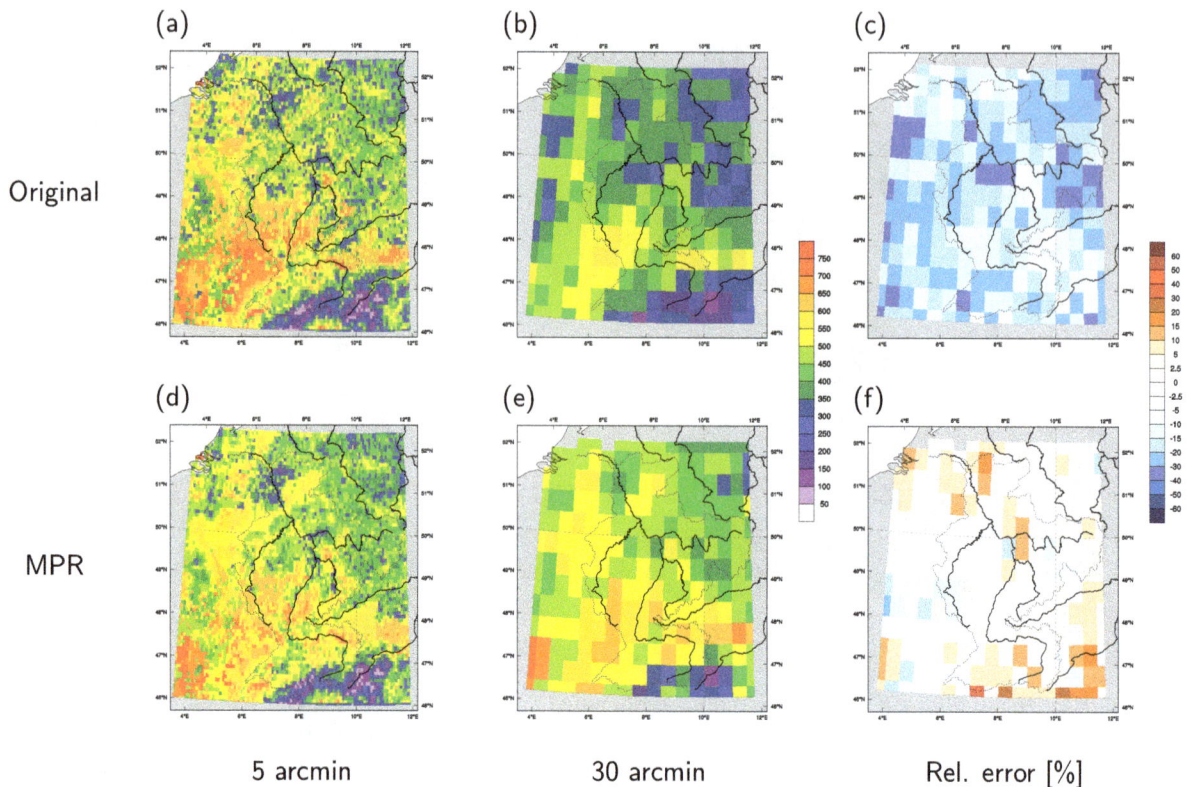

Figure 9. Annual ET fields in 2003 of PCR-GLOBWB before **(a, b)** and after implementing MPR **(c, d)** for two spatial resolutions of 5 and 30 arcmin. Dotted lines denote the Rhine Basin and the continuous line is the main EU river basin network. The relative errors in percentage of the coarse field estimates with respect to the finer ones are shown in panels **(c)** and **(f)**, respectively.

could potentially lead to better performance and transferability.

6 Conclusions

Hyper-resolution modeling initiatives (Wood et al., 2011; Bierkens et al., 2014) challenge the hydrological community to intensify efforts to make water (quantity and quality) and energy flux predictions "everywhere" and for these predictions to be "locally relevant." The predictions should have small uncertainties to be useful for the end users. These grand challenges also imply that the next generation of land surface and hydrologic models must incorporate probabilistic descriptions of the subgrid variability of geophysical land surface properties – such as POLARIS (Chaney et al., 2016b) and SoilGrids (Hengl et al., 2017) – to cope with the large uncertainties that characterize the related process below the representative elementary area (REA) scale. Consequently, great efforts should be made in hyper-resolution monitoring at the global scale in improving the computational efficiency of LSMs/HMs and in the development of scale-invariant parameterizations for these models. In this study, we have shown that the state-of-the-art parameterizations need to be

improved to address this grand challenge, especially with respect to better fulfill the flux-matching condition.

We revisited a technique called MPR (Samaniego et al., 2010b), originally available only in mHM but recently implemented in PCR-GLOBWB as a part of this study. Moreover, we proposed a "parameterization protocol" as a guideline to apply MPR and to retrofit existing LSMs/HMs to ease the implementation of MPR in the latter. We also discuss the advantages and limitations of MPR which should be considered while applying this concept to other LSMs/HMs.

This study has shown that two models that use ad hoc parameterizations can have reasonable efficiency with respect to simulated streamflow but poor performance with respect to distributed fluxes such as evapotranspiration. The implementation of this protocol in PCR-GLOBWB in this study increased the model efficiency by almost 6 % and improved the consistency of simulated ET fields across scales. For example, the estimation of evapotranspiration without MPR at 5 and 30 arcmin spatial resolutions for the Rhine River basin resulted in a difference of approximately 29 %. Applying MPR reduced this difference to 9 %. For total soil water, the differences without and with MPR are 25 and 7 %, respectively. We have also shown that the PCR-GLOBWB global parameters can be transferred across scales with consistent ET patterns and model efficiency.

In general, it can be concluded that the estimation of global parameters is feasible with MPR and that these scalars are transferable across scales and locations. The successful application of MPR implies that the averaging procedure of geophysical properties matters and that having the right physics with incorrect "effective" parameters leads to inconsistent fluxes and states. Consequently, MPR is a step forward to quasi-scale-invariant parameterizations and is feasible to implement in existing LSMs/HMs whose goal should be seamless parameter fields across scales that do not exhibit artificial spatial "discontinuities" such as calibration imprints, and that lead to consistent predictions across scales. We consider that this feature is the key for the next generation of LSM and NWP models such as the model for prediction across scales (MPAS) (www.mmm.ucar.edu) and the nested-domain ICON (www.earthsystemcog.org/projects/dcmip-2012/icon-mpi-dwd). Furthermore, a proper implementation of MPR in process-based (conceptual) models may contribute to recent efforts towards identifying their "effective" parameters through observational datasets at the scale of interest (Savenije and Hrachowitz, 2017).

Finally, we would like to reiterate that a flux obtained from a land surface/hydrologic model should always be evaluated with local observations when available and across scales. If "it disagrees with the experiment, it's wrong."

Competing interests. The authors declare that they have no conflict of interest.

Special issue statement. This article is part of the special issue "Observations and modeling of land surface water and energy exchanges across scales: special issue in Honor of Eric F. Wood". It is a result of the Symposium in Honor of Eric F. Wood: Observations and Modeling across Scales, Princeton, New Jersey, USA, 2–3 June 2016.

Acknowledgements. We kindly acknowledge our data providers: Noah-MP: Kirsten Warrach-Sagi (University of Hohenheim), PCR-GLOBWB: Niko Wanders (Princeton University), WaterGAP: Hannes Müller Schmied (University of Frankfurt), JULES: Anne Verhoef (The University of Reading), LISFLOOD: Peter Salamon (JRC), CABLE: Matthias Cuntz (formerly UFZ, now INRA), CLM: David Lawrence (UCAR) and Edwin H. Sutanudjaja and Marc F. P. Bierkens, Hannes Müller Schmied, Stephanie Eisner, Oldrich Rakovec for providing results from the HyperHydro WG1 Workshop 9–12 June 2015 in Utrecht. This study was carried out within the Helmholtz Association climate initiative REKLIM (www.reklim.de). This work has been partly funded by Helmholtz Alliance EDA – Remote Sensing and Earth System Dynamics, through the Initiative and Networking Fund of the Helmholtz Association, Germany. This study was partially performed under a contract for the Copernicus Climate Change Service (http://edge.climate.copernicus.eu). ECMWF implements this service and the Copernicus Atmosphere Monitoring Service on behalf of the European Commission. We thank Martin Schrön for kindly contributing to the artwork of Fig. 2. Rens van Beek is acknowledged for providing support with the MPR implementation in PCR-GLOBWB. Finally, we would like to thank the editor (Murugesu Sivapalan, Ross Woods) and two anonymous reviewers for their insightful and helpful comments.

Edited by: Murugesu Sivapalan

References

Abdulla, F. and Lettenmaier, D.: Development of regional parameter estimation equations for a macroscale hydrologic model, J. Hydrol., 197, 230–257, 1997.

Addor, N., Rössler, O., Köplin, N., Huss, M., Weingartner, R., and Seibert, J.: Robust changes and sources of uncertainty in the projected hydrological regimes of Swiss catchments, Water Resour. Res., 50, 7541–7562, 2014.

Andréassian, V., Bourgin, F., Oudin, L., Mathevet, T., Perrin, C., Lerat, J., Coron, L., and Berthet, L.: Seeking genericity in the selection of parameter sets: Impact on hydrological model efficiency, Water Resour. Res., 50, 8356–8366, 2014.

Ball, J. T., Woodrow, I. E., and Berry, J. A.: A Model Predicting Stomatal Conductance and its Contribution to the Control of Photosynthesis under Different Environmental Conditions, in: Progress in Photosynthesis Research, Springer, the Netherlands, Dordrecht, 221–224, 1987.

Barrios, M. and Francés, F.: Spatial scale effect on the upper soil effective parameters of a distributed hydrological model, Hydrol. Process., 26, 1022–1033, 2011.

Batjes, N. H.: Development of a world data set of soil water retention properties using pedotransfer rules, Geoderma, 71, 31–52, 1996.

Beck, H. E., van Dijk, A. I. J. M., de Roo, A., Miralles, D. G., McVicar, T. R., Schellekens, J., and Bruijnzeel, L. A.: Global-scale regionalization of hydrologic model parameters, Water Resour. Res., 52, 3599–3622, 2016.

Beldring, S., Engeland, K., Roald, L. A., Sælthun, N. R., and Voksø, A.: Estimation of parameters in a distributed precipitation-runoff model for Norway, Hydrol. Earth Syst. Sci., 7, 304–316, https://doi.org/10.5194/hess-7-304-2003, 2003.

Best, M. J., Pryor, M., Clark, D. B., Rooney, G. G., Essery, R. L. H., Ménard, C. B., Edwards, J. M., Hendry, M. A., Porson, A., Gedney, N., Mercado, L. M., Sitch, S., Blyth, E., Boucher, O., Cox, P. M., Grimmond, C. S. B., and Harding, R. J.: The Joint UK Land Environment Simulator (JULES), model description – Part 1: Energy and water fluxes, Geosci. Model Dev., 4, 677–699, https://doi.org/10.5194/gmd-4-677-2011, 2011.

Beven, K.: Linking parameters across scales: Subgrid parameterizations and scale dependent hydrological models, Hydrol. Process., 9, 507–525, 1995.

Bierkens, M. F. P.: Global hydrology 2015: State, trends, and directions, Water Resour. Res., 51, 4923–4947, 2015.

Bierkens, M. F. P., Bell, V. A., Burek, P., Chaney, N., Condon, L. E., David, C. H., de Roo, A., Döll, P., Drost, N., Famiglietti, J. S., Flörke, M., Gochis, D. J., Houser, P., Hut, R., Keune, J., Kollet,

S., Maxwell, R. M., Reager, J. T., Samaniego, L., Sudicky, E., Sutanudjaja, E. H., van de Giesen, N., Winsemius, H., and Wood, E. F.: Hyper-resolution global hydrological modelling: what is next? "Everywhere and locally relevant", Hydrol. Process., 29, 310–320, https://doi.org/10.1002/hyp.10391, 2014.

Blöschl, G., Reszler, C., and Komma, J.: A spatially distributed flash flood forecasting model, Environ. Modell. Softw., 23, 464–478, 2008.

Blöschl, G., Sivapalan, M., Wagener, T., Viglione, A., and Savenije, H. (Eds.): Runoff Prediction in Ungauged Basins: Synthesis Across Processes, Places and Scales, Cambridge University Press, ISBN: 978-1107028180, 2013.

Bock, A. R., Hay, L. E., McCabe, G. J., Markstrom, S. L., and Atkinson, R. D.: Parameter regionalization of a monthly water balance model for the conterminous United States, Hydrol. Earth Syst. Sci., 20, 2861–2876, https://doi.org/10.5194/hess-20-2861-2016, 2016.

Bonan, G. B., Lawrence, P. J., Oleson, K. W., Levis, S., Jung, M., Reichstein, M., Lawrence, D. M., and Swenson, S. C.: Improving canopy processes in the Community Land Model version 4 (CLM4) using global flux fields empirically inferred from FLUXNET data, J. Geophys. Res.-Atmos., 116, GB1008, https://doi.org/10.1029/2010JG001593, 2011.

Brynjarsdottir, J. and O'Hagan, A.: Learning about physical parameters: the importance of model discrepancy, Inverse Probl., 30, 114007, https://doi.org/10.1088/0266-5611/30/11/114007, 2014.

Burnash, R. J. C., Ferral, R. L., and McGuire, R. A.: A generalized streamflow simulation system: Conceptual modeling for digital computers, US Dept. of Commerce, National Weather Service, 1973.

Chaney, N. W., Metcalfe, P., and Wood, E. F.: HydroBlocks: a field-scale resolving land surface model for application over continental extents, Hydrol. Process., 30, 3543–3559, 2016a.

Chaney, N. W., Wood, E. F., McBratney, A. B., Hempel, J. W., Nauman, T. W., Brungard, C. W., and Odgers, N. P.: POLARIS: A 30-meter probabilistic soil series map of the contiguous United States, Geoderma, 274, 54–67, 2016b.

Christensen, N. S. and Lettenmaier, D. P.: A multimodel ensemble approach to assessment of climate change impacts on the hydrology and water resources of the Colorado River Basin, Hydrol. Earth Syst. Sci., 11, 1417–1434, https://doi.org/10.5194/hess-11-1417-2007, 2007.

Clapp, R. B. and Hornberger, G. M.: Empirical equations for some soil hydraulic properties, Water Resour. Res., 14, 601–604, 1978.

Clark, M. P., Nijssen, B., Lundquist, J. D., Kavetski, D., Rupp, D. E., Woods, R. A., Freer, J. E., Gutmann, E. D., Wood, A. W., Brekke, L. D., Arnold, J. R., Gochis, D. J., and Rasmussen, R. M.: A unified approach for process-based hydrologic modeling: 1. Modeling concept, Water Resour. Res., 51, 2498–2514, https://doi.org/10.1002/2015WR017198, 2015.

Clark, M. P., Schaefli, B., Schymanski, S. J., Samaniego, L., Luce, C. H., Jackson, B. M., Freer, J. E., Arnold, J. R., Dan Moore, R., Istanbulluoglu, E., and Ceola, S.: Improving the theoretical underpinnings of process-based hydrologic models, Water Resour. Res., 52, 2350–2365, https://doi.org/10.1002/2015WR017910, 2016.

Clark, M. P., Bierkens, M. F. P., Samaniego, L., Woods, R. A., Uijlenhoet, R., Bennett, K. E., Pauwels, V. R. N., Cai, X., Wood, A. W., and Peters-Lidard, C. D.: The evolution of process-based

hydrologic models: historical challenges and the collective quest for physical realism, Hydrol. Earth Syst. Sci., 21, 3427–3440, https://doi.org/10.5194/hess-21-3427-2017, 2017.

Cosby, B. J., Hornberger, G. M., Clapp, R. B., and Ginn, T. R.: A Statistical Exploration of the Relationships of Soil Moisture Characteristics to the Physical Properties of Soils, Water Resour. Res., 20, 682–690, 1984.

Crawford, N. H. and Linsley, R. K.: Digital simulation in hydrology: Stanford Watershed Model IV, Tech. Rep. 39, Stanford Univ. Dept. of Civil Engineering, 1966.

Cuntz, M., Mai, J., Zink, M., Thober, S., Kumar, R., Schäfer, D., Schrön, M., Craven, J., Rakovec, O., Spieler, D., Prykhodko, V., Dalmasso, G., Musuuza, J., Langenberg, B., Attinger, S., and Samaniego, L.: Computationally inexpensive identification of noninformative model parameters by sequential screening, Water Resour. Res., 51, 6417–6441, 2015.

Cuntz, M., Mai, J., Samaniego, L., Clark, M., Wulfmeyer, V., Branch, O., Attinger, S., and Thober, S.: The impact of standard and hard-coded parameters on the hydrologic fluxes in the Noah-MP land surface model, J. Geophys. Res.-Atmos., 121, 10676–10700, 2016.

Dagan, G.: Flow and transport in porous media, Springer Verlag, New York, 1989.

De Roo, A. and Wesseling, C. G.: Physically based river basin modelling within a GIS: the LISFLOOD model, Hydrol. Process., 14, 1981–1992, 2000.

Dooge, J.: Parameterization of hydrologic processes, in: Proceedings of the Greenbelt Study Conference, edited by: Eagleson, P., Cambridge University Press, new York, N.Y., 243–288, 1982.

Duckstein, L. and Opricovic, S.: Multiobjective optimization in river basin development, Water Resour. Res., 16, 14–20, 1980.

ECMWF: IFS DOCUMENTATION – Cy41r2 Operational implementation 8 March 2016, Tech. rep., European Centre for Medium-Range Weather Forecasts, http://www.ecmwf.int/search/elibrary/part?solrsort=sort_labe%lasc&title=part&secondary_title=41r1&f[0]=ts_biblio_year%3A2016, (last access: 2 February 2017, 2016.

Edijatno, de Oliveira Nascimento, N., Yang, X., Makhlouf, Z., and Michel, C.: GR3J: a daily watershed model with three free parameters, Hydrolog. Sci. J., 44, 263–277, 1999.

Edwards, P. N.: A Vast Machine: Computer Models, Climate Data, and the Politics of Global Warming, The MIP Press, 2010.

Famiglietti, J. and Wood, E.: Multiscale modeling of spatially variable water and energy balance processes, Water Resour. Res., 30, 3061–3078, 1994.

Famiglietti, J. S. and Wood, E. F.: Effects of Spatial Variability and Scale on Areally Averaged Evapotranspiration, Water Resour. Res., 31, 699–712, 1995.

Fenicia, F., Kavetski, D., and Savenije, H. H. G.: Elements of a flexible approach for conceptual hydrological modeling: 1. Motivation and theoretical development, Water Resour. Res., 47, W11510, https://doi.org/10.1029/2010WR010174, 2011.

Fernandez, W., Vogel, R., and Sankarasubramanian, A.: Regional calibration of a watershed model, Hydrolog. Sci. J., 45, 689–707, 2000.

Fisher, J. B., Huntzinger, D. N., Schwalm, C. R., and Sitch, S.: Modeling the Terrestrial Biosphere, Ann. Rev. Environ. Resour., 39, 91–123, 2014.

Flügel, W. A.: Delineating hydrological response units by geographical information system analyses for regional hydrological modelling using PRMS/MMS in the drainage basin of the river Bröl, Germany, Hydrol. Process., 9, 423–436, 1995.

Gelhar, L. W.: Stochastic Subsurface Hydrology, Prentice Hall, 1993.

Goehler, M., Mai, J., and Cuntz, M.: Use of eigendecomposition in a parameter sensitivity analysis of the Community Land Model, J. Geophys. Res.-Biogeo., 118, 904–921, 2013.

Gotzinger, J. and Bárdossy, A.: Comparison of four regionalisation methods for a distributed hydrological model, J. Hydrol., 333, 374–384, 2007.

Gupta, H. V., Perrin, C., Blöschl, G., Montanari, A., Kumar, R., Clark, M., and Andréassian, V.: Large-sample hydrology: a need to balance depth with breadth, Hydrol. Earth Syst. Sci., 18, 463–477, https://doi.org/10.5194/hess-18-463-2014, 2014.

Haddeland, I., Clark, D. B., Franssen, W., Ludwig, F., Voss, F., Arnell, N. W., Bertrand, N., Best, M., Folwell, S., Gerten, D., Gomes, S., Gosling, S. N., Hagemann, S., Hanasaki, N., Harding, R., Heinke, J., Kabat, P., Koirala, S., Oki, T., Polcher, J., Stacke, T., Viterbo, P., Weedon, G. P., and Yeh, P.: Multimodel Estimate of the Global Terrestrial Water Balance: Setup and First Results, J. Hydrometeorol., 12, 869–884, 2011.

Haughton, N., Abramowitz, G., Pitman, A. J., Or, D., Best, M. J., Johnson, H. R., Balsamo, G., Boone, A., Cuntz, M., Decharme, B., Dirmeyer, P. A., Dong, J., Ek, M., Guo, Z., Haverd, V., van den Hurk, B. J. J., Nearing, G. S., Pak, B., Santanello Jr., J. A., Stevens, L. E., and Vuichard, N.: The Plumbing of Land Surface Models: Is Poor Performance a Result of Methodology or Data Quality?, J. Hydrometeorol., 17, 1705–1723, 2016.

Hengl, T., Mendes de Jesus, J., Heuvelink, G. B. M., Ruiperez Gonzalez, M., Kilibarda, M., Blagotić, A., Shangguan, W., Wright, M. N., Geng, X., Bauer-Marschallinger, B., Guevara, M. A., Vargas, R., MacMillan, R. A., Batjes, N. H., Leenaars, J. G. B., Ribeiro, E., Wheeler, I., Mantel, S., and Kempen, B.: SoilGrids250m: Global gridded soil information based on machine learning, PLOS ONE, 12, 1–40, https://doi.org/10.1371/journal.pone.0169748, 2017.

Hundecha, Y. and Bárdossy, A.: Modeling of the effect of land use changes on the runoff generation of a river basin through parameter regionalization of a watershed model, J. Hydrol., 292, 281–295, 2004.

Hundecha, Y., Arheimer, B., Donnelly, C., and Pechlivanidis, I.: A regional parameter estimation scheme for a pan-European multi-basin model, J. Hydrol. Regional Studies, 6, 90–111, 2016.

Intsiful, J. and Kunstmann, H.: Upscaling of Land-Surface Parameters Through Inverse Stochastic SVAT-Modelling, Bound.-Lay. Meteorol., 129, 137–158, 2008.

Kavetski, D., Kuczera, G., and Franks, S. W.: Semidistributed hydrological modeling: A "saturation path" perspective on TOPMODEL and VIC, Water Resour. Res., 39, 1246, https://doi.org/10.1029/2003WR002122, 2003.

Kitanidis, P. K. and Vomvoris, E. G.: A geostatistical approach to the inverse problem in groundwater modeling (steady state) and one-dimensional simulations, Water Resour. Res., 19, 677–690, 2010.

Koren, V., Smith, M., and Duan, Q.: Use of a Priori Parameter Estimates in the Derivation of Spatially Consistent Parameter Sets of Rainfall-Runoff Models, American Geophysical Union, 239–254, https://doi.org/10.1002/9781118665671.ch18, 2013.

Kowalczyk, E. A., Wang, Y. P., and Law, R. M.: The CSIRO Atmosphere Biosphere Land Exchange (CABLE) model for use in climate models and as an offline model, CSIRO. Marine and Atmospheric Research, 13, 1–37, http://www.cmar.csiro.au/e-print/open/kowalczykea_2006a.pdf (last access: August 2017) 2006.

Kumar, R., Samaniego, L., and Attinger, S.: The effects of spatial discretization and model parameterization on the prediction of extreme runoff characteristics, J. Hydrol., 392, 54–69, 2010.

Kumar, R., Livneh, B., and Samaniego, L.: Toward computationally efficient large-scale hydrologic predictions with a multiscale regionalization scheme, Water Resour. Res., 49, 5700–5714, 2013a.

Kumar, R., Samaniego, L., and Attinger, S.: Implications of distributed hydrologic model parameterization on water fluxes at multiple scales and locations, Water Resour. Res., 49, 360–379, 2013b.

Kumar, R., Mai, J., Rakovec, O., Cuntz, M., Thober, S., Zink, M., Attinger, S., Schaefer, D., Schrön, M., and Samaniego, L. E.: Regionalized Hydrologic Parameters Estimates for a Seamless Prediction of Continental scale Water Fluxes and States, AGU Fall Meeting Abstracts, 2015.

Le Treut, H., Somerville, R., Cubasch, U., Ding, Y., Mauritzen, C., Mokssit, A., Peterson, T., and Prather, M.: Historical Overview of Climate Change, in: Climate Change 2007: The Physical Science Basis, Contribution of Working Group I to the Fourth Assessment Report of the Intergovernmental Panel on Climate Change, edited by: Solomon, S., Qin, D., Manning, M., Chen, Z., Marquis, M., Averyt, K., Tignor, M., and Miller, H. L., chap. 1, Cambridge University Press, Cambridge, United Kingdom and New York, NY, USA, 1–36, 2007.

Leavesley, G. H., Lichty, R. W., Troutman, B. M., and Saindon, L. G.: Precipitation-Runoff Modeling System: User's Manual, U.S. Geological Survey Water-Resources Investigations, Denver, Colorado, 83-4238 Edn., 1983.

Lerat, J., Andréassian, V., Perrin, C., Vaze, J., Perraud, J.-M., Ribstein, P., and Loumagne, C.: Do internal flow measurements improve the calibration of rainfall-runoff models?, Water Resour. Res., 48, W02511, https://doi.org/10.1029/2010WR010179, 2012.

Li, D., Bou-Zeid, E., Barlage, M., Chen, F., and Smith, J. A.: Development and evaluation of a mosaic approach in the WRF-Noah framework, J. Geophys. Res.-Atmos., 118, 11918–11935, 2013.

Li, H., Sivapalan, M., and Tian, F.: Comparative diagnostic analysis of runoff generation processes in Oklahoma DMIP2 basins: The Blue River and the Illinois River, J. Hydrol., 418–419, 90–109, 2012.

Liang, X., Lettenmaier, D., Wood, E., and Burges, S.: A Simple Hydrologically Based Model of Land-Surface Water and Energy Fluxes for General-Circulation Models, J. Geophys. Res.-Atmos., 99, 14415–14428, 1994.

Liang, X., Lettenmaier, D. P., and Wood, E. F.: One-dimensional statistical dynamic representation of subgrid spatial variability of precipitation in the two-layer variable infiltration capacity model, J. Geophys. Res.-Atmos., 101, 21403–21422, 1996.

Lindstrom, G., Johansson, B., Persson, M., Gardelin, M., and Bergström, S.: Development and test of the distributed HBV-96 hydrological model, J. Hydrol., 201, 272–288, 1997.

Lindström, G., Pers, C., Rosberg, J., Strömqvist, J., and Arheimer, B.: Development and testing of the HYPE (Hydrological Predictions for the Environment) water quality model for different spatial scales, Hydrol. Res., 41, 295–26, 2010.

Livneh, B. and Lettenmaier, D. P.: Regional parameter estimation for the unified land model, Water Resour. Res., 49, 100–114, 2013.

Livneh, B., Kumar, R., and Samaniego, L.: Influence of soil textural properties on hydrologic fluxes in the Mississippi river basin, Hydrol. Process., 29, 4638–4655, 2015.

Ludovicus, P. H. (Rens), van Beek, Sutanudjaja, E. H., Wada, Y., Bosmans, J. H. C., Drost, N., de Graaf, I. E. M., de Jong, K., Lopez Lopez, P., Pessenteiner, S., Schmitz, O., Straatsma, M. W., Wanders, N., Wisser, D., and Bierkens, M. F. P.: PCR-GLOBWB, https://doi.org/10.1029/2010WR009792, data available at: https://github.com/UU-Hydro/PCR-GLOBWB_model/blob/develop/README.txt, last access: 2 August 2017.

Martina, M. L. V., Todini, E., and Liu, Z.: Preserving the dominant physical processes in a lumped hydrological model, J. Hydrol., 399, 121–131, 2011.

Mendoza, P. A., Clark, M. P., Barlage, M., Rajagopalan, B., Samaniego, L., Abramowitz, G., and Gupta, H.: Are we unnecessarily constraining the agility of complex process-based models?, Water Resour. Res., 51, 716–728, https://doi.org/10.1002/2014WR015820, 2015.

Merz, R. and Blöschl, G.: Regionalisation of catchment model parameters, J. Hydrol., 287, 95–123, 2004.

Miller, E. E. and Miller, R. D.: Physical Theory for Capillary Flow Phenomena, J. Appl. Phys., 27, 324–332, 1956.

Mizukami, N., Clark, M., Newman, A. J., Wood, A. W., Gutmann, E., Nijssen, B., Rakovec, O., and Samaniego, L.: Towards seamless large domain parameter estimation for hydrologic models, Water Resour. Res., accepted, https://doi.org/10.1002/2017WR020401, 2017.

Morris, M. D.: Factorial Sampling Plans for Preliminary Computational Experiments, Technometrics, 33, 161–174, 1991.

Mueller, B., Hirschi, M., Jimenez, C., Ciais, P., Dirmeyer, P. A., Dolman, A. J., Fisher, J. B., Jung, M., Ludwig, F., Maignan, F., Miralles, D. G., McCabe, M. F., Reichstein, M., Sheffield, J., Wang, K., Wood, E. F., Zhang, Y., and Seneviratne, S. I.: Benchmark products for land evapotranspiration: LandFlux-EVAL multi-data set synthesis, Hydrol. Earth Syst. Sci., 17, 3707–3720, https://doi.org/10.5194/hess-17-3707-2013, 2013.

Müller Schmied, H., Eisner, S., Franz, D., Wattenbach, M., Portmann, F. T., Flörke, M., and Döll, P.: Sensitivity of simulated global-scale freshwater fluxes and storages to input data, hydrological model structure, human water use and calibration, Hydrol. Earth Syst. Sci., 18, 3511–3538, https://doi.org/10.5194/hess-18-3511-2014, 2014.

Nearing, G. S., Tian, Y., Gupta, H. V., Clark, M. P., Harrison, K. W., and Weijs, S. V.: A philosophical basis for hydrological uncertainty, Hydrolog. Sci. J., 61, 1666–1678, 2016.

Neuman, S. P.: Universal scaling of hydraulic conductivities and dispersivities in geologic media, Water Resour. Res., 26, 1749–1758, 2010.

Nijzink, R. C., Samaniego, L., Mai, J., Kumar, R., Thober, S., Zink, M., Schäfer, D., Savenije, H. H. G., and Hrachowitz, M.: The importance of topography-controlled sub-grid process heterogeneity and semi-quantitative prior constraints in distributed hydrological models, Hydrol. Earth Syst. Sci., 20, 1151–1176, https://doi.org/10.5194/hess-20-1151-2016, 2016.

Niu, G.-Y.: THE COMMUNITY NOAH LAND-SURFACE MODEL (LSM) WITH MULTI-PHYSICS OPTIONS, Tech. rep., National Centers for Environmental Prediction (NCEP), Oregon State University, Air Force, and Hydrology Lab – NWS, https://www.jsg.utexas.edu/noah-mp/users-guide/, (last access: 2 February 2017), 2011.

Niu, G.-Y., Yang, Z.-L., Mitchell, K. E., Chen, F., Ek, M. B., Barlage, M., Kumar, A., Manning, K., Niyogi, D., Rosero, E., Tewari, M., and Xia, Y.: The community Noah land surface model with multiparameterization options (Noah-MP): 1. Model description and evaluation with local-scale measurements, J. Geophys. Res.-Atmos., 116, D12109, https://doi.org/10.1029/2010JD015139, 2011.

Oleson, K., Lawrence, D., Bonan, G., Drewniak, B., Huang, M., Koven, C., Levis, S., Li, F., Riley, W., Subin, Z., Swenson, S., Thornton, P., Bozbiyik, A., Fisher, R., Kluzek, E., Lamarque, J.-F., Lawrence, P., Leung, L., Lipscomb, W., Muszala, S., Ricciuto, D., Sacks, W., Sun, Y., Tang, J., and Yang, Z.-L.: Technical Description of version 4.5 of the Community Land Model (CLM), Tech. rep., Ncar Technical Note NCAR/TN-503+STR, National Center for Atmospheric Research, Boulder, CO, http://www.cesm.ucar.edu/models/cesm1.2/clm/, (last access: 2 February 2017), 2013.

Peters-Lidard, C. D., Clark, M., Samaniego, L., Verhoest, N. E. C., van Emmerik, T., Uijlenhoet, R., Achieng, K., Franz, T. E., and Woods, R.: Scaling, similarity, and the fourth paradigm for hydrology, Hydrol. Earth Syst. Sci., 21, 3701–3713, https://doi.org/10.5194/hess-21-3701-2017, 2017.

Pielke Sr., R.: Mesoscale meteorological modeling, Academic Press, Elsevier, International Geophysics, 3 Rev Edn., 2013.

Pokhrel, P. and Gupta, H. V.: On the use of spatial regularization strategies to improve calibration of distributed watershed models, Water Resour. Res., 46, W01505, https://doi.org/10.1029/2009WR008066, 2010.

Rakovec, O., Hill, M. C., Clark, M. P., Weerts, A. H., Teuling, A. J., and Uijlenhoet, R.: Distributed Evaluation of Local Sensitivity Analysis (DELSA), with application to hydrologic models, Water Resour. Res., 50, 1–18, https://doi.org/10.1002/2013WR014063, 2014.

Rakovec, O., Kumar, R., Attinger, S., and Samaniego, L.: Improving the realism of hydrologic model functioning through multivariate parameter estimation, Water Resour. Res., 52, 7779–7792, 2016a.

Rakovec, O., Kumar, R., Mai, J., Cuntz, M., Thober, S., Zink, M., Attinger, S., Schäfer, D., Schrön, M., and Samaniego, L.: Multiscale and Multivariate Evaluation of Water Fluxes and States over European River Basins, J. Hydrometeorol., 17, 287–307, 2016b.

Reggiani, P., Sivapalan, M., and Majid Hassanizadeh, S.: A unifying framework for watershed thermodynamics: balance equations for mass, momentum, energy and entropy, and the second law of thermodynamics, Adv. Water Resour., 22, 367–398, 1998.

Samaniego, L. and Bárdossy, A.: Robust parametric models of runoff characteristics at the mesoscale, J. Hydrol., 303, 136–151, 2005.

Samaniego, L., Bárdossy, A., and Kumar, R.: Streamflow

prediction in ungauged catchments using copula-based dissimilarity measures, Water Resour. Res., 46, W02506, https://doi.org/10.1029/2008WR007695, 2010a.

Samaniego, L., Kumar, R., and Attinger, S.: Multiscale parameter regionalization of a grid-based hydrologic model at the mesoscale, Water Resour. Res., 46, W05523, https://doi.org/10.1029/2008WR007327, 2010b.

Samaniego, L., Kumar, R., and Jackisch, C.: Predictions in a datasparse region using a regionalized grid-based hydrologic model driven by remotely sensed data, Hydrol. Res., 42, 338–355, 2011.

Samaniego, L. E., Warrach-Sagi, K., Zink, M., and Wulfmeyer, V.: Verification of High Resolution Soil Moisture and Latent Heat in Germany, AGU Fall Meeting Abstracts, http://adsabs.harvard.edu/abs/2012AGUFM.H23G..02S, last access: 2 August 2017, provided by the SAO/NASA Astrophysics Data System, 2012.

Samaniego, L., Brenner, J., Cuntz, M., Demirel, C. M., Kumar, R., Langenberg, B., Mai, J., Rakovec, O., Schäfer, D., Schrön, M., Stisen, S., Thober, S., and Zink, M.: mHM, https://doi.org/10.1029/2008WR007327, data available at: http://www.ufz.de/index.php?en=40114, last access: 2 August 2017.

Savenije, H. H. G. and Hrachowitz, M.: HESS Opinions "Catchments as meta-organisms – a new blueprint for hydrological modelling", Hydrol. Earth Syst. Sci., 21, 1107–1116, https://doi.org/10.5194/hess-21-1107-2017, 2017.

Seibert, J.: Regionalisation of parameters for a conceptual rainfall-runoff model, Agr. Forest Meteorol., 98–99, 279–293, 1999.

Sellers, P. J., Dickinson, R. E., Randall, D. A., Betts, A. K., Hall, F. G., Berry, J. A., Collatz, G. J., Denning, A. S., Mooney, H. A., Nobre, C. A., Sato, N., Field, C. B., and Henderson-Sellers, A.: Modeling the exchanges of energy, water, and carbon between continents and the atmosphere, Science, 275, 502–509, 1997.

Singh, R., Archfield, S. A., and Wagener, T.: Identifying dominant controls on hydrologic parameter transfer from gauged to ungauged catchments – A comparative hydrology approach, J. Hydrol., 517, 985–996, 2014.

Singh, S. K., Bárdossy, A., Götzinger, J., and Sudheer, K. P.: Effect of spatial resolution on regionalization of hydrological model parameters, Hydrol. Process., 26, 3499–3509, 2012.

Sobol', I. M.: Global sensitivity indices for nonlinear mathematical models and their Monte Carlo estimates, Math. Comput. Simulat., 55, 271–280, https://doi.org/10.1016/S0378-4754(00)00270-6, 2001.

Sutanudjaja, E., Bosmans, J., Chaney, N., Clark, M. P., Condon, L. E., David, C. H., De Roo, A. P. J., Doll, P. M., Drost, N., Eisner, S., Famiglietti, J. S., Floerke, M., Gilbert, J. M., Gochis, D. J., Hut, R., Keune, J., Kollet, S. J., Maxwell, R. M., Pan, M., Rakovec, O., Reager, II, J. T., Samaniego, L. E., Mueller Schmied, H., Trautmann, T., Van Beek, L. P., Van De Giesen, N., Wood, E. F., Bierkens, M. F., and Kumar, R.: The HyperHydro (H^2) experiment for comparing different large-scale models at various resolutions, AGU Fall Meeting Abstracts, http://adsabs.harvard.edu/abs/2015AGUFM.H23E1622S (last access: 2 August 2017), 2015.

Sutanudjaja, E., van Beek, R., Wada, Y., Bosmans, J., Drost, N., de Graaf, I., de Jong, K., Lopez Lopez, P., Pessenteiner, S., Schmitz, O., Straatsma, M., Wanders, N., Wisser, D., and Bierkens, M.: PCR-GLOBWB_model: PCR-GLOBWB version v2.1.0_alpha, https://doi.org/10.5281/zenodo.60764, 2016.

Troy, T. J., Wood, E. F., and Sheffield, J.: An effi-

cient calibration method for continental-scale land surface modeling, Water Resour. Res., 44, W09411, https://doi.org/10.1029/2007WR006513, 2008.

van Beek, L. P. H., Wada, Y., and Bierkens, M. F. P.: Global monthly water stress: 1. Water balance and water availability, Water Resour. Res., 47, W07517, https://doi.org/10.1029/2010WR009791, 2011.

Viterbo, P. and Beljaars, C. M.: An improved land surface parameterization scheme in the ECMWF model and its validation, J. Climate, 8, 2716–2748, 1995.

Viviroli, D., Zappa, M., Gurtz, J., and Weingartner, R.: An introduction to the hydrological modelling system PREVAH and its pre- and post-processing-tools, Environ. Modell. Softw., 24, 1209–1222, 2009.

Wada, Y. and Bierkens, M. F. P.: Sustainability of global water use: past reconstruction and future projections, Environ. Res. Lett., 9, 104003, https://doi.org/10.1088/1748-9326/9/10/104003, 2014.

Wagener, T. and Wheater, H. S.: Parameter estimation and regionalization for continuous rainfall-runoff models including uncertainty, J. Hydrol., 320, 132–154, 2006.

Wanders, N. and Wada, Y.: Human and climate impacts on the 21st century hydrological drought, J. Hydrol., 526, 208–220, 2015.

Wöhling, T., Samaniego, L., and Kumar, R.: Evaluating multiple performance criteria to calibrate the distributed hydrological model of the upper Neckar catchment, Environmental Earth Sciences, 69, 453–468, 2013.

Wood, A. and Mizukami, N.: Project Summary Report: CMIP5 1/8 Degree Daily Weather and VIC Hydrology Datasets for CONUS, Tech. rep., B. o. R. U.S. Department of the Interior, Technical Services Center, Denver, Colorado, http://www.corpsclimate.us/docs/cmip5.hydrology.2014.final.re%port.pdf (last access: 24 Januar 2017), 2014.

Wood, E. (Ed.): Land Surface, atmosphere interactions for climate modelling: observations. models, and analysis, Kluwer, 1990.

Wood, E.: Effects of soil moisture aggregation on surface evaporative fluxes, J. Hydrol., 190, 397–412, 1997.

Wood, E. F., Sivapalan, M., Beven, K., and Band, L.: Effects of Spatial Variability and Scale with Implications to Hydrologic Modeling, J. Hydrol., 102, 29–47, 1988.

Wood, E. F., Lettenmaier, D. P., Liang, X., Lohmann, D., Boone, A., Chang, S., Chen, F., Dai, Y., Dickinson, R. E., Duan, Q., Ek, M., Gusev, Y. M., Habets, F., Irannejad, P., Koster, R., Mitchel, K. E., Nasonova, O. N., Noilhan, J., Schaake, J., Schlosser, A., Shao, Y., Shmakin, A. B., Verseghy, D., Warrach, K., Wetzel, P., Xue, Y., Yang, Z.-L., and Zeng, Q. C.: The project for intercomparison of land-surface parameterization schemes (PILPS) phase 2(c) Red-Arkansas River basin experiment: 1. Experiment description and summary intercomparisons, Global Planet. Change, 19, 115–135, 1998.

Wood, E. F., Roundy, J. K., Troy, T. J., van Beek, L. P. H., Bierkens, M. F. P., Blyth, E., de Roo, A., Doell, P., Ek, M., Famiglietti, J., Gochis, D., van de Giesen, N., Houser, P., Jaffé, P. R., Kollet, S., Lehner, B., Lettenmaier, D. P., Peters-Lidard, C., Sivapalan, M., Sheffield, J., Wade, A., and Whitehead, P.: Hyperresolution global land surface modeling: Meeting a grand challenge for monitoring Earth's terrestrial water, Water Resour. Res., 47, W05301, https://doi.org/10.1029/2010WR010090, 2011.

Wösten, J. H. M., Pachepsky, Y. A., and Rawls, W. J.: Pedotransfer functions: bridging the gap between available basic soil data and missing soil hydraulic characteristics, J. Hydrol., 251, 123–150, 2001.

Yadav, M., Wagener, T., and Gupta, H.: Regionalization of constraints on expected watershed response behavior for improved predictions in ungauged basins, Adv. Water Res., 30, 1756–1774, 2007.

Zink, M., Kumar, R., Cuntz, M., and Samaniego, L.: German data set, https://doi.org/10.5194/hess-21-1769-2017, data available at: https://www.ufz.de/drp/de/index.php?drp_data%5Bmvc% 5D=Search%2Fsearch&drp_data%5Bfilter%5D%5Barchive% 5D=on&drp_data%5Bkeywords%5D=Water+fluxes+and+ states+dataset+for+Germany+from+1951+to+2010), last access: 2 August 2017.

Zehe, E., Ehret, U., Pfister, L., Blume, T., Schröder, B., Westhoff, M., Jackisch, C., Schymanski, S. J., Weiler, M., Schulz, K., Allroggen, N., Tronicke, J., van Schaik, L., Dietrich, P., Scherer, U., Eccard, J., Wulfmeyer, V., and Kleidon, A.: HESS Opinions: From response units to functional units: a thermodynamic reinterpretation of the HRU concept to link spatial organization and functioning of intermediate scale catchments, Hydrol. Earth Syst. Sci., 18, 4635–4655, https://doi.org/10.5194/hess-18-4635-2014, 2014.

4

Coupled hydro-meteorological modelling on a HPC platform for high-resolution extreme weather impact study

Dehua Zhu[1,a], **Shirley Echendu**[1], **Yunqing Xuan**[1], **Mike Webster**[1], **and Ian Cluckie**[1]

[1]College of Engineering, Swansea University Bay Campus, Swansea, SA1 8EN, UK
[a]now at: School of Hydrometeorology, Nanjing University of Information Science and Technology. Nanjing, 210044, China

Correspondence to: Dehua Zhu (d.zhu@nuist.edu.cn)

Abstract. Impact-focused studies of extreme weather require coupling of accurate simulations of weather and climate systems and impact-measuring hydrological models which themselves demand larger computer resources. In this paper, we present a preliminary analysis of a high-performance computing (HPC)-based hydrological modelling approach, which is aimed at utilizing and maximizing HPC power resources, to support the study on extreme weather impact due to climate change. Here, four case studies are presented through implementation on the HPC Wales platform of the UK mesoscale meteorological Unified Model (UM) with high-resolution simulation suite UKV, alongside a Linux-based hydrological model, Hydrological Predictions for the Environment (HYPE). The results of this study suggest that the coupled hydro-meteorological model was still able to capture the major flood peaks, compared with the conventional gauge- or radar-driving forecast, but with the added value of much extended forecast lead time. The high-resolution rainfall estimation produced by the UKV performs similarly to that of radar rainfall products in the first 2–3 days of tested flood events, but the uncertainties particularly increased as the forecast horizon goes beyond 3 days. This study takes a step forward to identify how the online mode approach can be used, where both numerical weather prediction and the hydrological model are executed, either simultaneously or on the same hardware infrastructures, so that more effective interaction and communication can be achieved and maintained between the models. But the concluding comments are that running the entire system on a reasonably powerful HPC platform does not yet allow for real-time simulations, even without the most complex and demanding data simulation part.

1 Introduction

Extreme precipitation with great intensity and the subsequent flash flooding events arising from rivers and mountainous watersheds often lead to considerable economic damage and casualties, because water levels can react extremely quickly within rather limited warning lead time (flash flooding). Therefore, the evaluation of potential flooding risks in extreme weather conditions, and the corresponding protection measures required, demands accurate short-term flood forecasting, and more often very short lead-time forecasting – termed "nowcasting" (Cloke and Pappenberger, 2009).

Understandably, hydrological models together with hydraulic models play a key role in predicting runoff, river flow, and possible inundations. However, the lead time, which is crucial for hazard mitigation and evacuation, is often highly limited in such a classic model chain configuration, since, the lead time is basically then the travelling time of flood water. It is therefore other means of providing rainfall estimates with extra lead time (e.g. weather radar observations) which have become increasingly essential in flood forecasting under extreme weather conditions (Zhu et al., 2014). However, it has also been recognized, for example by Golding (1998) and Smith and Austin (2000), that the performance of radar-based rainfall nowcasting deteriorates rapidly when the lead time goes beyond 0.5 h. Then, the combination of radar nowcasting and hydrological forecasting is reduced to that of a normal model, or even worse. In fact, early attempts, whilst using the NIMROD radar rainfall product, already introduced a rainfall forecast from numerical weather prediction models to compensate for this shortcoming.

The fast development of HPC (high-performance computing), as well as that of NWP (numerical weather prediction) models, has since given rise to the use of NWP, either directly or indirectly in hydrological simulations, in an effort to push hydrological forecasting beyond the limit of the rainfall-observation time horizon. This link between two different modelling disciplines is often referred to as model coupling. The resulting coupled meteorological–hydrological models appeared from the beginning of the 21st century, being initially focused on flash-flood forecasting, and later extended to handle climatic–hydrological coupling. This has facilitated many climate-change impact studies on water resources that rely heavily on the use of climate projections or simulations. Nevertheless, the linkage between the meteorological and hydrological models is scientifically challenging due to differences in model structures and issues of incompatible units (use of different scales in time and space). This is encapsulated, in particular, in the task of how best to transform and regionalize global climate scenarios, with spatial resolutions of $1000–10\,000\,km^2$, to hydrological mesoscale catchments of $10–1000\,km^2$.

Simulation with meteorological–hydrological coupling in high spatial and temporal resolution is a comparatively new field of hydrological research, yet some pioneering work has recently appeared. In order to analyse the prediction of selected events characterized by peak flows, Westrick et al. (2002) proposed a hydrometeorological forecasting system for mountainous watersheds by coupling the Penn State–NCAR Mesoscale Meteorological Model (also known as MM5 for brevity; Dudhia, 1993; Grell et al., 1994) in $4 \times 4\,km^2$ resolution and the distributed hydrological model DHVSM (Wigmosta et al., 1994). Jasper et al. (2002) compared the hydrological performance of radar and gauge measurements with five different high-resolution NWP models and grid-cell sizes between 2 and 14 km. This work covered the prediction of peak flows on the alpine Ticino–Toce watershed, using the distributed hydrological model WaSiM (Schulla and Jasper, 2000). The results suggest that the accuracy and consistency of NWP rainfall in hydrological applications heavily depend on their process modelling at all scales of model nesting. Particularly so, as inaccuracies introduced by downscaling of precipitation from NWP models can lead to large differences in the predicted hydrological results, especially during extreme convective storm periods. Kunstmann and Stadler (2005) coupled (in a one-way manner) the mesoscale meteorological model MM5 with the distributed hydrological model WaSiM. The meteorological re-analysis data were dynamically downscaled with MM5 grid-cell sizes from 100 to 2 km using four nests. Findings show that the MM5-based interpolation of precipitation yielded 21 % less total yearly precipitation in the catchment area, compared to the station-based interpolation. Yarnal et al. (2000) linked a high-resolution meteorological model (MM5 at 4 km resolution) and a suite of coupled hydrological models in the Susquehanna River Basin Experiment

(SRBEX). This work points out that the coupled model has to confront several issues, such as physics and parameterizations, for a mesoscale atmospheric model to match the timescales of climate coupled to the hydrological, meteorological and climatological process models with different scales, and accordingly the immense computational needs. Xuan et al. (2009) also indicated that the inaccuracies and uncertainties in NWP could propagate to the downstream hydrological models, and they proposed to use an ensemble-based approach, together with effective bias correction, to mitigate this problem.

The majority of the studies cited above have been relying on the use of the so-called downscaling of large-scale NWP results using regional meteorological model such as MM5. These studies are often conducted in an offline manner where hydrological modellers have hardly any control of NWP except the meso-scale one used for downscaling. However, the work presented in this paper not only focuses on the performance of coupled high-resolution meteorological–hydrological simulations for extreme storm events on a HPC platform. It is also aimed at exploring the potential of building and running fully coupled NWP-hydrological forecasts on a single computer platform, and therefore being able to obtain first-hand knowledge on fully integrated hydro-meteorological modelling. As such, we did not apply the meteorological model in forecasting mode, but used hindcasting mode instead, to test the model performance and benchmarking over several selected historical events.

One of the main challenges faced in coupled NWP-hydrological model simulation, or operational forecasting, is their reliance on computationally implementing NWP. In turn, this necessitates the use of HPC, a procedure which can be performed in two different fashions: firstly, through an offline approach, where the hydrological model receives data that are generated from NWP beforehand (e.g. the data disseminated from various national meteorological centres), and secondly through an online mode, where both NWP and the hydrological model are executed, either simultaneously or on the same hardware infrastructures, so that more effective interaction and communication can be achieved and maintained between the models. Most existing studies have adopted the former approach to ease technical demands on HPC as well as on NWP.

In contrast, this study takes a step forward to identify how the latter approach can be used, once HPC installation has been resolved. Moreover, it is worth noting that this experiment of a fully coupled NWP-hydrological forecast is preliminarily designed to be a one-way coupling system in this study, which will form the basis for extension into a two-way coupling system, which will be developed further in the future.

2 Materials and methods

In this study, the principal goal of the experiment has been to simulate the river basin response to extreme storm events, by linking a semi-distributed hydrological HYPE model to the UK Met Office Unified Model (UM) at a much finer spatial scale (1.5 km). The combined high-resolution one-way driven model experiments generate runoff hydrographs for three extreme flood events, which occurred in the Upper Medway catchment (220 km^2) located south of London in the UK (see Fig. 1).

The catchment elevation varies between 30 and 220 m above mean sea level and the majority of slopes range from 2 to 8°, which makes up around 70 % of the whole catchment. This suggests that the main topography of the Upper Medway catchment is made up of small hills surrounding the flat, low-relief, low-lying area without much variation of elevation. The land use in the catchment can be simplified and described as permanent grass (over 95 %). The major soil types in the Upper Medway catchment can be categorized as silty loam and clayey silt, according to the National Soil Resources Institute (NSRI) data. The geology of the catchment is a mixture of permeable (chalk) and impermeable (clay) and the dominant aquifers consist of the Ashdown Formation and the Tunbridge Wells Formation of the Hastings Group. The saturation-excess mechanism is the major runoff generation process in the catchment.

In such model experiments, two different sets of meteorological input data were used: (1) surface observation data from station measurements and from weather radar estimation, and (2) forecast rainfall data from high-resolution UM simulation suite, UKV, with grid-cell sizes of 1.5 km. The experiments were designed as follows: (1) selecting representative storms and hydrographs for simulation, (2) simulating these storms using the high-resolution UKV simulation model, and (3) modelling the river-basin response to the simulated storm events using the HYPE hydrological model.

One notes that Met Office has used the operational high-resolution UK 1.5 km model (UKV) under the New Dynamics algorithm specification. This introduces nested operations, through parallel suites PS30 and the time periods of interest. As such, this consists of a global 25 km simulation, followed by a North Atlantic–European 12 km simulation, and finally, a UKV 1.5 km simulation. Each such simulation stage provides the necessary lateral (spatio-temporal) boundary conditions for the regionally refined subsequent stage.

Rainfall observations from weather radars were also introduced in this study to check the UKV output, since rain gauges are point-based and the radar rainfall can provide well represented rainfall distribution. Moreover, the comparison with UKV input through a hydrological model can be drawn, in terms of streamflow differences.

The rain-gauge measurements are collected from nine real-time, tipping-bucket rain gauges (TBRs) operated by the UK Environment Agency (EA). Figure 1 shows the locations of the rain gauges (circles) and the flow gauges (triangles) on the catchment. And all the flow comparisons in this study were carried out at the Chafford flow gauge close to the catchment outlet.

The radar rainfall estimates used in this study are extracted from the UK NIMROD composite data set. This has been provided and quality-controlled by the UK Met Office using the lowest available scan. It has been adjusted against available rain-gauge measurement and undergone extensive processing to correct for various sources of radar error. Such radar error would include noise, clutter, anomalous propagation, attenuation, occultation, and "bright band" and orographic enhancement. Therefore, these high-resolution radar composite rainfall estimates incorporate the latest UK Met Office processing algorithms to account for the different sources of errors in the estimation of precipitation using weather radars (Harrison et al., 2000). This implies that this data set is the best possible estimate of rainfall available at the ground level in the UK (i.e. it is the most error-free).

More details in regards to the properties of this catchment and data description used in this study, such as topography, vegetation, and soil types, as well as the availability of a hydrological data set, have been detailed in Zhu and Cluckie (2012) and Zhu et al. (2014).

2.1 UKV model configuration and implementation

The Unified Model is an atmospheric predictive numerical modelling software, offered by the UK Met Office and written in FORTRAN. Here, its output is coupled with a hydrological model for the purpose of accurate flood and extreme storm prediction. The UM was built on Archer hardware, with specification as a Cray XC30 MPP supercomputer and with up to 4920 compute nodes, each having a two 12-core Intel Ivy Bridge series processor, providing a total of 118 080 processing cores. Each node has 64 GB memory, with a subset of large memory nodes possessing 128 GB.

Further to the successful build and implementation of the UM, output from the various implementations has been validated against results derived from other HPC architectures. The UM features a new dynamics algorithm, which is based on a semi-implicit, semi-Lagrangian formulation, that uses a common finite-difference scheme for the fully compressible, non-hydrostatic Euler equations, discretized on a latitude–longitude grid. The algorithm is designed around a matrix-bound approach that is used to solve the semi-implicit aspects of the scheme.

The high-resolution simulation was achieved using the UKV suite which is a regional configuration of the UM, derived from operations through Parallel Suites (PS30), which consist of three nested domain simulations: a global 25 km simulation, a North Atlantic–European (NAE) 12 km simulation, and a UKV 1.5 km simulation. The Global N512L70 problem suite is discretized into approximately 25 km mid-latitudes, upon a 1024 × 769 grid. There are 70 model lev-

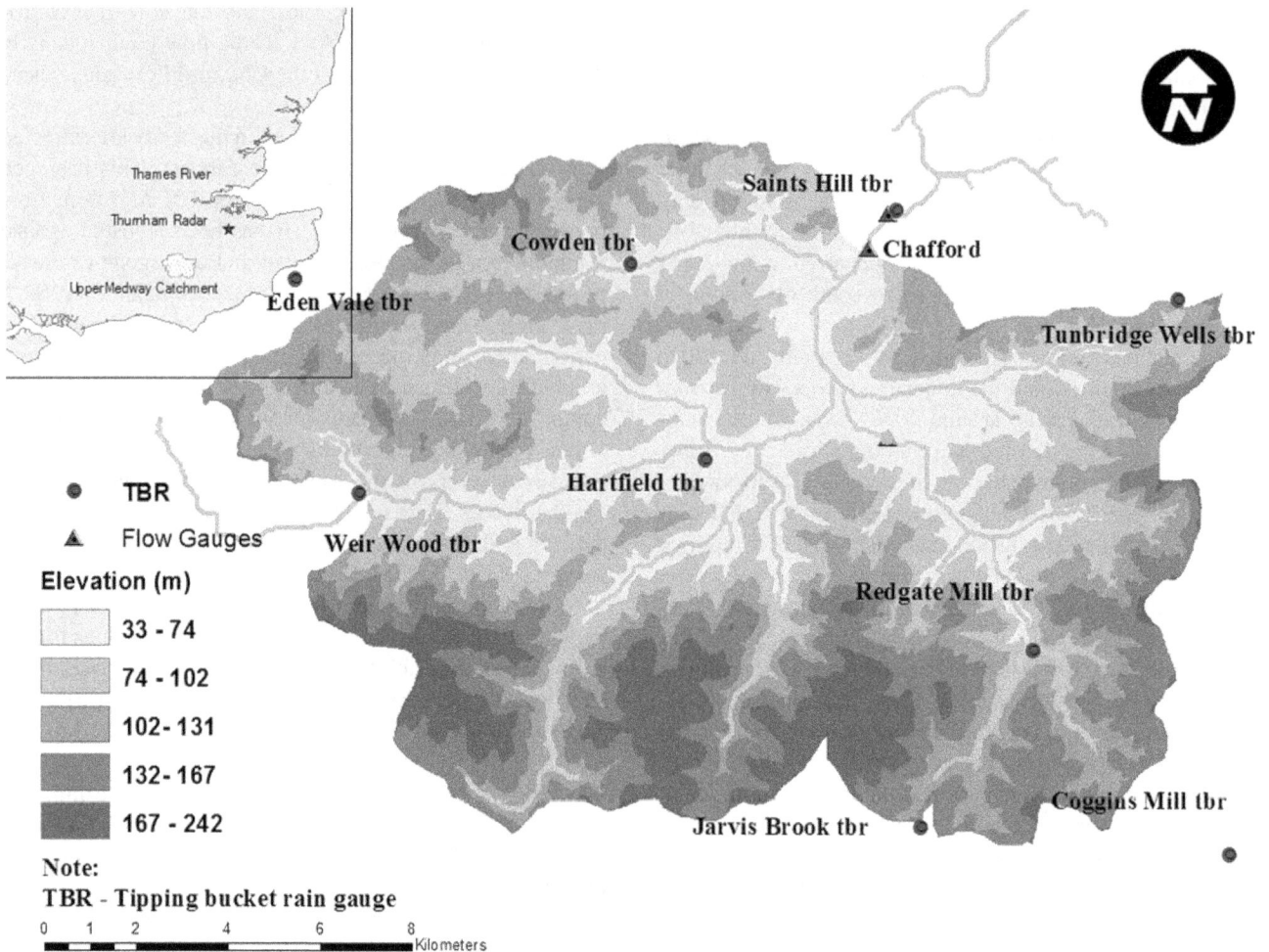

Figure 1. Map of rain gauges and flow gauge locations on the Upper Medway catchment (source: Zhu and Cluckie, 2012).

els vertically and a time step of 10 min is used. The regional NAE problem suite has a resolution of 12 km, across a 600 × 360 grid. The NAE suite also has 70 vertical levels but the time-step choice is 5 min. Finally, the regional UKV is set at 1.5 km resolution over a 622 × 810 grid with a time step of 50 s.

UKV model implementation requires a few events for model run. This includes an initialization date and a number of subsequent time-duration periods, i.e. 3, 6, 8, and 12 days. A selection of 8-day start dumps was used in this study, requested from ECMWF or the Met Office. The Met Office holds start dumps to a back-date of up to 5 years only; prior to that, start dumps would need to be obtained from other sources.

The steps of the UKV process in the overall procedure are to run as follows: first, the global reconfiguration and forecast; second, the European reconfiguration and forecast; and finally, the UKV reconfiguration and forecast. These independent simulation steps are all dynamically linked through lateral boundary conditions (LBCs), and regionalization of a

start dump. With the start dump reconfigured for a UM input file format (global region), this is then utilized to initialize the global, European and UKV reconfiguration and to obtain an additional start dump for the forecasting stage. In turn, the global forecast is run to obtain lateral boundary conditions for the European stage, whilst the European forecast provides lateral boundary conditions for the UKV.

The UKV model outputs were also on a rotated longitude–latitude grid, whose resolution is not constant, with small deviations from 1.5 km depending on the locations. The data were further projected onto the National Grid Reference Grid to become comparable with other sources of data, such as the weather radar rainfall observation from the NIMROD system. A nearest-neighbour interpolation was used to produce the evenly distributed grid data after projecting.

2.2 The configuration and calibration of hydrological model HYPE

Whilst many hydrological models could have been selected (for example see Zhu et al. (2014)), an open-source model –

HYPE (Hydrological Predictions for the Environment) – has been selected in this study to avoid reliance on commercial modelling packages. HYPE is developed at Swedish Meteorological and Hydrological Institute (SMHI), with a focus on integrating water and water quality throughout the model compartments and predictions in ungauged catchments with large model set-ups, e.g. across Europe. It is a dynamical model forced with time series of precipitation and air temperature, typically on a daily time step. Forcing in the form of nutrient loads is not dynamical. Examples of HYPE applications include atmospheric deposition, fertilizers, and waste water.

The HYPE model is able to predict water and nutrient concentrations in the landscape at the catchment scale. Its spatial division is related to sub-catchments and corresponding characteristics, including land use, vegetation, soil type, and elevation. Within a particular catchment, the model will simulate water content in different compartments, including soil moisture, shallow groundwater, rivers, and lakes.

The default time step in HYPE is daily, but it can be reduced to hourly, which is normally specified in the input data set, such as precipitation. Since there is no 2-D surface runoff algorithm built in the HYPE model, it is in principle a lumped model. However, spatial variations can be accounted for by portioning the catchment into smaller sub-catchments. In this respect, the simulated precipitation was processed as the catchment average rainfall before being fed to the HYPE model.

The winter flood event, which took place from 6 December 2003 to 28 February 2004, was used for model calibration, carried out using 1 h time step rain-gauge measurements and parameterized with the streamflow observation at the catchment outlet. In order to achieve the best fit between observed and modelled flow, the model parameters were calibrated in simulation mode using a mixture of manual and automatic parameter adjustment, according to their functionalities in the model.

First, all the parameters went through an initial manual sensitivity analysis, to choose those worthy of further automatic parameterization. In this study, the maximum amount of percolation (mperc1, mperc2) in soil layers needs to be calibrated for percolation to occur. In addition, the soil-type-related parameters, such as the available storage of water in the soil and the runoff coefficient of the topsoil layer (rrcs1), are sensitive in the model. And the peak velocity of flow in rivers (rivvel) determines the peak flow delay in the model, which also needs to be calibrated. After the sensitive parameters are selected, the progressive Monte Carlo simulation was employed to reduce the parameter space in stages and finally determine the calibrated parameters for later rainfall–runoff comparisons.

The soil properties setting is critical in HYPE model. Figure 2 shows the model calibration performance with single soil type (SST) and multiple soil type (MST) settings. The soil types and the corresponding properties for the Upper

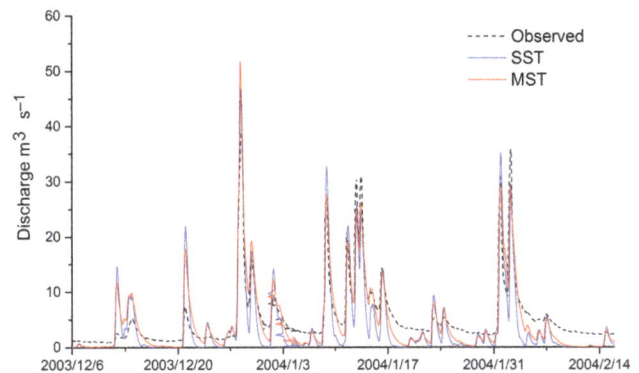

Figure 2. The comparison of model calibration with different soil settings.

Table 1. Soil properties for corresponding HOST number.

HOST	Water content at saturated condition	Field capacity	Wilting point	Infiltration rate (m s^{-1})
9	0.501	0.418	0.244	3.4E-06
18	0.474	0.367	0.162	1.04E-06
16	0.46	0.378	0.219	1.9E-06
33	0.472	0.35	0.144	1.04E-06
3	0.441	0.295	0.117	3.6E-05
25	0.473	0.408	0.255	6.9E-08

Medway catchment are derived from the Hydrology of Soil Types (HOST, see Table 1), provided by the National Soil Resources Institute (NSRI) in the UK.

These data clearly indicate that the recessions period with SST setting was much faster than the observation, possibly due to less resilience from a single-soil-type setting and the shallow depth of soil layer in the model. Consequently, multiple soil types and the increment depth of the soil layer were introduced to the model while the recession of the flood was improved. Additionally, the most critical performance criterion for the model, the Nash–Sutcliffe efficiency (NSE), increases from 0.68 (SST) to 0.82 (MST).

2.3 The settings of a coupled UKV-HYPE case study

The UKV model is set to make 36 h forecasts with a high-resolution inner domain (1.5 km grid boxes) over the area of forecast interest, separated from a coarser grid (4 km) near the boundaries by a variable resolution transition zone. This variable resolution approach allows the boundaries to be moved further away from the region of interest, reducing unwanted boundary effects on the forecasts.

Part of the motivation of using such resolution is to improve forecasts of convective rainfall. The variable resolution model with 1.5 km grid length over the UK (although increasing to 4 km at the edges of the domain) enables the

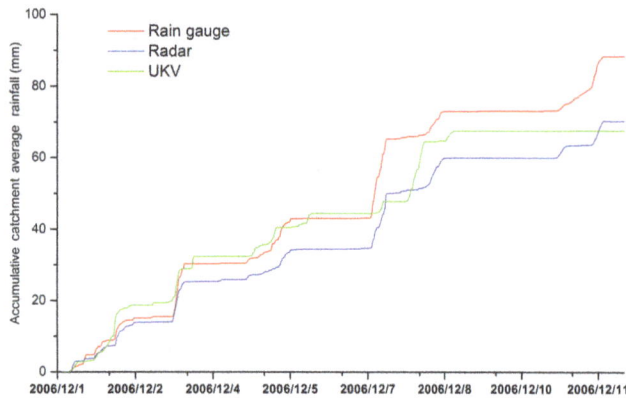

Figure 3. The comparison of accumulative catchment average rainfall (event December 2006).

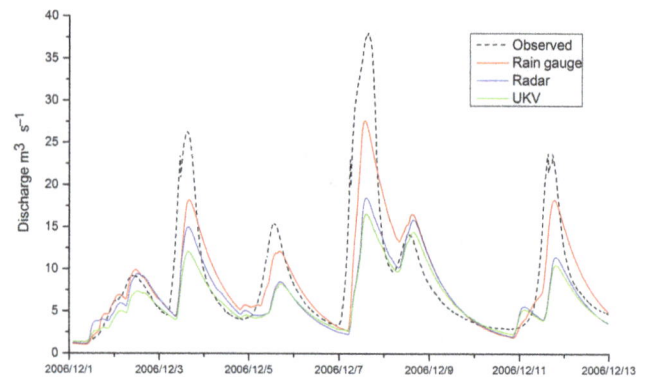

Figure 4. The comparison of flow simulation in HYPE (event December 2006).

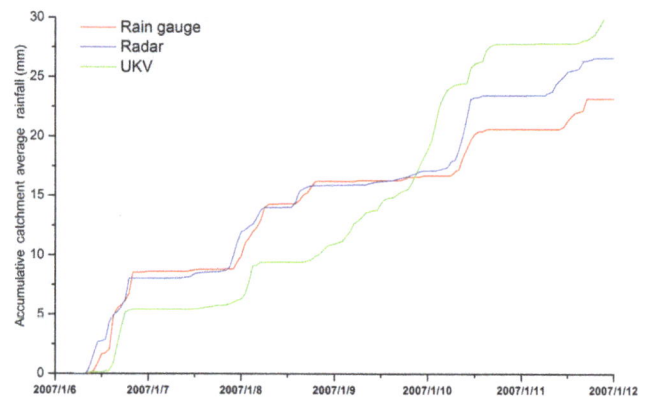

Figure 5. The comparison of accumulative catchment average rainfall (event January 2007).

boundaries of the model to be pushed further away from the area of interest at lower cost, and also enables reduction of the resolution mismatch with the driving (12 km) model. The UKV rainfall estimation produced by the Unified Model is used as the input for the HYPE model, which provides the cornerstone to the coupled UKV-HYPE model.

3   Results and discussions

Four flood events were selected to evaluate the performance of UKV rainfall products through HYPE hydrological model, by comparing the simulated streamflow driven by rain-gauge measurements, NIMROD radar rainfall estimates, and UKV rainfall data. For the first flood event, there was around 100 mm depth of precipitation over the Upper Medway catchment during 1 to 13 December 2006, according to the rain-gauge rainfall record.

Figure 3 shows the rainfall comparison on the accumulation of catchment average rainfall over the flood period. The UKV rainfall products had quite a good agreement with rain-gauge measurements before the high peak flow occurred on 7 December 2006, in terms of the accumulative catchment average rainfall. However, the hydrological simulations illustrated in Fig. 4 indicate that the rain-gauge measurements outperform the UKV rainfall product, especially on peak-flow simulations.

Figure 4 shows the comparison of the hydrological model performances driven by three different rainfall products in the entire event. The NIMROD radar rainfall estimates and UKV rainfall products were underpredicted on all the peak flows, especially on the highest peak flow that occurred around 8 December 2006, compared with the rain-gauge measurements. However, the UKV rainfall products have very similar performance with radar rainfall estimates, on the peak-flow volume and the time of the peak, which implies that the high-resolution NWP rainfall products are as good as the radar rainfall estimates in this flood event.

For the second flood event, the comparison of accumulative catchment rainfall is shown in Fig. 5. The trends on the rainfall data are reasonably good across all three data sets. The UKV rainfall data do however pick up some exaggerated noisy peaks over the period between 9 and 10 January 2007.

Figure 5 also shows that the NIMROD radar data produced more rainfall depth over the catchment than rain-gauge measurements, but less than the UKV rainfall. In addition, it shows similar rising cumulative rainfall for this event between all three data sets, and particularly between rain-gauge measurements and radar rainfall estimation up to 10 January. In contrast, one notes that the UKV rainfall underestimates rain-gauge and radar data sets before 10 January, but with a similar rising trend. Departure arises subsequently between all three data sets, with UKV rainfall providing the extreme outcome.

The performance of UKV rainfall in the HYPE model simulation for the January 2007 flood event of Fig. 6 shows that the peaks and troughs are reasonably well represented against the observed data up to 10 January, after which the fourth peak is overestimated, and thus so is the final peak. The radar

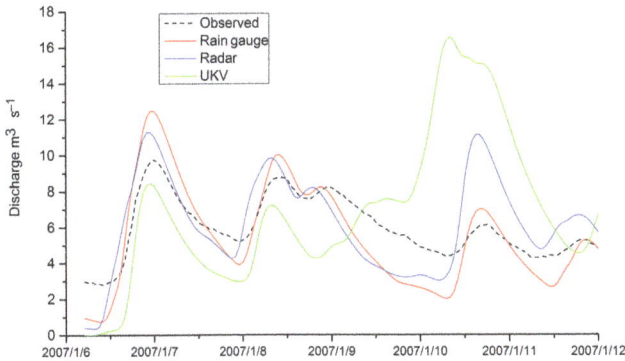

Figure 6. The comparison of flow simulation in HYPE (event January 2007).

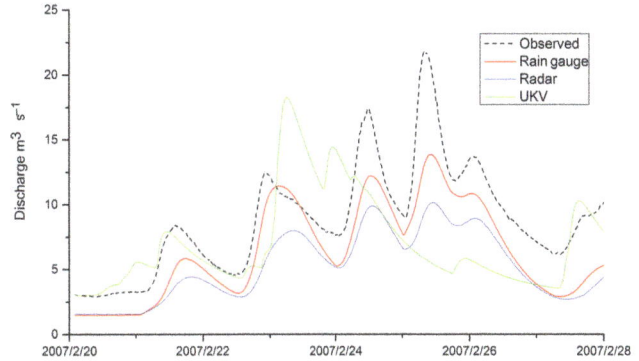

Figure 8. The comparison of flow simulation in HYPE (event February 2007).

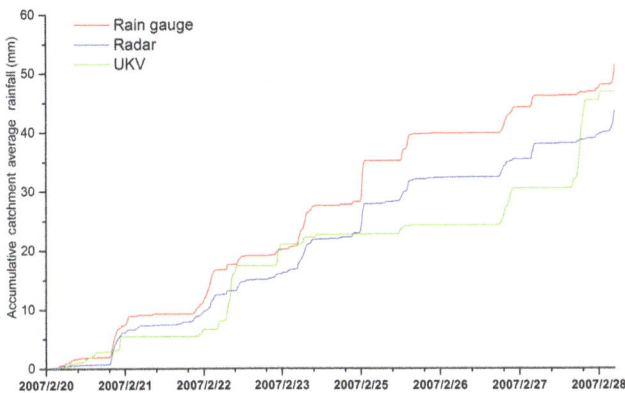

Figure 7. The comparison of accumulative catchment average rainfall (event February 2007).

data suffer likewise, over the final two peaks, which are better captured by the rain-gauge data. The rain-gauge data do however underestimate the observed data output over this period.

During the third flood event, there was a 50 mm rainfall depth in total over the catchment, recorded by the rain gauges, which triggered the highest discharge at the catchment outlet of about $25\,\mathrm{m}^3\,\mathrm{s}^{-1}$ during the flood period. In terms of the cumulative catchment rainfall, the rain-gauge measurement produced more precipitation than the UKV rainfall, followed by the radar rainfall estimation.

However, Fig. 7 shows that the UKV rainfall product did not capture the trend of accumulative rainfall over the catchment, and therefore totally missed the two flow peaks after 24 February 2007 (illustrated in Fig. 8) compared with the rain-gauge measurement and radar rainfall estimates. The rain-gauge data outperform the radar data in this whole event, of which all the peak flows are better captured. However, the rain-gauge data do underestimate the observed data output over this period.

During the final event of July 2007, in terms of the flood magnitude, there was around 80 mm precipitation recorded by the rain gauges over 4 days which caused over $40\,\mathrm{m}^3\,\mathrm{s}^{-1}$

discharge at the catchment outlet. It can be regarded as a similar case to the first flood event on December 2006, where the recorded streamflow was also around $40\,\mathrm{m}^3\,\mathrm{s}^{-1}$, triggered by around 100 mm of precipitation in the catchment over 12 days. However, there were no other peaks before the highest flow appeared in this event and the peak only lasted 1 day, which implied that this was a flash flood (sudden high peak flow over a short period). It can also be identified from Fig. 9, which clearly showed that there was a significant increase (over 40 mm difference) on 20 July for the accumulative catchment precipitation calculated from all rainfall measurements and rainfall estimation products, especially during the period from 20 July at 08:00 LT to 20 July 2007 at 11:00 LT, when over 30 mm of precipitation fell on the catchment in 3 h, detected from the rain-gauge network.

Considering the differences between the rain-gauge measurements and radar rainfall estimates, the precipitation estimated from radar reflectivity could be heavily attenuated. After being converted to Cartesian format, the details of the signal were further smoothed by the averaging process, which could explain the reason that the radar rainfall estimates underestimated a lot more than rain-gauge measurements. Additionally, because the model rainfall input for HYPE is the catchment average precipitation, the rainfall distribution and heterogeneities are not simulated, so that all the modelled flow was not comparable with the observation in this extreme rainfall flood event.

The flow simulation shown in Fig. 10 appears to pick up an exaggerated peak in the UKV rainfall through HYPE model simulation after the first day (16 July 2007), which is not reflected in the other data sets. This early disturbance influences the early undershoot of the observed-data first peak (at 21 July 2007), and the overshoot of the observed-data second peak (before 22 July 2007). Notably, rain-gauge data output overshoots the observed-data first peak, whilst NIMROD radar data output provides an undershoot; both undershoot the observed-data second peak. This is rather a testing event with only one single main flood event to sharply capture. Clearly, one would need to investigate further in this instance

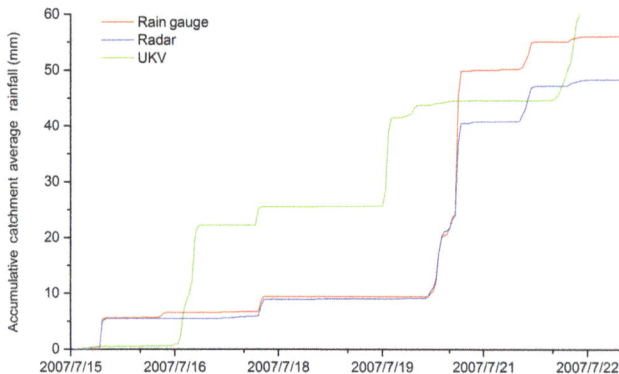

Figure 9. The comparison of accumulative catchment average rainfall (event July 2007).

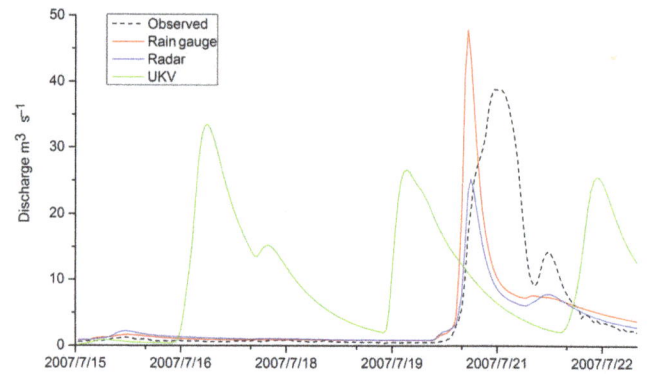

Figure 10. The comparison of flow simulation in HYPE (event July 2007).

as to why the early disturbance has arisen for UKV output in this case, and provide more data evidence to prove or refute this particular finding. Further case study events would help clarify this issue, as the January 2007 event did not show this.

4 Conclusions

This paper describes a recent effort to integrate both the driver NWP models and the impact analyser–hydrological model on a single HPC platform to support better and more refined studies on extreme weather impacts. What distinguishes this study from others is it is first time that modellers are able to simulate the entire system, ranging from the global circulation down to a target catchment, for an impact study. This study also explores the feasibility of building weather and climate services together with the impact-oriented analysis on a single platform, and what can be done if this is not feasible: for example, how computing resources can be re-arranged to deal with the issues.

The initial idea of this study was in fact to include a whole system model including the UKV, data processing, and hydrological models on an HPC platform. This differs from the usual approach of getting the IC/BC (initial condition/boundary condition) from a weather centre and then running a limited area model such as WRF and then some hydrological simulations. Further, it aimed to fully integrate a hydrological model component into the UKV, which is not revealed by this paper yet, but it will form the basis for extension into a two-way coupling system, which will be developed further in the future. Therefore, this study aimed to construct and run both NWP and the hydrological model, either simultaneously or on the same hardware infrastructures, to achieve more effective interaction and communication, so that the potential of fully coupled NWP-hydrological forecasts on a single computer platform is explored, and first-hand knowledge on fully integrated hydro-meteorological modelling can be obtained.

The study finds that when running the entire system on a reasonably powerful HPC platform, the overall time frame does not yet allow for a real-time simulation even without the most complex and demanding data simulation part. It is therefore suggested that the components responsible for large scale simulation, such as global and European areas, should remain at national weather service centres where dedicated HPC resources can deal with the demand as they already have been doing. However, it is still possible to have a high-resolution version with less geographical coverage running on a general-purpose HPC platform together with the impact analysing model such as a hydrological model and further inundation models. This configuration also allows for finer control and/or tuning of the models to fit various purposes.

The other main purpose of this study is to gain sight of how a common hydrological model can utilize the high-resolution precipitation (among others) forecast and simulation in an impact study of extreme weather events. It is encouraging to find that even without fine-tuning, such as using various parameterization schemes, the coupled hydro-meteorological was still able to capture the major flood peaks with much longer lead time compared with the conventional gauge- or radar-driven forecast (2–3 days vs. 2–3 h). The high-resolution UKV rainfall shows some promising agreement with rain-gauge measurements and radar estimation in the first 2–3 days in this flood event, both in the average catchment rainfall amount and hydrological simulation in HYPE.

The study also identified uncertainties associated with precipitation forecast, particularly that it will increase as the forecast horizon goes beyond 3 days. For example, the latter part of the flood event was not represented well by the HYPE model simulation using the UKV rainfall, compared with those using other sources of rainfall, e.g. radar and rain gauges. This is, however, understandable and consistent with our previous studies using other models; see, for example, Seyoum et al. (2013). Apparently, other more complicated uncertainty-aware techniques need to be applied in this model coupling configuration, which, in fact, is the key research topic for further studies.

Overall, the primary value of this study is in making sure that the high-resolution UKV can be used to drive hydrological models like HYPE; and secondly, the study demonstrates that it is possible that with a moderate resources requirement, a fully integrated system can be established to benefit from independent data assimilation/NWP coupled with hydrological models. Although many deficiencies (such as uncertainty) still exist in such a system, as they do in many WRF-related studies, the ability to simulate from source (atmospheric observations) to end (flow simulation) means that these issues will be addressed more efficiently in the future.

Consequently, the following recommendations for future work are made:

1. The study needs to be repeated and extended, as more data sets become available from UKV.

2. The impact of the high-resolution new radar data needs to be explored in the context of distributed hydrological modelling.

3. The UKV rainfall needs to be fully assessed by various lead times and ensemble simulations, which encapsulate uncertainty generation and propagation through complex "cloud to catchment" or "Whole Systems Modelling" concepts.

Acknowledgements. This study was partly funded under the Knowledge Transfer Partnership scheme, between Fujitsu Europe Laboratories Ltd and Swansea University, acting through support from Innovate UK (KTP009201) and the Welsh Government (GON CFI 468). The authors D. Zhu and S. Echendu would like to gratefully acknowledge this support.

The study is also partly supported under the ongoing research initiative, "Welsh Extreme Weather Study". The HPC resources were provided by HPC Wales (HPCW179). Access to the UM/UKV was supported by UK Met Office, under license agreement UM-L0030, which is also gratefully acknowledged. The NIMROD data were provided by the UK Met Office. The catchment soil data were provided by National Soil Resources Institute. The HYPE model and continuous modelling technical support were provided by the Swedish Meteorological and Hydrological Institute. Finally, our thanks go to the National Centre for Atmospheric Sciences for their technical support and access to HPC resources.

Edited by: Q. Chen

References

Cloke, H. L. and Pappenberger, F.: Ensemble flood forecasting: A review, J. Hydrol., 375, 613–626, 2009.

Dudhia, J.: A non-hydrostatic version of the Penn State/NCAR mesoscale model: validation tests and simulation of an Atlantic cyclone and cold front, Mon. Weather Rev., 121, 1493, doi:10.1175/1520-0493(1993)121<1493:ANVOTP>2.0.CO;2, 1993.

Golding B. W.: NIMROD: a system for generating automated very short range forecasts, Met. Appl., 5, 1–16, 1998.

Grell, G., Dudhia, J., and Stauffer, D.: A description of the fifth generation Penn State/NCAR Mesoscale Model (MM5), NCAR Technical Note, NCAR/TN-398CSTR, 117 pp., 1994.

Jasper, K., Gurtz, J., and Lang, H.: Advanced flood forecasting in Alpine watersheds by coupling meteorological observations and forecasts with a distributed hydrological model, J. Hydrol., 267, 40–52, 2002.

Kunstmann, H. and Stadler, C.: High resolution distributed atmospheric-hydrological modelling for Alpine catchments., J. Hydrol., 314, 105–124, 2005.

Schulla, J. and Jasper, K.: Model Description WASIM-ETH (Water Balance Simulation Model ETH), ETH-Zurich, Zurich, 2000.

Seyoum, M., van Andel, S., Xuan, Y., and Amare, K.: Precipitation Forecasts for Rainfall Runoff Predictions: A case study in poorly gauged Ribb and Gumara catchments, upper Blue Nile, Ethiopia, Phys. Chem. Earth, 61–62, 43–51, doi:10.1016/j.pce.2013.05.005, 2013.

Smith, K. T. and Austin, G. L.: Nowcasting precipitation – a proposal for a way forward, J. Hydrol., 239, 34–45, 2000.

Westrick, K., Storck, P., and Mass, C.: Description and evaluation of a hydrometeorological forecast system for mountainous watersheds, Weather Forecast, 17, 250–262, 2002.

Wigmosta, M. S., Vail, L. W., and Lettenmaier, D. P.: A distributed hydrology-soil-vegetation model for complex terrain, Water Resour. Res., 30, 1665–1679, 1994.

Xuan, Y., Cluckie, I. D., and Wang, Y.: Uncertainty analysis of hydrological ensemble forecasts in a distributed model utilising short-range rainfall prediction, Hydrol. Earth Syst. Sci., 13, 293–303, doi:10.5194/hess-13-293-2009, 2009.

Yarnal, B., Lakhtakia, M. N., Yu, Z., White, R. A., Pollard, D., Miller, D. A., and Lapenta, W. M.: A linked meteorological and hydrological model system: the Susquehanna River Basin Experiment (SRBEX), Global Planet. Change, 25, 149–161, 2000.

Zhu, D. and Cluckie, I. D.: A preliminary appraisal of Thurnham Dual Polarisation Radar in the context of hydrological modelling structure, J. Hydrol. Res., 43, 736–752, 2012.

Zhu, D., Xuan, Y., and Cluckie, I.: Hydrological appraisal of operational weather radar rainfall estimates in the context of different modelling structures, Hydrol. Earth Syst. Sci., 18, 257–272, doi:10.5194/hess-18-257-2014, 2014.

5

Event-based stochastic point rainfall resampling for statistical replication and climate projection of historical rainfall series

Søren Thorndahl[1], **Aske Korup Andersen**[2], **and Anders Badsberg Larsen**[2]

[1]Department of Civil Engineering, Aalborg University, Aalborg, 9220, Denmark
[2]Niras A/S, Aalborg, 9000, Denmark

Correspondence to: Søren Thorndahl (st@civil.aau.dk)

Abstract. Continuous and long rainfall series are a necessity in rural and urban hydrology for analysis and design purposes. Local historical point rainfall series often cover several decades, which makes it possible to estimate rainfall means at different timescales, and to assess return periods of extreme events. Due to climate change, however, these series are most likely not representative of future rainfall. There is therefore a demand for climate-projected long rainfall series, which can represent a specific region and rainfall pattern as well as fulfil requirements of long rainfall series which includes climate changes projected to a specific future period.

This paper presents a framework for resampling of historical point rainfall series in order to generate synthetic rainfall series, which has the same statistical properties as an original series. Using a number of key target predictions for the future climate, such as winter and summer precipitation, and representation of extreme events, the resampled historical series are projected to represent rainfall properties in a future climate. Climate-projected rainfall series are simulated by brute force randomization of model parameters, which leads to a large number of projected series. In order to evaluate and select the rainfall series with matching statistical properties as the key target projections, an extensive evaluation procedure is developed.

1 Introduction

In design of new and analysis of existing storm water drainage systems valid rainfall statistics are crucial. With climate changes anticipated to impact precipitation patterns, the historical rainfall statistics upon which the traditional design is based, is no longer valid for future design. There is therefore a need for climate projection of the rainfall statistics in order for these to represent the future loads on storm water drainage systems.

Traditionally many simple urban drainage systems are designed with intensity–duration–frequency (IDF) relationships, or types of design storms (e.g. Unit Hydrograph: Sherman, 1932; Chicago Design Storm, CDS: Keifer and Chu, 1957; SCS: NRCS, 1986) which represent statistics for rain with specific return periods. Climate projection of these types of design methods can be relatively simple, e.g. by multiplying the design rain by a bias climate factor (e.g. Semadeni-Davies et al., 2008; Olsson et al., 2009; Willems et al., 2012a; Willems, 2013b; Shahabul Alam and Elshorbagy, 2015), assuming that extreme rainfall events for a specific return period will be increased linearly with a given factor as a function of time. The most recognized approach for estimating climate factors is the downscaling of global circulation models (GCMs) and/or regional climate models (RCMs) (e.g. Wilby and Wigley, 1997; Fowler et al., 2007).

In general, statistical downscaling determines a statistical relationship between a large- and a local-scale climate variable based on historical records. The relationship can be used in a GCM/RCM to obtain local variables for a specific domain in a given time frame of climate projection (e.g. Wilby et al., 2002; Nguyen et al., 2007; Willems and Vrac, 2011; Willems et al., 2012b; Arnbjerg-Nielsen, 2012; Sunyer et al., 2015). The statistical downscaling approach requires long historical records of observations in order to establish the necessary statistical relationships. Based on various types of statistical downscaling assumptions and methods, climate factors for urban drainage design purposes (e.g for multiplication on IDF relationships) can be derived by statistically comparing contemporary climate conditions with projected future rainfall with regards to specific return periods, and aggregation levels (durations) or rainfall (e.g. Mailhot et al., 2007; Larsen et al., 2009; Madsen et al., 2009; Nguyen et al., 2009, 2010; Willems and Vrac, 2011; Olsson et al., 2012; Willems, 2013b).

Whereas a large proportion of the recent research described above has been conducted on estimating climate factors for design purposes, there is also a significant need, not only to describe future extremes (e.g. in the form of IDF relationships) but also to be able to project climate changes to continuous rainfall time series. Basically, simple design methods assume agreement between the return period of the rain intensity (for a given duration), and on the other hand the return period of the critical load in the drainage system (water level, flow, basin storage, etc.). Multiplication of climate factors to design storms, e.g. IDF relationships, is sufficient for many applications of urban drainage design; however, for more complex drainage systems with non-linear rainfall runoff response the simple design methods falls short. That is, for complex systems the return periods of the rainfall duration and intensity are not in agreement with the return periods of the corresponding drainage system state. Therefore, historical rainfall series (or climate-projected rainfall series) are required for complex systems in order to estimate maximum water levels in manholes, flooding, to estimate the return periods, and other loads on the drainage system such as outlet to recipient, inlet to wastewater treatment plants, combined sewer overflow, outlet flow, and pollutants loads in the future climate (e.g. Schaarup-Jensen et al., 2009; Thorndahl, 2009; Thorndahl et al., 2015).

According to Willems et al., (2012a, b) there are generally two methods that produce continuous climate-projected time series either by (1) stochastic rainfall generators which generate locally representative synthetic rainfall conditioned on climate variables in present and future climate or (2) statistical approaches to downscaling such as change factor, resampling or weather typing methods, in which future local rainfall is sought in historical rainfall records under equivalent historical climate conditions as projected in the future, or modified to represent future climate conditions.

In the literature, the most acknowledged methods for stochastically generating synthetic rainfall series are based on Poisson cluster processes and rectangular pulse models such as Bartlett–Lewis (Koutsoyiannis and Onof, 2001; Onof and Wheater, 1994, 1993; Segond et al., 2007; Onof and Arnbjerg-Nielsen, 2009; Paschalis et al., 2014; Kossieris et al., 2016) or Neyman–Scott (e.g. Entekhabi et al., 1989; Cowpertwait, 1991, 2010; Cowpertwait et al., 2002; Fowler et al., 2005; Burton et al., 2008; Paschalis et al., 2014; Sørup et al., 2016). Calibration of the generators is typically performed by comparing generated series to observed series and adjusting relevant parameters prior to climate projection. Methods for estimating point rainfall (e.g. Cowpertwait et al.,

1996; Marani and Zanetti, 2007; Onof and Arnbjerg-Nielsen, 2009) and spatially distributed rainfall or multi-site generators with spatial dependency (e.g. Kilsby et al., 2007; Burton et al., 2008; Sørup et al., 2016) have been applied. These methods have been shown to provide valid results for hourly or daily time steps but also have significant shortcomings in terms of modelling rainfall at a finer temporal resolution. For urban hydrological applications with fast rainfall response, a temporal resolution of input data down to 1–10 min is required (e.g. Schilling, 1991; Willems, 2000; Thorndahl et al., 2008, 2016, 2017). Because we are interested in maintaining the fine temporal resolution of observed rainfall series, generation of synthetic rainfall series using Poisson clusters is rejected here as an applicable method.

Change factor, resampling or weather typing methods (Willems et al., 2012a, b) of statistical downscaling outcomes of RCMs/GCMs can provide data in the required temporal resolution, since directly based upon historical records. Arnbjerg-Nielsen (2012) applied historical rain series originating from another geographical region, which had a climate analogue to the projected climate in order to obtain continuous representative rainfall series for future climate conditions. Zorita and Von Storch (1999), Olsson et al. (2009), Willems and Vrac (2011), and Ntegeka et al. (2014) used historical records of rain and modified these records to represent climate-representative continuous climate-projected rain series. Ntegeka et al. (2014) alternated the number of dry and wet days and used *quantile perturbation* (an advanced delta change method) to modify rainfall intensities. Olsson et al. (2009) applied the *delta change method* to multiply historical records with bias climate factors depending on rainfall intensity levels in order to fit projections of extreme, seasonal, and annual precipitation. This approach, however, was implemented without alternating the temporal variability and the seasonal distribution of events of the rain series and maintaining the chronology of the original series. This particular shortcoming might be problematic in order to project the frequency of extreme events sufficiently.

The approach presented in this paper is different from the methods presented above, although it can be considered as a variation of *resampling* combined with *stochastic generation*. Whereas other methods use other climate variables, e.g. pressure and temperature, as climate predictors, this approach aims at fitting statistical properties of climate-projected precipitation directly. In this case, these properties are derived from other studies of RCM projection (see Sect. 2 for details). The validity of the method therefore depends on whether the climate-projected target variables are comprehensive and detailed enough to project the future rainfall upon. The aim is to develop a generally adaptive method which can be applied to an arbitrary rainfall series and with different climate scenarios and projection period. in contrast to the studies described above, climate-projected time series are generated directly for urban drainage modelling purposes. The objective has been to develop a generally applica-

Table 1. The calculated Danish climate changes in annual and seasonal precipitation as well as extremes. The values are expressed as a multiplicative climate factor describing the difference between the reference period 1961–1990 and 2071–2100. The A1B scenario is presented in Olesen et al. (2014) and represents 14 regional climate model runs from the ENSEMBLES project. The climate factors from the two RCP scenarios are previously unpublished, but derived from the Euro-CORDEX-11 database (Jacob et al., 2014) and processed statistically for this paper. Standard deviation is listed in parentheses. The indices marked with bold are the ones used in this paper.

Parameter	Climate factors for the period 2071–2100		
	Scenario A1B (Olesen et al., 2014)	Scenario RCP4.5 (unpublished)	Scenario RCP8.5 (unpublished)
Annual precipitation	1.14 (±0.06)	1.08 (±0.06)	1.14 (±0.07)
Winter precipitation (DJF)	1.25 (±0.06)	1.12 (±0.06)	1.24 (±0.07)
Spring precipitation (MAM)	1.13 (±0.06)	1.13 (±0.08)	1.23 (±0.11)
Summer precipitation (JJA)	1.05 (±0.08)	1.06 (±0.18)	1.03 (±0.21)
Fall precipitation (SON)	1.13 (±0.06)	1.05 (±0.07)	1.09 (±0.13)
Events above 10 mm	1.37 (±0.12)	1.20 (±0.13)	1.35 (±0.14)
Events above 20 mm	2.50 (±0.14)	1.41 (±0.30)	1.80 (±0.40)
Max. daily precipitation	1.16 (±0.12)	1.12 (±0.09)	1.24 (±0.11)

ble method that can be used directly by practitioners and scientists within the field of urban drainage, who do not necessarily have detailed knowledge of climate projection, RCM's, downscaling, etc.

The procedure is divided into two major parts: (1) resampling of a historical point rainfall time series ("Method development": Sect. 3.1; "Results and evaluation": Sect. 4.1); and (2) climate projection of resampled time series ("Method development": Sect. 3.2; "Results and evaluation": Sect. 4.2).

The essential concept of the method is to stochastically generate a large number of either resampled historical series or climate-projected series, and to evaluate the statistical properties of the generated series against a number of key target variables. Rather than optimizing for the best parameter fit, the basic concept is to sample parameters from broad uniform distribution functions for each parameter and to either accept or reject each stochastically simulated series using a specified criterion. Repeating this procedure for a large number of realizations of rainfall series, it is possible to select a number of rainfall series which has a satisfying statistical representativeness in comparison with historical series or climate projection targets. The evaluation procedure is inspired by the generalized likelihood uncertainty estimation (GLUE) method (Beven and Binley, 1992; Thorndahl et al., 2008) and is presented in detail in Sect. 3.3.

The method assumptions and subjectivity are discussed in Sect. 5 and in Sect. 6 conclusions on this approach to climate projection of single-point historical rainfall series are provided.

2 Data

The development of the model is based on rain gauge data from Denmark and projection of Danish climate conditions,

but could easily be extended to other regions/countries of interest.

Specific statistical properties for the future precipitation in Denmark are necessary in order to climate project the resampled rainfall series. In Olesen et al. (2014) the Danish Meteorological Institute has collected and processed data from the ENSEMBLES project (http://www.ensembles-eu.org/, http://ensemblesrt3.dmi.dk/; Van der Linden and Mitchell, 2009; Boberg et al., 2010; Maule et al., 2013). The report includes projection of weather extremes (including precipitation) using the SRES A1B scenario (IPCC, 2007) and is produced from an ensemble of 14 regional climate models in the ENSEMBLES project. The RCMs are simulated for 1961–1990, 2021–2050, and 2071–2100, but in this case only the first and last time interval are applied. Table 1 presents annual and seasonal precipitation increment (expressed as a climate change factor) in 2071–2100 compared to the reference in 1961–1990. Furthermore, the report specifies changes in other climate indices. In the context of precipitation, the variables *number of events above 10 mm*, *number of events above 20 mm*, and *max. daily precipitation* are relevant (Table 1). In this paper these three variables are used to climate-project the resampled rainfall series, as they are considered important with regards to urban drainage modelling. Because the data from Olesen et al. (2014) represent the SRES scenarios (IPCC, 2007), new data representing the representative concentration pathway (RCP) scenarios (IPCC, 2013; Christensen et al., 2015) are developed for this paper. Daily RCM simulations from an ensemble of 14 models has been derived over Denmark from the Euro-CORDEX database (Casanueva et al., 2016; Jacob et al., 2014; Prein et al., 2016) and statistically processed by the same variables as in Olesen et al. (2014). Derived values are provided in Table 1. For the climate projections in this paper the RCP4.5 scenario is

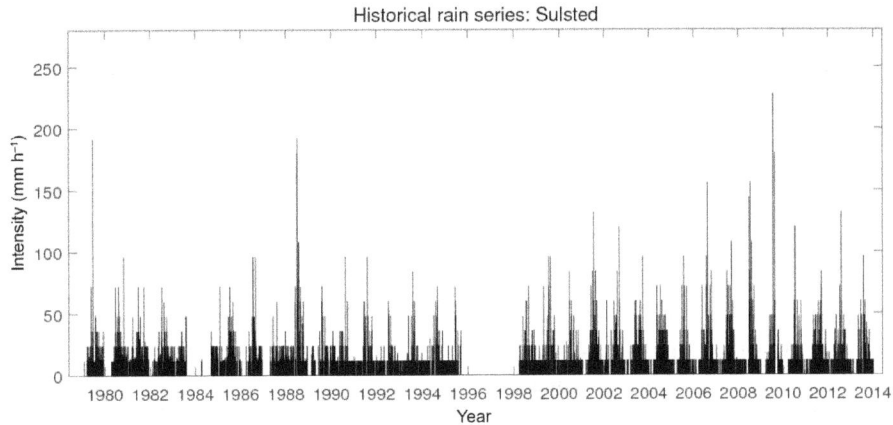

Figure 1. Measured time series of the Sulsted rain gauge. The temporal resolution of rainfall data is 1 min.

Table 2. Recommended climate factors for design of drainage systems in Denmark according to WPC (2008, 2014) and Gregersen et al. (2014b). The climate factors are valid for a duration of 1 h but also recommended for other durations up to 3 h. The indices marked with bold are the ones used in this paper. The standard deviations are not provided directly in the references, but estimated from tables and figures.

Return period (years)	Climate factors for the period 2071–2100		
	Scenario A2 (WPC, 2008)	Scenario RCP4.5 (WPC, 2014)	RCP8.5 (WPC, 2014)
2	1.20 (±0.1)	1.20 (±0.1)	1.45 (±0.1)
10	1.30 (±0.2)	1.30 (±0.2)	1.70 (±0.2)
100	1.40 (±0.3)	1.40 (±0.3)	2.00 (±0.3)

chosen throughout, but the paper could easily have been presented with other SRES or RCP scenarios.

The Water Pollution Committee of the Society of Danish Engineers has published reports (guidelines nos. 29 and 30) with recommendations for design of drainage systems considering climate change (WPC, 2008, 2014, background report: Gregersen et al., 2014b). Based also on the climate simulations of the ENSEMBLES project, the climate factors for drainage system design in Denmark are recommended (Table 2). Design rainfall, e.g. IDF relationships, with a specified return period is recommended to be multiplied by these climate factors. The values are derived for rainfall intensities over 1 h but also recommended for other durations (up to 3 h). In this paper these values are used to certify a correct representation of extreme events.

The rainfall series which are applied in this study has its origin in the rain gauge network of the Water Pollution Committee (WPC) of the Society of Danish Engineers. At present, the network consists of 145 tipping bucket rain gauges (DMI, 2014). The rain gauge no. 5047 located in Sulsted, North Jutland (lat 57.17, long 9.96), is applied since this is a station with a long recording time and few errors compared to other gauge records. The gauge has been in operation over a period of 34 years from 1979 to 2014, but due to minor interruptions in the dataset, the effective length of the series is 32 full years. The interruptions do not affect the statistical calculations as these are excluded from the data before the calculations are performed. The time series of 1 min. values for the Sulsted rain gauge is shown in Fig. 1.

In the WPC rain gauge network the temporal resolution of data is 1 min. The start time of an event is determined at the minute of the first tip of 0.2 mm. All events therefore have initial values equivalent to a multiple of 0.2 mm min^{-1} (12 mm h^{-1}). These initial values are easily identifiable in Fig. 1. The end of an event is specified when there is no registered tip within 1 h. Using this definition of events, the minimum *inter-event time* (time between events) will be 1 h.

Using Danish rainfall data on a daily scale Gregersen et al. (2014a) have been able to identify multidecadal climate oscillations (Ntegeka and Willems, 2008; Willems, 2013a) as well as climate-related changes in precipitation patterns over the past 140 years. Nevertheless, since this paper is based on evidently shorter rainfall series, it is assumed that no significant trends or climate changes in this period are present. The historical records from the Sulsted series are therefore assumed to be stationary in terms of climate properties.

3 Method development

The procedure of the method is presented in two sections: the *resampling of historical rainfall series* (Sect. 3.1) and the *stochastic climate projection of resampled historical rainfall series* (Sect. 3.2). Since both methods involve random selection of events and brute force randomization of parameters there is a need for a unique method to evaluate the generated series against target values. This evaluation method is inspired by the GLUE methodology (Beven and Binley, 1992). The basic concept is to generate a large number of rainfall series and evaluate whether each generated series should be accepted or rejected based on an empirical likelihood (performance) measure based on individual criteria for each target value. For the accepted generated rainfall series a combined performance measure for each realization is calculated in order to find the rainfall series realization which in general fits the target values the best. This method is described in detail in Sect. 3.3.

3.1 Historical rainfall series resampling

The objective is to create synthetic rainfall series resampled stochastically from a historical series such that the synthetic and the historical series have the same statistical properties. The first step is to divide the historical rainfall series into smaller parts in order to describe variability of intensities, event duration, and time between events over the year. We chose to divide the series into four seasons (winter: DJF; spring: MAM; summer: JJA; autumn: SON), although a finer division (e.g. monthly) could have been implemented. Because the target projections (Table 1) are implemented in seasons, this is the one used. The summer precipitation in the synthetic rainfall series is thus generated based on statistics calculated for every summer period's precipitation in the historical rainfall series and correspondingly for the other seasons.

The stochastic generation (resampling) is based on the following:

1. Statistics of the inter-event time (also referred to as *rainfall intermittency*, e.g. by Molini et al., 2001, and Schleiss et al., 2011) using the definition of events presented in Sect. 2.

2. Sampling of rainfall events including original event durations and intensities randomly from the pool of historical rain events for each season.

The concept is outlined in Fig. 2.

The inter-event times(t_{ie}) for each season are approximated by a *two-component mixed exponential probability density function*:

$$\lambda f(t_{ie}) = p\left[\lambda_{a,ie}\exp\left(\lambda_{a,ie}t_{ie}\right)\right] \\ + (1-p)\left[\lambda_{b,ie}\exp\left(\lambda_{b,ie}t_{ie}\right)\right], \quad (1)$$

where $\lambda_{a,ie}$ and $\lambda_{b,ie}$ are the rate parameters for two populations, "a" and "b", with different exponential distributions and p is the weight of population "a". This mixed distribution function was also applied by Rossi et al. (1984) and Willems (2000). Willems (2000) applied the distribution for fitting rainfall intensities arguing that the two distributions originated from two different types of storms (convective thunder storms and frontal storms respectively). The same rationale is applied here. The approximation to inter-event times for each season thus require approximation of three parameters, p, $\lambda_{a,ie}$, and $\lambda_{b,ie}$.

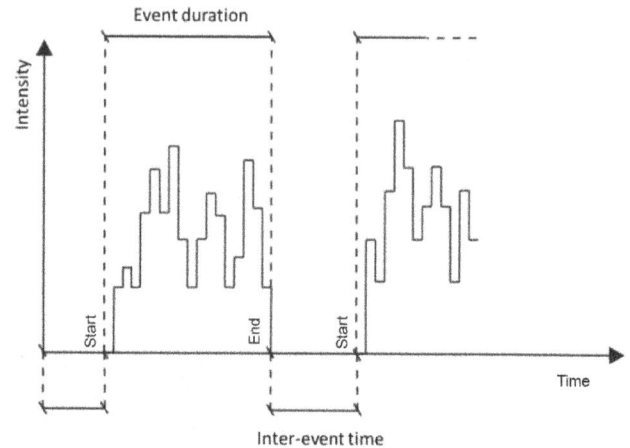

Figure 2. Diagram of the construction of the synthetic (resampled) rainfall series.

Molini et al. (2001) applied a Weibull distribution to describe the inter-event time of rainfall events. The Weibull distribution, along with exponential, gamma, and generalized Pareto distributions was also investigated for this paper, but was however outperformed by the mixed exponential distribution, especially in fitting both ends of the distribution.

As opposed to other rainfall generators which use a fixed timescale (e.g. Furrer and Katz, 2008), the time is sampled discontinuously in this case.

The sampling of the events is an automated process with random selection of events from the pool of historical rainfall events for each season. When sampling a specific event, the intensity sequence and consequently also the duration is maintained. Synthetic resampled time series are, therefore produced by random alternating sampling of the inter-event times and historical events from a specific season. It is possible to sample the same event more than once. The procedure is repeated until the length of the generated series corresponds to the length of the historical series or any other specified length shorter than the total length of the original series. The number and the chronology of events are therefore different from season to season and from year to year.

A vital assumption here is that events from the historical series can be sampled independently. Depending on the meteorological conditions at the time of a specific event there might potentially be some correlation to prior and posterior events due to short inter-event times. Extreme event statistics and development of IDF relationships from partial duration series in Denmark is also produced assuming independent events (Mikkelsen et al., 1998; Madsen et al., 2009), so in order to preserve this methodology, no inter-correlation between events has been implemented in the presented approach.

3.2 Climate projection and stochastic resampling of rainfall series

The climate-projected rainfall series is generated in three steps:

1. The inter-event time for each season is sampled using the same procedure as described in the previous section; however, the parameters of the mixed exponential distribution for each season are implemented as stochastic variables and thus sampled randomly from a uniform distribution with fixed upper and lower boundaries. This allows for different distributions of inter-event times than the ones used in the resampling of historical series. In the climate-projected series, it is thus possible to accommodate for climate changes in seasonal precipitation and the distribution between small and large events, by changing the number of events per season. As an example the method is able to accommodate a moderate increase of total summer precipitation, and at the same time a considerable increase in frequency and intensity of extreme events, with generally a lower number of total events in summer as a result.

2. Rainfall events are sampled from the pool of historical events for each season in the same way as described in Sect. 3.1. The duration of each event is not alternated under impact of climate change, since there is presently no evidence that single events will become shorter or longer in the future. This is obviously a crucial assumption, but nonetheless the best current estimate, which also has been applied by, for example, Olsson et al. (2009). The sampling of events is therefore done without alternating the events from the pool, other than multiplying by different change factors as presented below.

3. The climate projection of the generated time series is inspired by the delta change method. However unlike Olsson, the change factors are implemented as random variables. The change factor for a given rainfall intensity, i, is derived using the probability, $F(i)$, of that the intensity being less than or equal to i. For each season, the rainfall intensities from the original historical rain series are fitted to the same type of mixed exponential distribution (Willems, 2000a) as applied for fitting the inter-event times (Eq. 1):

$$F(i) = p\left[1 - \exp(\lambda_a i)\right] + (1 - p)\left[1 - \exp(\lambda_b i)\right], \quad (2)$$

where λ_a and λ_b are rate parameters for two populations "a" and "b", and p is the weight given to population "a". $F(i)$ has a range from 0 to 1.

For each season change factors are multiplied by intensities on the minute scale. The change factor as a function of intensity, $c(i)$, is thus calculated for each season

by a linear function:

$$c(i) = \alpha F(i) + \beta, \quad (3)$$

where α and β are random variables sampled from uniform distributions with fixed limits.

For each projected rainfall series there is a different value of α and β for each season. During the development of the procedure, the limits of the uniform distribution of α and β for each season were empirically selected starting with broad intervals which were reduced by discarding non-accepted runs (see below).

The total number of random variables for generating climate-projected stochastic rain series in the current setup with four yearly seasons is 20 (2×4 for the change factor plus 3×4 for the mixed exponential distributions).

3.3 Evaluation and optimization procedure

The governing assumption behind the resampling procedure is that the resampled rainfall series should have the equivalent statistical characteristics as the historical series on a number of key target variables. The climate-projected resampled series should therefore also have the equivalent statistical characteristics by means of a number of key target climate projections (as the ones presented in Tables 1–2). It is not a necessity that the same target variables are used to evaluate resampled historical rainfall series and the climate-projected series, but we chose to do so in this paper in order to keep the evaluation procedures the same regardless of generating series which should statistically represent historical series or climate-projected series. The key target variables are described in detail below:

1. Annual precipitation (ap). This target variable is included as it is a measure of the total "mass" balance. Since the individual years of the resampled and historical series are not directly comparable year by year, the mean of all years is applied as target variable.

2. Seasonal precipitation (sp). The mean seasonal precipitation is applied as a target variable in order to ensure same distribution between seasons in the resampled series. The four target parameters are labelled spwi, spsp, spsu, spau corresponding to winter, spring, summer, and autumn precipitation respectively.

3. Number of events above 10 mm per day (n10mm). This target variable provides a measure of the representation of extreme events.

4. Number of events above 20 mm per day (n20mm); same procedure as for no. 3.

5. Maximum daily precipitation (mdp, as a mean of the maximum day for all years). This target variable also certifies the representation of extreme events.

6. IDF relationships. The IDF relationships are traditionally applied in design of urban drainage systems and are therefore relevant to include as a target variable. In accordance with Table 2, it is chosen to use the mean rain intensity over a duration of 60 min for return periods of 2 and 10 years respectively as a target value. The two values are labelled d60T2 and d60T10 respectively.

The performance of each individual target variable is estimated using a simple ratio measure between the target value and the corresponding modelled value:

$$P_{i,j} = 1 - \frac{|T_i - M_{i,j}|}{T_i}. \tag{4}$$

Here $P_{i,j}$ is the individual performance parameter for target variable i (as presented above) corresponding to realization j, T_i is the target value, and $M_{i,j}$ is the modelled value of the target variable of the jth realization. For the evaluation of the resampled series against the historical series, $T_i = H_i$, where H_i is the value of the target variable of the historical series. With respect to the evaluation of the climate-projected rainfall series, where the target value is given by a climate factor (cf) multiplied by the target variable of the historical series,

$$T_i = \mathrm{cf}_i \cdot H_i. \tag{5}$$

Thus the performance measure is

$$P_{i,j} = 1 - \frac{|\mathrm{cf}_i \cdot H_i - M_{i,j}|}{\mathrm{cf}_i \cdot H_i}. \tag{6}$$

Here P can vary between 0 and 1, where $P = 1$ corresponds to a perfect fit.

In order for a simulated rainfall series to be accepted $P_{i,j}$ has to be larger than a specified threshold. For the resampled historical series the acceptance criterion for the individual performance measures is fixed and has been chosen as $P_{\mathrm{crit},i} = 0.90$, hence all 10 individual performance measures should exceed this value in order for the realization to be accepted (Table 4). This means that if a target value of just one of the 10 target values deviates more than 10 % from the value of the historical series, the realization is rejected.

For the climate-projected series, it is possible to estimate individual values of the performance using the standard deviations of the climate factors (cf) given in Tables 1 and 2:

$$P_{\mathrm{crit},i} = 1 - \frac{2 \cdot \sigma_{\mathrm{cf},i}}{\mathrm{cf}_i}. \tag{7}$$

Assuming Gaussian distributed target variables, we will thus accept values which are within the 95 % confidence intervals of the distribution of each target variable. The acceptance criteria of the performance measure will thus be different for each target variable depending on the uncertainty (standard deviation) related to that specific climate projection (see Tables 1 and 2). The acceptance criteria for the performance

of each target variable are presented in Table 6 along with climate factors and standard deviation for each variable.

The combined performance measure P_j of each realization series (j) is estimated as

$$P_j = \sum_{i=1}^{I} w_i P_{i,j}, \tag{8}$$

where w_i is the weights of the individual performance measures, $\sum w_i = 1$, and I is the total number of individual performance parameters.

The individual weights are presented in Sect. 4.2 and Table 6. One could argue that each season should be given the same weight; however, because summer precipitation tends to be more important in terms of extreme events in Denmark this is given a higher weight. Moreover, because winter precipitation might be associated with larger measurement errors due to poor measurement of solid precipitation, this is given a smaller weight.

4 Results and evaluation

4.1 Historical rainfall series resampler

The synthetic resampled series are generated with the same total length as the original historical series – in this case 32 years.

The inter-event times for each season are sampled from the mixed exponential distribution as detailed in Sect. 3.1. The estimated parameters are presented in Table 3. By comparing the parameters, it is evident that there is a significant difference for each season. Therefore, it is important that the inter-event times are sampled individually for each season to ensure a representative number of events in the resampled rainfall series compared to the historical rainfall series. Figure 3 exemplifies empirical cumulative distribution functions for summer inter-events times for the historical series and for the fitted mixed exponential distribution of summer inter-event times. Furthermore, the empirical distribution from the resampled series with the best combined performance measure is presented ($P_j = 0.98$). Using the mixed exponential distribution, there is small underestimation of inter-event times between 1 and 6 h and an equivalent overestimation between 6 and 24 h. This is, however, insignificant in comparison to other fitted distribution functions and thus not considered a problem in random sampling of inter-event times from these distributions.

There is a stringent dependency between inter-event times and number of events in the rainfall series. In order to generate a valid and representational resampled rain series, the number of events series should correspond somewhat to the number of events in the historical rainfall series. Table 4 therefore includes the mean and standard deviation of the number of events per year even though the number of events

Table 3. The fitted rate and weight parameters for the mixed exponential distribution specified for each season.

Parameter	Winter	Spring	Summer	Autumn
Rate, population a, $\lambda_{a,ie}$ (days)	0.38	0.33	0.24	0.26
Rate, population b, $\lambda_{b,ie}$ (days)	4.87	4.46	3.00	2.90
Weight population a, p (–)	0.69	0.56	0.55	0.64

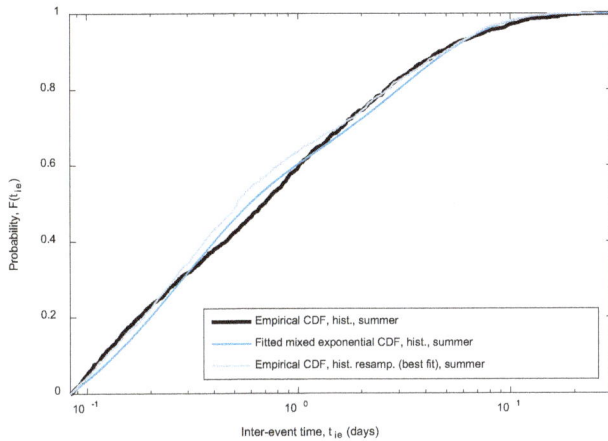

Figure 3. Example of cumulative distribution functions for summer inter-event times.

is not used as a target variable for estimating the individual performances.

The resampling of the observed rainfall series is performed generating 5000 different resampled rainfall series and assessing the performance of each generated series using the method described in Sect. 3.3. Out of the 5000 realizations of simulated series, 275 (5.5 %) are accepted using the criterion of a minimum individual performance measure ($P_{crit,i}$) of 0.90. The fact that all 10 individual performance measures have to be larger than the acceptance criteria has been shown to be a tough condition to fulfil. Often one or two of the 10 has a slightly lower value and the realization is thus rejected. On average the accepted realizations have a combined performance measure (P_j) of 0.95 (ranging between 0.92 and 0.98). Figure 4 presents a bar plot (blue shades) of each of the target variables for the historical series, the one resampled series with the highest combined performance measure, as well as the mean of the accepted resampled series (with uncertainty bounds indicating the minimum and maximum of the accepted series).

Generally there is a good agreement between the historical series and the accepted series on the target parameters with the highest weights, i.e. the seasonal precipitation. This is actually the case for the majority of the 5000 realizations; however, the performance measures becomes rather low if the extreme events are not represented correctly in the resampled series and they are in that case rejected. The variability between the resampled series is only due to the randomness

assembling events and inter-event times from the historical series because the mixed exponential parameters for each season are fixed corresponding to the fits (Table 3). The rejection of resampled series is therefore often due to either sampling of too few or too many "extreme" events within a season.

In many situations, only the one resampled series with the highest performance measure is of interest. Table 4, therefore, lists target values of the historical series and the resampled series with the highest performance measure (best fit). Besides the best combined performance measure of $P_j = 0.98$, the individual performance measures are given in the right column. In order not only to compare series on mean values, Table 4 also presents standard deviations describing the year-to-year variability over the total length of the series. Generally there is a satisfactory agreement (below 10 %) of both mean and standard deviations between the historical series and the "best" accepted resampled rainfall series.

To verify the representativeness of extreme rainfall, Fig. 5 (left) presents IDF relationships (from 10 to 360 min durations) for the historical and "best" resampled series for return periods of 2 and 10 years respectively. Grey areas represent the variability in all the accepted realizations. Generally, there is an acceptable agreement between the curves which verifies the resampling method. There is, however, a minor divergence for short durations of the 10-year return period. In general, the longer the return period the larger the divergence between the curves to be expected as a result of the random sampling of historical events in the generated series. Figure 6 shows the time series of the "best fit" resampled time series.

The overall assessment of the previous evaluation indicates that the rainfall resampler can represent the historical rainfall series well based on the selected performance parameters. Due to the stochasticity of the sampling of inter-event times and rainfall events, there is obviously some variability from year to year and from series to series, but because none of the target variables are significantly biased, the overall performance of the resampler is accepted. As it is possible to produce resampled rainfall series with the same statistics as the corresponding original historical series, the resampling algorithm will be applied to generate climate-projected rainfall series in the following section.

Table 4. Target variables (mean and standard deviation) and performance measures for the historical series and the one resampled series with the highest performance measure.

Target variable		Acceptance criteria and weights		Historical series (target)		"Best fit" resampled series		
		$P_{crit,i}$	w_i	Mean	SD	Mean	SD	P_i
Annual no. of events				200.1	39.4	218.2	45.7	
Annual precipitation	ap (mm)	0.90		576.1	122.3	586.5	140.6	0.96
Seasonal precipitation, winter	spwi (mm)	0.90	0.05	90.9	36.5	101.6	26.7	0.99
Seasonal precipitation, spring	spsp (mm)	0.90	0.10	86.5	43.1	84.9	29.3	0.98
Seasonal precipitation, summer	spsu (mm)	0.90	0.25	213.0	57.0	209.1	75.3	0.98
Seasonal precipitation, autumn	spau (mm)	0.90	0.10	185.7	53.6	190.9	63.9	0.97
Annual number of events above 10 mm per day	n10mm (#)	0.90	0.17	16.0	4.5	15.8	5.4	0.99
Annual number of events above 20 mm per day	n20mm (#)	0.90	0.08	3.3	2.1	3.3	2.3	0.99
Annual maximum daily precipitation	mdp (mm)	0.90	0.08	35.2	12.7	32.0	12.0	0.99
Rain intensity for 60 min, $T = 2$ years	d60T2 (mm h^{-1})	0.90	0.08	15.7		16.4		0.95
Rain intensity for 60 min, $T = 10$ years	d60T10 (mm h^{-1})	0.90	0.08	28.4		28.7		0.99
Combined performance measure	P							0.98

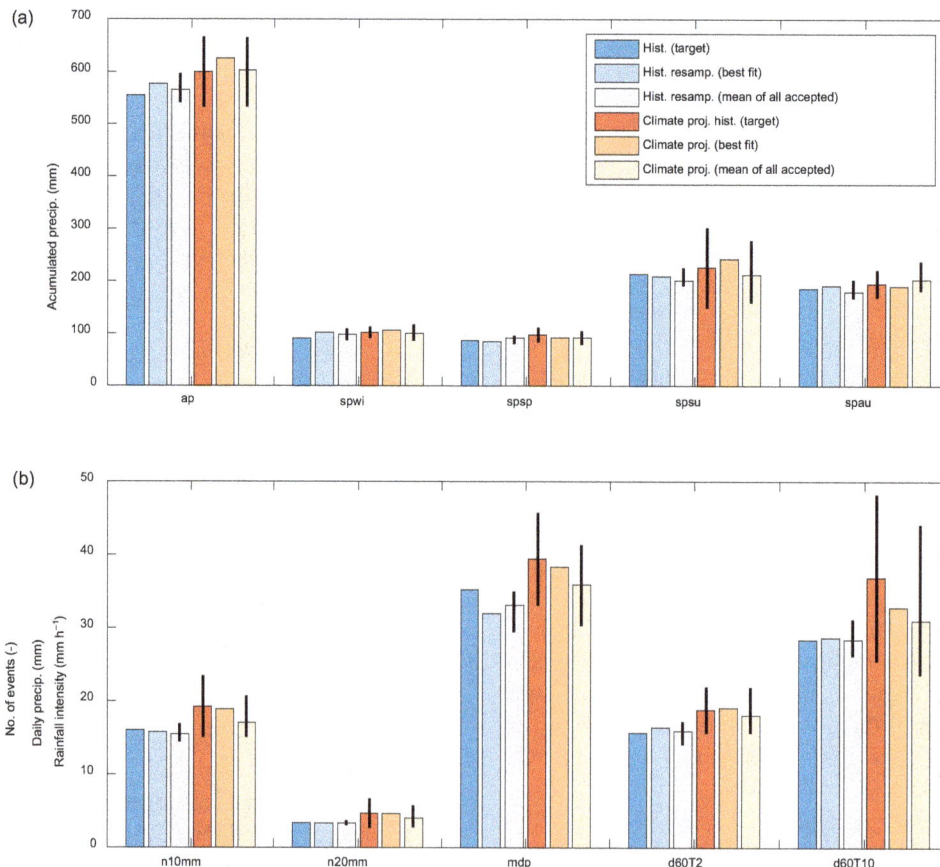

Figure 4. Target variables and their values for comparing historical series and resampled series (in blue shades) and climate-projected historical series and climate-projected and resampled series (in red shades). For the climate-projected target (deep red) the uncertainty bounds (black lines) represent 2 times the standard deviation of Tables 1 and 2. For the resampled series the uncertainty bounds represent the total range of the accepted realizations.

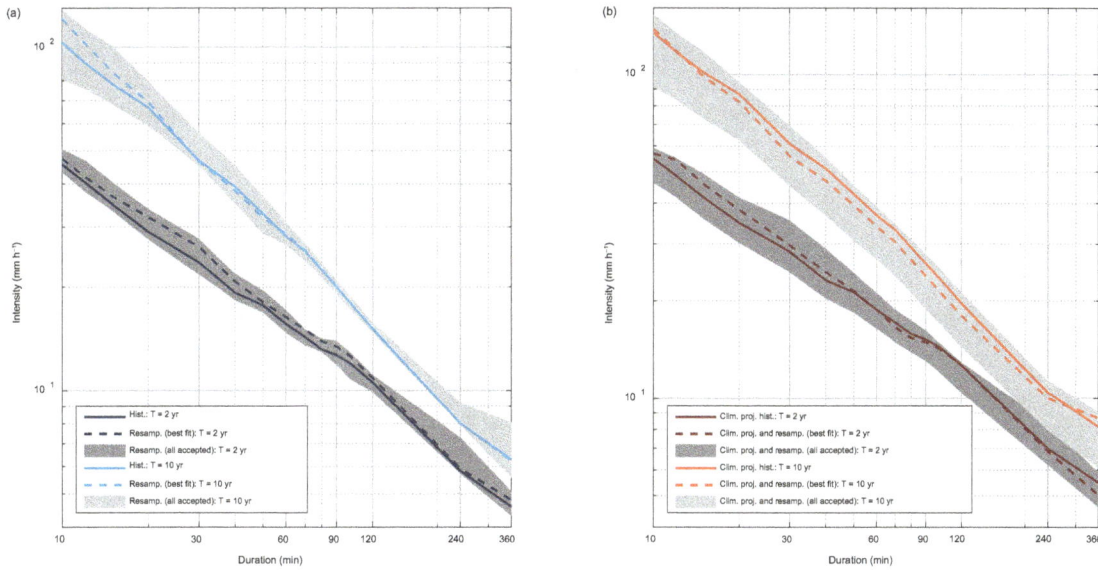

Figure 5. IDF curves for historical and resampled rainfall series **(a)** and climate-projected historical and resampled series **(b)**.

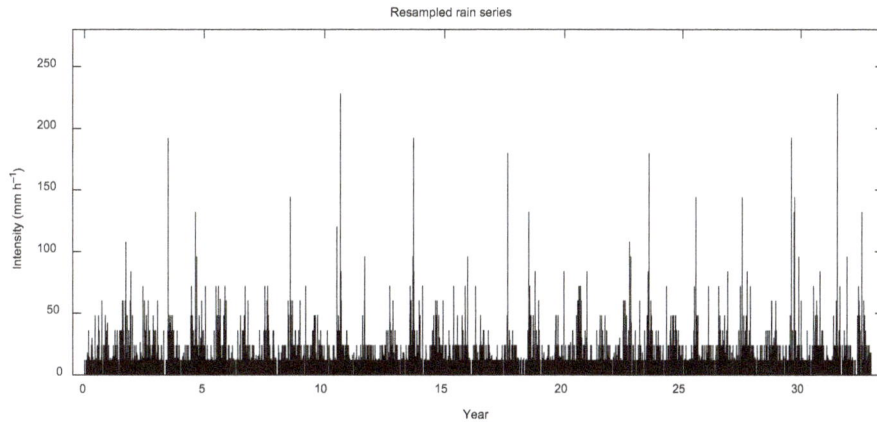

Figure 6. Time series example of resampled rainfall series. The temporal resolution of rainfall data is 1 min.

4.2   Climate-projected rainfall series

Figure 4 and Table 6 provides results for the climate-projected rainfall series. The target variables (climate-projected historical) are estimated using Eq. (5) and are thus the mean values of the historical series of Table 4 multiplied by the climate factors specified in Table 6. In addition Fig. 4 provides an uncertainty estimate on the target values obtained from the standard deviations of Tables 1 and 2.

Because the climate projection of rainfall series involves randomization of not only the event assembling but also randomization of *mixed exponential distribution* parameters and change factors as function of intensity for each season, the generation of rainfall series requires a larger quantity of realizations compared to the resampling of series described in the previous section. Therefore a total of 10 000 climate-projected rainfall series are generated. The acceptance cri-

teria implemented are, however, slightly different compared to the ones detailed in Sect. 4.1. In the evaluation of climate-projected rainfall series an individual acceptance criterion for each target variable is estimated using Eq. (7). For the 10 target variables the acceptance criteria range between 0.59 (n20mm) and 0.89 (spwi) as presented in Table 6. The total number of accepted realizations is 721 (7.2 %). The reason that a larger percentage is accepted here than in the previous section is that the acceptance criterion is somewhat softer encountering the uncertainty of climate factors. On average the accepted realizations have a combined performance measure (P_j) of 0.90 (ranging between 0.81 and 0.97).

Table 5 presents the range of mixed exponential distribution parameters as well as ranges of change factor parameters for the accepted climate-projected realizations for each season. Comparing with Table 3 (in which the parameter assessment is based on fitting the historical data) it is clear that

Table 5. Ranges of accepted parameter values for the mixed exponential distribution applied to sampling inter-event times and for the linear function applied to sample change factors for each season.

Parameter	Winter		Spring		Summer		Autumn	
	min	max	min	max	min	max	min	max
Rate, population a, $\lambda_{a,ie}$, (days)	0.32	0.44	0.27	0.39	0.20	0.28	0.23	0.29
Rate, population b, $\lambda_{b,ie}$, (days)	4.10	5.60	3.90	5.00	2.70	3.30	2.60	3.20
Weight population a, p (–)	0.63	0.74	0.50	0.61	0.51	0.60	0.60	0.68
Change factor slope, α (–)	0.000	0.050	0.000	0.050	0.000	0.025	0.000	0.049
Change factor intercept, β (–)	0.80	1.20	0.80	1.20	0.81	1.20	0.86	1.20

Table 6. Climate factors of target variables and minimum acceptance criteria of the individual performance parameters Pi, j, empirical combined performance measure weights (w_i), climate-projected target variables, and the corresponding values (±standard deviation) of the best-fit climate-projected series.

Target variable		Climate factors c_f	Acceptance criteria and weights $P_{crit,i}$	w_i	Climate proj. hist. series (target) Mean	"Best fit" climate proj. series Mean	SD	P_i
Annual no. of events						206.8	39.4	
Annual precipitation	ap (mm)	1.08 (±0.06)	0.89		599.6	629.8	147.3	0.96
Seasonal precipitation, winter	spwi (mm)	1.12 (±0.06)	0.89	0.05	101.8	105.8	28.7	0.93
Seasonal precipitation, spring	spsp (mm)	1.13 (±0.08)	0.86	0.10	97.7	92.2	39.8	0.99
Seasonal precipitation, summer	spsu (mm)	1.06 (±0.18)	0.66	0.25	225.8	242.1	89.3	0.99
Seasonal precipitation, autumn	spau (mm)	1.05 (±0.07)	0.87	0.10	195.0	189.8	65.9	0.90
Annual number of events above 10 mm per day	n10mm (#)	1.20 (±0.13)	0.78	0.17	19.2	18.9	5.5	0.98
Annual number of events above 20 mm per day	n20mm (#)	1.41 (±0.30)	0.57	0.08	4.7	4.6	2.7	0.99
Annual maximum daily precipitation	mdp (mm)	1.12 (±0.09)	0.84	0.08	39.4	38.3	14.3	0.92
Rain intensity for 60 min, $T = 2$ years	d60T2 (mm h^{-1})	1.20 (±0.10)	0.83	0.08	18.8	19.0		0.99
Rain intensity for 60 min, $T = 10$ years	d60T10 (mm h^{-1})	1.30 (±0.20)	0.69	0.08	36.9	32.8		0.96
Combined performance measure	P							0.97

the parameter values obtained by random sampling have a broader range, indicating that an accepted realization with a high performance value can be obtained from a broad range of parameter values. Scatter plotting the performance values as a function of parameter values (not shown) shows flat tops indicating that an equal performance can be obtained from low and high values within the range (uniform distribution). This means that there is a dependency between inter-event time parameters and chance factor parameters.

As seen in Table 5, the change factor is allowed to be both smaller and larger than 1. This allows for both decrease and increase in precipitation amounts in each seasons. The climate-projected precipitation can thus be obtained from an insignificant change in seasonal precipitation, but a rather large increase in extreme precipitation.

Generally there is an acceptable agreement of climate-projected target variables (*climate-projected historical*) and corresponding values for the climate-projected resampled series (red shades in Fig. 4). There is, however, slightly more deviation compared to the present-time simulations and larger ranges of target variable values. This is as expected since the climate projection includes more random parame-

ters and complexity as well as broader acceptance criteria. For the accepted realization with the highest performance measure, $P = 0.97$ (Fig. 4 and Table 6), there is a tendency for the target variables related to the extreme values to be marginally underestimated. This is inevitably a result of high weights given to the target values related to seasonal precipitation, especially summer precipitation. By changing the weights it would be possible to obtain more equal extreme values, however potentially at the expense of a poorer fit of the accumulated precipitation values.

In Fig. 5 (right) the IDF curves for the climate-projected series are shown. There is a slight underestimation of extremes for the 10-year return period, but an overestimation of the 2-year return periods on low durations. Since the total length of the series is 32 years, return periods larger than 10 years are not presented well, since they the associated with large uncertainties (see e.g. Thorndahl, 2009). The uncertainty bands (grey areas) however cover the climate-projected intensities. Figure 7 shows the time series of the "best fit" resampled time series.

The overall performance of the climate projection of re-sampled rainfall series is considered to be acceptable within

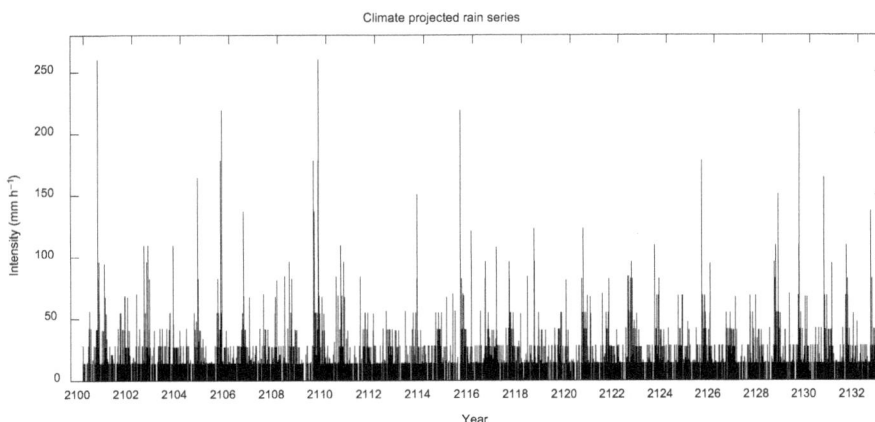

Figure 7. Time series example of climate-projected rainfall series. The temporal resolution of rainfall data is 1 min.

the range of uncertainties related to the climate projections. The introduction of 20 random variables and the random assembling of rain events obviously require many realizations in order to produce accepted rainfall series which have a satisfactory degree of agreement on all target parameters.

5   Discussion

The developed procedure obviously involves a large degree of subjectivity in the choice of processes and parameters to include. This section will discuss and argue for some of these choices.

The target variables have been chosen to represent both annual and seasonal precipitation as well as more extreme values. The choice of the 10 specific target variables is closely connected to the fact that this is what is currently available for Danish future climate conditions. However, when other, maybe more detailed, target variables becomes available, it would be possible to redo the generation of climate-projected rainfall series with other target variables. It was initially decided only to present values from the RCP4.5 climate scenario; however, the implementation of the method could just as well have been implemented with another RCP or SRES scenario. Another possibility could be to implement other durations and return periods than for 60 min durations for 2 and 10 years respectively in order to emphasize specific extremes further.

It is of utmost importance that the chosen target variables are representative of the future precipitation patterns and that they are comprehensive in the way that they cover both annual/seasonal variations and single events and the statistics related to these. In this paper, we chose only to include yearly mean values of target parameters (except for the target variables related to return periods), but it could also be relevant to apply the year-to-year variability as a target in itself in order to certify the correct representativeness of the resampled series in comparison with the original historical series.

The weights applied in estimating the overall performance of resampled series are chosen in order to emphasize the accumulated precipitation values but, on the other hand, not neglect the extremes. Other weights could have been applied. One could imagine that the weights were chosen according to the purpose of use of the resampled and climate-projected series. If, for example, the series were to be used as an input to an urban drainage model simulating overflow from combined sewer systems to a recipient, it would probably be most important to have a good representation of the precipitation (event) totals. On the other hand if the purpose was simulating surcharge or flooding of a drainage system, the representation of extremes would be more important.

In the present approach a linear function and the probability of a given rainfall intensity for a given season is applied to derive the change factor as a function of intensity. The choice of parameters allows change factors to be both smaller and larger than 1. This might entail that the lowest fraction of intensities is allowed to be smaller in a future climate while the highest fraction of intensities will increase. Other continuous functions, rather than the applied linear function, might be an objective of future studies.

The proposed method applies two major assumptions which are relevant to discuss here. The first assumption is that events are sampled independently for each season. With inter-event times down to 1 h, this might constitute a problem in hydrological applications where the response time of the system in question is larger than 1 h. Hence, coupled events might impact the hydrological system response. The second assumption is that the duration of events does not change under changed climate signals. It has presently not been possible to find evidence for this contention in the scientific literature on climate change. Both of the assumptions are subject to further investigations.

6 Conclusions

This paper presented a procedure to generate both statistically representative resampled rainfall series from original historical rainfall series as well as climate-projected rainfall series, which includes the advantages in local historical rainfall series as well as projections on changes in rain patterns in the future climate.

The simulated rainfall series can represent the climate-projected target variables and it is shown possible to produce rainfall series which project not only accumulated seasonal precipitation but also extremes in correspondence with the climate projection of the RCP4.5 scenario. The procedure is generic, so if other climate scenarios and potentially other target variables for further precipitation patterns are available, the method will be able to adapt to these as well.

The procedure for generating resampled and climate-projected rainfall series fulfils a need for having local representative rainfall series which are valid both for the present and future climate. The series can be applied directly as inputs to urban drainage models in order to analyse the loads on a drainage system, e.g. combined sewer overflow, surcharge, storage filling, and flooding in the present and future climate.

Competing interests. The authors declare that they have no conflict of interest.

Acknowledgements. The authors would like to acknowledge Cathrine Fox Maule at the Danish Meteorological Institute for providing RCM data from EURO-CORDEX database.

Edited by: Nadia Ursino

References

Arnbjerg-Nielsen, K.: Quantification of climate change effects on extreme precipitation used for high resolution hydrologic design, Urban Water Journal, 9, 57–65, https://doi.org/10.1080/1573062X.2011.630091, 2012.

Beven, K. and Binley, A.: The future of distributed models: Model calibration and uncertainty prediction, Hydrol. Process., 6, 279–298, https://doi.org/10.1002/hyp.3360060305, 1992.

Boberg, F., Berg, P., Thejll, P., Gutowski, W. J., and Christensen, J. H.: Improved confidence in climate change projections of precipitation further evaluated using daily statistics from ENSEMBLES models, Clim. Dynam., 35, 1509–1520, https://doi.org/10.1007/s00382-009-0683-8, 2010.

Burton, A., Kilsby, C. G., Fowler, H. J., Cowpertwait, P. S. P., and O'Connell, P. E.: RainSim: A spatial-temporal stochastic rainfall modelling system, Environ. Model. Softw., 23, 1356–1369, https://doi.org/10.1016/j.envsoft.2008.04.003, 2008.

Casanueva, A., Kotlarski, S., Herrera, S., Fernández, J., Gutiérrez, J. M., Boberg, F., Colette, A., Christensen, O. B., Goergen, K., Jacob, D., Keuler, K., Nikulin, G., Teichmann, C.,

and Vautard, R.: Daily precipitation statistics in a EURO-CORDEX RCM ensemble: added value of raw and bias-corrected high-resolution simulations, Clim. Dynam., 47, 719–737, https://doi.org/10.1007/s00382-015-2865-x, 2016.

Christensen, O. B., Yang, S., Boberg, F., Maule, C. F., Thejll, P., Olesen, M., Drews, M., Sørup, H. J. D., and Christensen, J. H.: Scalability of regional climate change in Europe for high-end scenarios, Climate Res., 64, 25–38, https://doi.org/10.3354/cr01286, 2015.

Cowpertwait, P. S. P.: Further developments of the neyman-scott clustered point process for modeling rainfall, Water Resour. Res., 27, 1431–1438, https://doi.org/10.1029/91WR00479, 1991.

Cowpertwait, P. S. P.: A spatial-temporal point process model with a continuous distribution of storm types, Water Resour. Res., 46, 1–12, https://doi.org/10.1029/2010WR009728, 2010.

Cowpertwait, P. S. P., O'Connell, P. E., Metcalfe, A. V., and Mawdsley, J. A.: Stochastic point process modelling of rainfall. I. Single-site fitting and validation, J. Hydrol., 175, 17–46, https://doi.org/10.1016/S0022-1694(96)80004-7, 1996.

Cowpertwait, P. S. P., Kilsby, C. G., and O'Connell, P. E.: A space-time Neyman-Scott model of rainfall: Empirical analysis of extremes, Water Resour. Res., 38, 6-1–6-14, https://doi.org/10.1029/2001WR000709, 2002.

Danish Meteorological Institute (DMI): The rain gauge network of the Danish Water Pollution Committee of the Society of Danish Engineers, available at: https://www.dmi.dk/erhverv/anvendelse-af-vejrdata/spildevandskomiteens-regnmaalersystem/, last access: 31 August 2017.

DMI: Drift af Spildevandskomitéens Regnmålersystem, Technical report 15-03, Rikke Sjølin Thomsen (ed.), Danish Meteorological Institute, Copenhagen, Denmark, 2014.

Entekhabi, D., Rodriguez-Iturbe, I., and Eagleson, P. S.: Probabilistic representation of the temporal rainfall process by a modified Neyman-Scott Rectangular Pulses Model: Parameter estimation and validation, Water Resour. Res., 25, 295–302, https://doi.org/10.1029/WR025i002p00295, 1989.

Fowler, H. J., Kilsby, C. G., O'Connell, P. E., and Burton, A.: A weather-type conditioned multi-site stochastic rainfall model for the generation of scenarios of climatic variability and change, J. Hydrol., 308, 50–66, https://doi.org/10.1016/j.jhydrol.2004.10.021, 2005.

Fowler, H. J., Blenkinsop, S., and Tebaldi, C.: Linking climate change modelling to impacts studies: Recent advances in downscaling techniques for hydrological modelling, Int. J. Climatol., 27, 1547–1578, https://doi.org/10.1002/joc.1556, 2007.

Furrer, E. M. and Katz, R. W.: Improving the simulation of extreme precipitation events by stochastic weather generators, Water Resour. Res., 44, 1–13, https://doi.org/10.1029/2008WR007316, 2008.

Gregersen, I. B., Madsen, H., Rosbjerg, D., and Arnbjerg-Nielsen, K.: Long term variations of extreme rainfall in Denmark and southern Sweden, Clim. Dynam., 44, 3155–3169, https://doi.org/10.1007/s00382-014-2276-4, 2014a.

Gregersen, I. B., Sunyer, M. A. P., Madsen, H., Funder, S., Luchner, J., Rosbjerg, D., and Arnbjerg-Nielsen, K.: Past, present, and future variations of extreme precipitation in Denmark, DTU Environment, Kgs. Lyngby, Denmark, 2014b.

IPCC: Climate change 2007: the physical science basis summary for policymakers, Energ. Environ., 18, 433–440, https://doi.org/10.1260/095830507781076194, 2007.

IPCC: Climate Change 2013: The Physical Science Basis. Summary for Policymakers, IPCC, 1–29, https://doi.org/10.1017/CBO9781107415324, 2013.

Jacob, D., Petersen, J., Eggert, B., Alias, A., Christensen, O. B., Bouwer, L. M., Braun, A., Colette, A., Déqué, M., Georgievski, G., Georgopoulou, E., Gobiet, A., Menut, L., Nikulin, G., Haensler, A., Hempelmann, N., Jones, C., Keuler, K., Kovats, S., Kröner, N., Kotlarski, S., Kriegsmann, A., Martin, E., van Meijgaard, E., Moseley, C., Pfeifer, S., Preuschmann, S., Radermacher, C., Radtke, K., Rechid, D., Rounsevell, M., Samuelsson, P., Somot, S., Soussana, J. F., Teichmann, C., Valentini, R., Vautard, R., Weber, B., and Yiou, P.: EURO-CORDEX: New high-resolution climate change projections for European impact research, Regional Environmental Change, 14, 563–578, https://doi.org/10.1007/s10113-013-0499-2, 2014.

Keifer, C. J. and Chu, H. H.: Synthetic Storm Pattern for Drainage Design, Journal of the Hydraulics Division, 83, 1–25, 1957.

Kilsby, C. G., Jones, P. D., Burton, A., Ford, A. C., Fowler, H. J., Harpham, C., James, P., Smith, A., and Wilby, R. L.: A daily weather generator for use in climate change studies, Environ. Model. Softw., 22, 1705–1719, https://doi.org/10.1016/j.envsoft.2007.02.005, 2007.

Kossieris, P., Makropoulos, C., Onof, C., and Koutsoyiannis, D.: A rainfall disaggregation scheme for sub-hourly time scales: Coupling a Bartlett-Lewis based model with adjusting procedures, J. Hydrol., https://doi.org/10.1016/j.jhydrol.2016.07.015, in press, 2016.

Koutsoyiannis, D. and Onof, C.: Rainfall disaggregation using adjusting procedures on a Poisson cluster model, J. Hydrol., 246, 109–122, https://doi.org/10.1016/S0022-1694(01)00363-8, 2001.

Larsen, A. N., Gregersen, I. B., Christensen, O. B., Linde, J. J., and Mikkelsen, P. S.: Potential future increase in extreme one-hour precipitation events over Europe due to climate change, Water Sci. Technol., 60, 2205–2216, https://doi.org/10.2166/wst.2009.650, 2009.

Madsen, H., Arnbjerg-Nielsen, K., and Mikkelsen, P. S.: Update of regional intensity-duration-frequency curves in Denmark: Tendency towards increased storm intensities, Atmos. Res., 92, 343–349, https://doi.org/10.1016/j.atmosres.2009.01.013, 2009.

Mailhot, A., Duchesne, S., Caya, D., and Talbot, G.: Assessment of future change in intensity-duration-frequency (IDF) curves for Southern Quebec using the Canadian Regional Climate Model (CRCM), J. Hydrol., 347, 197–210, https://doi.org/10.1016/j.jhydrol.2007.09.019, 2007.

Marani, M. and Zanetti, S.: Downscaling rainfall temporal variability, Water Resour. Res., 43, 1–7, https://doi.org/10.1029/2006WR005505, 2007.

Maule, C. F., Thejll, P., Christensen, J. H., Svendsen, S. H., and Hannaford, J.: Improved confidence in regional climate model simulations of precipitation evaluated using drought statistics from the ENSEMBLES models, Clim. Dynam., 40, 155–173, https://doi.org/10.1007/s00382-012-1355-7, 2013.

Mikkelsen, P. S., Madsen, H., Arnbjerg-Nielsen, K., Jorgensen, H. K., Rosbjerg, D., and Harremoes, P.: A rationale for using local and regional point rainfall data for design and analysis of urban storm drainage systems, Water Sci. Technol., 37, 7–14, https://doi.org/10.1016/S0273-1223(98)00310-2, 1998.

Molini, A., La Barbera, P., Lanza, L. G., and Stagi, L.: Rainfall intermittency and the sampling error of tipping-bucket rain gauges, Physics and Chemistry of the Earth, Part C: Solar, Terrestrial & Planetary Science, 26, 737–742, https://doi.org/10.1016/S1464-1917(01)95018-4, 2001.

Nguyen, V. T. V., Nguyen, T. D., and Cung, A.: A statistical approach to downscaling of sub-daily extreme rainfall processes for climate-related impact studies in urban areas, in: Water Science and Technology: Water Supply, vol. 7, pp. 183–192, 2007.

Nguyen, V.-T.-V., Desramaut, N., and Nguyen, T.-D.: Estimation of urban design storms in consideration of GCM-based climate change scenarios, in Water and Urban Development Paradigms: Towards an Integration of Engineering, Design and Management Approaches – Proceedings of the International Urban Water Conference, 347–356, 2009.

Nguyen, V. T. V., Desramaut, N., and Nguyen, T. D.: Optimal rainfall temporal patterns for urban drainage design in the context of climate change, Water Sci. Technol., 62, 1170–1176, https://doi.org/10.2166/wst.2010.295, 2010.

NRCS: Urban Hydrology for Small Watersheds TR-55, USDA Natural Resource Conservation Service Conservation Engeneering Division Technical Release 55, 164, Technical Release 55, 1986.

Ntegeka, V. and Willems, P.: Trends and multidecadal oscillations in rainfall extremes, based on a more than 100-year time series of 10 min rainfall intensities at Uccle, Belgium, Water Resour. Res., 44, 1–15, https://doi.org/10.1029/2007WR006471, 2008.

Ntegeka, V., Baguis, P., Roulin, E., and Willems, P.: Developing tailored climate change scenarios for hydrological impact assessments, J. Hydrol., 508, 307–321, https://doi.org/10.1016/j.jhydrol.2013.11.001, 2014.

Olesen, M., Madsen, K. S., Ludwigsen, C. A., Boberg, F., Christensen, T., Cappelen, J., Christensen, O. B., Andersen, K. K., and Hesselbjerg Christensen, J.: Fremtidige klimaforandringer i Danmark, 2014.

Olsson, J., Berggren, K., Olofsson, M., and Viklander, M.: Applying climate model precipitation scenarios for urban hydrological assessment: A case study in Kalmar City, Sweden, Atmos. Res., 92, 364–375, https://doi.org/10.1016/j.atmosres.2009.01.015, 2009.

Olsson, J., Willen, U., and Kawamura, A.: Downscaling extreme short-term regional climate model precipitation for urban hydrological applications, Hydrol. Res., 43, 341–351, https://doi.org/10.2166/nh.2012.135, 2012.

Onof, C. and Arnbjerg-Nielsen, K.: Quantification of anticipated future changes in high resolution design rainfall for urban areas, Atmos. Res., 92, 350–363, https://doi.org/10.1016/j.atmosres.2009.01.014, 2009.

Onof, C. and Wheater, H. S.: Modelling of British rainfall using a random parameter Bartlett-Lewis Rectangular Pulse Model, J. Hydrol., 149, 67–95, https://doi.org/10.1016/0022-1694(93)90100-N, 1993.

Onof, C. and Wheater, H. S.: Improvements to the modelling of British rainfall using a modified Random Parameter Bartlett-Lewis Rectangular Pulse Model, J. Hydrol., 15, 177–195, https://doi.org/10.1016/0022-1694(94)90104-X, 1994.

Paschalis, A., Molnar, P., Fatichi, S., and Burlando, P.: On temporal stochastic modeling of precipitation, nesting models across scales, Adv. Water Resour., 63, 152–166, https://doi.org/10.1016/j.advwatres.2013.11.006, 2014.

Prein, A. F., Gobiet, A., Truhetz, H., Keuler, K., Goergen, K., Teichmann, C., Fox Maule, C., van Meijgaard, E., Déqué, M., Nikulin, G., Vautard, R., Colette, A., Kjellström, E., and Jacob, D.: Precipitation in the EURO-CORDEX 0.11 and 0.44 simulations: high resolution, high benefits?, Clim. Dynam., 46, https://doi.org/10.1007/s00382-015-2589-y, 2016.

Rossi, F., Fiorentino, M., and Versace, P.: Two-Component Extreme Value Distribution for Flood Frequency Analysis, Water Resour. Res., 20, 847–856, https://doi.org/10.1029/WR020i007p00847, 1984.

Schaarup-Jensen, K., Rasmussen, M. R., and Thorndahl, S.: To what extent does variability of historical rainfall series influence extreme event statistics of sewer system surcharge and overflows?, Water Sci. Technol., 60, 87–95, https://doi.org/10.2166/wst.2009.290, 2009.

Schilling, W.: Rainfall data for urban hydrology: what do we need?, Atmos. Res., 27, 5–21, https://doi.org/10.1016/0169-8095(91)90003-F, 1991.

Schleiss, M., Jaffrain, J., and Berne, A.: Statistical analysis of rainfall intermittency at small spatial and temporal scales, Geophys. Res. Lett., 38, L18403, https://doi.org/10.1029/2011GL049000, 2011.

Segond, M.-L., Neokleous, N., Makropoulos, C., Onof, C., and Maksimovic, C.: Simulation and spatio-temporal disaggregation of multi-site rainfall data for urban drainage applications, Hydrol. Sci. J., 52, 917–935, https://doi.org/10.1623/hysj.52.5.917, 2007.

Semadeni-Davies, A., Hernebring, C., Svensson, G., and Gustafsson, L. G.: The impacts of climate change and urbanisation on drainage in Helsingborg, Sweden: Combined sewer system, J. Hydrol., 350, 100–113, https://doi.org/10.1016/j.jhydrol.2007.05.028, 2008.

Shahabul Alam, M. and Elshorbagy, A.: Quantification of the climate change-induced variations in Intensity–Duration–Frequency curves in the Canadian Prairies, J. Hydrol., 527, 990–1005, https://doi.org/10.1016/j.jhydrol.2015.05.059, 2015.

Sherman: Streamflow from rainfall by the unit-graph method, Engineering News Record, 108, 1932.

Sørup, H. J. D., Christensen, O. B., Arnbjerg-Nielsen, K., and Mikkelsen, P. S.: Downscaling future precipitation extremes to urban hydrology scales using a spatio-temporal Neyman-Scott weather generator, Hydrol. Earth Syst. Sci., 20, 1387–1403, https://doi.org/10.5194/hess-20-1387-2016, 2016.

Sunyer, M. A., Hundecha, Y., Lawrence, D., Madsen, H., Willems, P., Martinkova, M., Vormoor, K., Bürger, G., Hanel, M., Kriauciuniene, J., Loukas, A., Osuch, M., and Yücel, I.: Intercomparison of statistical downscaling methods for projection of extreme precipitation in Europe, Hydrol. Earth Syst. Sci., 19, 1827–1847, https://doi.org/10.5194/hess-19-1827-2015, 2015.

Thorndahl, S.: Stochastic long term modelling of a drainage system with estimation of return period uncertainty, Water Sci. Technol., 59, 2331–2339, https://doi.org/10.2166/wst.2009.305, 2009.

Thorndahl, S., Beven, K. J., Jensen, J. B., and Schaarup-Jensen, K.: Event based uncertainty assessment in urban drainage modelling, applying the GLUE methodology, J. Hydrol., 357, 421–437, https://doi.org/10.1016/j.jhydrol.2008.05.027, 2008.

Thorndahl, S., Schaarup-Jensen, K., and Rasmussen, M. R.: On hydraulic and pollution effects of converting combined sewer catchments to separate sewer catchments, Urban Water Journal, 12, 120–130, https://doi.org/10.1080/1573062X.2013.831915, 2015.

Thorndahl, S., Balling, J. D., and Larsen, U. B. B.: Analysis and integrated modelling of groundwater infiltration to sewer networks, Hydrol. Process., 30, 3228–3238, https://doi.org/10.1002/hyp.10847, 2016.

Thorndahl, S., Einfalt, T., Willems, P., Nielsen, J. E., ten Veldhuis, M.-C., Arnbjerg-Nielsen, K., Rasmussen, M. R., and Molnar, P.: Weather radar rainfall data in urban hydrology, Hydrol. Earth Syst. Sci., 21, 1359–1380, https://doi.org/10.5194/hess-21-1359-2017, 2017.

Van Der Linden, P. and Mitchell, J. F. B.: ENSEMBLES: Climate Change and its Impacts: Summary of research and results from the ENSEMBLES project, Met Office Hadley Centre, Exeter, UK, 2009.

Wilby, R. L. and Wigley, T. M. L.: Downscaling general circulation model output: a review of methods and limitations, Prog. Phys. Geogr., 21, 530–548, https://doi.org/10.1177/030913339702100403, 1997.

Wilby, R. L., Dawson, C. W., and Barrow, E. M.: SDSM – a decision support tool for the assessment of regional climate change impacts, Environ. Model. Softw., 17, 145–157, https://doi.org/10.1016/S1364-8152(01)00060-3, 2002.

Willems, P.: Compound intensity/duration/frequency-relationships of extreme precipitation for two seasons and two storm types, J. Hydrol., 233, 189–205, https://doi.org/10.1016/S0022-1694(00)00233-X, 2000a.

Willems, P.: Compound intensity/duration/frequency-relationships of extreme precipitation for two seasons and two storm types, J. Hydrol., 233, 189–205, https://doi.org/10.1016/S0022-1694(00)00233-X, 2000b.

Willems, P.: Multidecadal oscillatory behaviour of rainfall extremes in Europe, Climatic Change, 120, 931–944, https://doi.org/10.1007/s10584-013-0837-x, 2013a.

Willems, P.: Revision of urban drainage design rules after assessment of climate change impacts on precipitation extremes at Uccle, Belgium, J. Hydrol., 496, 166–177, https://doi.org/10.1016/j.jhydrol.2013.05.037, 2013b.

Willems, P. and Vrac, M.: Statistical precipitation downscaling for small-scale hydrological impact investigations of climate change, J. Hydrol., 402, 193–205, https://doi.org/10.1016/j.jhydrol.2011.02.030, 2011.

Willems, P., Arnbjerg-Nielsen, K., Olsson, J., and Nguyen, V. T. V: Climate change impact assessment on urban rainfall extremes and urban drainage: Methods and shortcomings, Atmos. Res., 103, 106–118, https://doi.org/10.1016/j.atmosres.2011.04.003, 2012a.

Willems, P., Olsson, J., Arnbjerg-Nielsen, K., Beecham, S., Pathirana, A., Gregersen, I. B., Madsen, H., and Nguyen, V. T. V.: Impacts of Climate Change on Rainfall Extremes and Urban Drainage Systems, IWA publishing, London, 2012b.

WPC: Forventede ændringer i ekstremregn som følge af klimaændringer, Skrift nr. 29 (Anticipated changes in extrem precipitation as a result of climate change, Guideline no. 29), The Water Pollution Committee of the Society of Danish Engineers, Copenhagen, Denmark, 2008 (in Danish).

6

Towards simplification of hydrologic modeling: identification of dominant processes

Steven L. Markstrom[1], **Lauren E. Hay**[1], **and Martyn P. Clark**[2]

[1]US Geological Survey, P.O. Box 25046, MS 412, Denver Federal Center, Denver, Colorado, 80225, USA
[2]National Center for Atmospheric Research, P.O. Box 3000, Boulder, Colorado, 80307, USA

Correspondence to: Steven L. Markstrom (markstro@usgs.gov)

Abstract. The Precipitation–Runoff Modeling System (PRMS), a distributed-parameter hydrologic model, has been applied to the conterminous US (CONUS). Parameter sensitivity analysis was used to identify: (1) the sensitive input parameters and (2) particular model output variables that could be associated with the dominant hydrologic process(es). Sensitivity values of 35 PRMS calibration parameters were computed using the Fourier amplitude sensitivity test procedure on 110 000 independent hydrologically based spatial modeling units covering the CONUS and then summarized to process (snowmelt, surface runoff, infiltration, soil moisture, evapotranspiration, interflow, baseflow, and runoff) and model performance statistic (mean, coefficient of variation, and autoregressive lag 1). Identified parameters and processes provide insight into model performance at the location of each unit and allow the modeler to identify the most dominant process on the basis of which processes are associated with the most sensitive parameters.

The results of this study indicate that: (1) the choice of performance statistic and output variables has a strong influence on parameter sensitivity, (2) the apparent model complexity to the modeler can be reduced by focusing on those processes that are associated with sensitive parameters and disregarding those that are not, (3) different processes require different numbers of parameters for simulation, and (4) some sensitive parameters influence only one hydrologic process, while others may influence many.

1 Introduction

It has long been recognized that distributed-parameter hydrology models (DPHMs) are complex because of the subtlety and diversity of the hydrologic cycle which they aim to simulate (Freeze and Harlan, 1969; Amorocho and Hart, 1964). In this study, two different aspects of this complexity are addressed:

1. DPHMs have too many input parameters (Jakeman and Hornberger, 1993; Kirchner et al., 1996; Brun et al., 2001; Perrin et al., 2001; McDonnell et al., 2007). In this article, distributed parameters are defined as model inputs that remain constant through time, but can vary spatially across the landscape. Those who apply these models often have difficulty with understanding what these parameters are and how they are used in the model. Regularly, there are several parameters that may have similar effect on the computations or may constrain the model in unintended ways (Hrachowitz et al., 2014). Despite the developer's claims that these DPHMs are more or less physically based, often there are not measurements or data sources available for reliable development of all of the input parameters. Duan et al. (2006) describes "a gap in our understanding of the links between model parameters and the land surface characteristics". These unmeasured parameters, ostensibly tangible, are really empirical coefficients when it comes to application and calibration (Samaniego et al., 2010).

2. The output produced by DPHMs is difficult to interpret (Schaefli and Gupta 2007; Gupta et al., 2009, 2012;

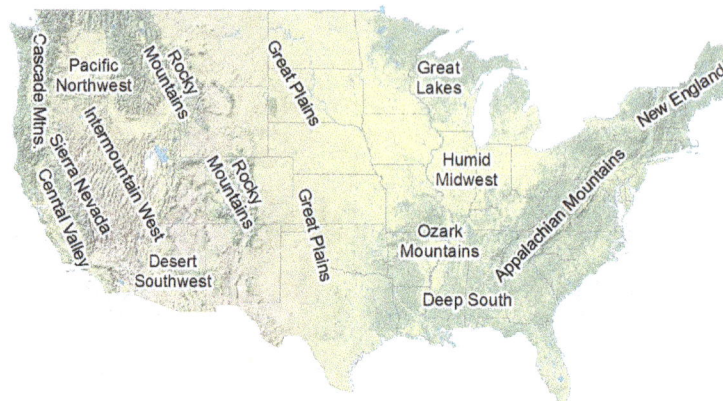

Figure 1. Location map of the conterminous US showing the different geographic regions referred to this study.

Mayer and Butler, 1993; Ewan, 2011). Often, the meaning of output variables is not always intuitive and results sometimes can seem contradictory (e.g., when streamflow does not seem to correlate with climate information). The result of these complex issues has led to the study of parameter interaction (Clark and Vrugt, 2006) and equifinality (Beven, 2006).

Developing effective DPHM applications require that the modeler address these two aspects of complexity at the same time (i.e., the uncertainty problem: "If I am uncertain when estimating input parameters, due to either incomplete or inaccurate information, what effect does it have on the output?", and the calibration problem: "I know the output I want, which parameters should I change and how much should I change them?") (Chaney et al., 2015; Reusser and Zehe, 2011). While the user of a DPHM can do nothing about the complexity of the model's internal structure, the apparent complexity can be reduced by limiting the parameters and the affected output under consideration (as described by Jakeman and Hornberger, 1993; Hay et al., 2006).

Global parameter sensitivity analysis can determine the degree to which different values of parameters can affect the simulation of certain model outputs (Sanadhya et al., 2013). Furthermore, parameter sensitivity can be evaluated with respect to selected output variables, each representing a different aspect of the hydrologic cycle (hereafter referred to as *processes*). Sensitivity analysis of this form can be used to identify both the input parameters that are the most sensitive (i.e., the parameters that affect the simulation the most) and the dominant process(es) (i.e., those processes which are affected most, by the most sensitive parameters) according to the DPHM.

Any particular DPHM must necessarily be able to simulate any and all hydrological processes that may occur anywhere on the landscape. However, with the application of a DPHM to a specific site, it can become much less complex when the dominant hydrological process(es) are identified, as not all processes are active to the same degree. The mod-

eling problem becomes less complex to the modeler when hydrological processes not relevant to the modeled domain or watershed are removed from consideration (Wagener et al., 2003; Reusser et al., 2011; Guse et al., 2014; Bock et al., 2016). Related to this, various methods have been developed that will group similar watersheds together for purposes of study (Wolock et al., 2004; Winter, 2001; Ali et al., 2012) or for parameter regionalization (He et al., 2011; Merz and Blöschl, 2004; Seibert, 1999; Vogel, 2005). In addition, dominant process concepts have been explored by several researchers as a way to classify watersheds and natural hydrologic systems for the purpose of simplifying DPHMs (Sivakumar and Singh, 2012; Sivakumar et al., 2007). Some have suggested this approach for use as a possible classification framework (e.g., Woods, 2002; Sivakumar, 2004). Pfannerstill et al. (2015) developed a framework for identification and verification of hydrologic processes in simulation models on the basis of temporal sensitivity analysis. Cuntz et al. (2015) describe a method of identifying only informative parameters as a screening step in order to reduce the effort required to perform global sensitivity analysis on the full parameter space. McDonnell et al. (2007) discuss the possibility of simplifying hydrologic modeling by identifying "fundamental laws" so that over-parameterized models are not needed. However, in our opinion we have not made much progress on that front and DPHMs are, in many ways and for many reasons, more complex than ever.

This article describes an approach for identification of sensitive parameters and processes for a modeling application of the conterminous US (CONUS, Fig. 1). Identification and simulation of regional CONUS sub-watersheds are determined by the resolution of the available information and how the DPHM responds to geophysical (e.g., topography, vegetation and soils) and climatological variation. Specifically, we propose to identify the sensitive parameters and dominant hydrologic process(es), thereby reducing the amount of parameter input and number of output variables to consider

(Chaney et al., 2015) and address the two aspects of complexity as described above.

2   Methods

2.1   Distributed-parameter hydrology model

The US Geological Survey's Precipitation–Runoff Modeling System (PRMS) is the DPHM used in this study. PRMS is a modular, deterministic, distributed-parameter, physical-process watershed model used to simulate and evaluate the effects of various combinations of precipitation, climate, and land use on watershed response. Each hydrologic process simulated by PRMS is encoded in a modular piece of source code (i.e., a "module") and is represented by an algorithm that is based on a physical law (e.g., balance of energy required to melt the ice in a snowpack) or empirical relation with measured or estimated characteristics (e.g., a tank model used to simulate interflow). The reader is referred to Markstrom et al. (2015) for a complete description of PRMS.

A fundamental assumption of this study is that PRMS is able to simulate and differentiate hydrologic signals from all the different processes at the scale of the CONUS. Two possible ways to evaluate this are: (1) an analysis of PRMS's internal structure, and (2) the history of PRMS applications. A detailed analysis of PRMS's structure is beyond the scope of this article (see Markstrom et al., 2015); however, PRMS is implemented in a very linear fashion. Each parameter is clearly identified with an equation that is related to simulation of a specific process. Equations are solved sequentially, generally in the order that is defined by water moving through the hydrologic cycle, starting from the atmosphere as precipitation and moving through the rivers as streamflow. The outputs of one equation may be used as inputs to subsequent equations. All of the inputs for a particular equation are required before that equation can be solved. This interdependency in equations can lead to parameter interaction in the simulation of subsequent processes (as described by Beven, 1989; Grayson et al., 1992; Yilmaz et al., 2008; Pfannerstill et al., 2015). For example, parameters related to distribution of temperature and solar radiation may show correlation with each other when evaluated with respect to simulation of evapotranspiration, despite these parameters not being explicit terms in the evapotranspiration equations. Past studies indicate that PRMS has been very useful in water-resource and research studies across the CONUS (Battaglin et al., 2011; Boyle et al., 2006; Hay et al., 2011; Markstrom et al., 2012) and is capable of matching measured data (Bower, 1985; Cary, 1991; Dudley, 2008; Koczot et al., 2011) in a variety of geophysical and climatological settings.

To define the spatial domain for the CONUS application of PRMS, the locations of major river confluences, water bodies, and stream gages have been geo-referenced. Approximately 56 000 stream segments are used to connect these lo-

cations. Using these stream segments, the left and right bank areas that contribute runoff directly to each segment have been identified, resulting in approximately 110 000 irregularly shaped hydrologic response units (HRUs) of various sizes ($500 \, \mathrm{m}^2$ to $14\,000 \, \mathrm{km}^2$) (Viger and Bock, 2014). These HRUs are derived by their geographic and topographic location, affecting their extent and resolution. The CONUS application is forced with values of daily precipitation and daily maximum and minimum air temperature from the DAYMET data set (Thornton et al., 2014). The climate information covers a time period from 1980 to 2013 on a daily time step, but a shorter period (1987–1989 used for warmup, and 1990–2000 used for evaluation) was used in this study.

2.2   Calibration parameters

The version of PRMS used in this study has 108 input parameters. A parameter is defined as an input value that does not change over the course of a simulation run. Of these parameters, most would never be modified from their initial values (hereafter referred to as *non-calibration parameters*, see Viger, 2014) because they are (1) computed directly from digital data sets through the use of a geographic information system (e.g., land–surface characterization parameters), (2) boundary conditions (e.g., parameters to adjust daily precipitation and daily air temperature forcings), or (3) model configuration options (e.g., unit conversions and model output options). This leaves 35 parameters under consideration for improved model performance, hereafter referred to as *calibration parameters* (Table 1). Each parameter is used within a PRMS code module that simulates a single hydrologic process in PRMS. The output variables of one module may be used as input variables to other modules. It is through these connections that calibration parameters associated with a PRMS module may affect the results of other modules.

2.3   Hydrologic processes

PRMS produces more than 200 output variables that indicate the simulated hydrologic response of a watershed through time (Markstrom et al., 2015, see Table 5 in Appendix 1). In this study, eight of these output variables have been selected to represent the response of major hydrologic processes at the HRU resolution. These processes are: (1) snowmelt (PRMS output variable snowmelt) – the amount of water that has changed from ice to liquid and becomes either surface runoff or infiltrates into the soil zone of the HRU; (2) surface runoff (sroff) – water from a rainfall or snowmelt event that travels quickly over the land surface from the HRU to the connected stream segment; (3) infiltration (infil) – the sum of rain and snowmelt that passes into the soil zone of the HRU; (4) soil moisture (soil_moist) – the storage state that represents the amount of soil water in the soil zone above wilting point and below total saturation in the HRU; (5) evapotranspiration (hru_actet) – the total actual evapotranspira-

Table 1. Precipitation Runoff Modeling System (PRMS) calibration parameters used in this study. The values in the column labeled "PRMS module" identify the module type equation(s) from the PRMS source code (see Markstrom et al., 2015).

Parameter name	Description	PRMS module	Range
adjmix_rain	Factor to adjust rain proportion in a mixed rain/snow event	climate	0.6–1.4
tmax_allrain	Maximum air temperature above which precipitation is rain	climate	−8.0–60.0
tmax_allsnow	Maximum air temperature below which precipitation is snow	climate	−10.0–40.0
dday_intcp	Intercept in degree–day equation	solar radiation	−60.0–10.0
dday_slope	Slope in degree–day equation	solar radiation	0.2–0.9
ppt_rad_adj	Solar radiation adjustment threshold for precipitation days	solar radiation	0.0–0.5
radj_sppt	Solar radiation adjustment on summer precipitation days	solar radiation	0.0–1.0
radj_wppt	Solar radiation adjustment on winter precipitation days	solar radiation	0.0–1.0
radmax	Maximum solar radiation due to atmospheric effects	solar radiation	0.1–1.0
tmax_index	Temperature to determine precipitation adjustments to solar radiation	solar radiation	−10.0–110.0
jh_coef	Coefficient used in Jensen–Haise potential ET computations	Potential ET	0.005–0.06
jh_coef_hru	Coefficient used in Jensen–Haise potential ET computations	Potential ET	5.0–25.0
srain_intcp	Summer rain interception storage capacity	interception	0.0–1.0
wrain_intcp	Winter rain interception storage capacity	interception	0.0–1.0
cecn_coef	Convection condensation energy coefficient	snow	2.0–10.0
emis_noppt	Average emissivity of air on days without precipitation	snow	0.757–1.0
freeh2o_cap	Free-water holding capacity of snowpack	snow	0.01–0.2
potet_sublim	Snow sublimation fraction of potential ET	snow	0.1–0.75
carea_max	Maximum area contributing to surface runoff	surface runoff	0.0–1.0
smidx_coef	Non-linear contributing area coefficient	surface runoff	0.001–0.06
smidx_exp	Exponent in non-linear contributing area coefficient	surface runoff	0.1–0.5
fastcoef_lin	Linear coefficient in equation to route preferential-flow	soil-zone	0.001–0.8
fastcoef_sq	Non-linear coefficient in equation to route preferential-flow	soil-zone	0.001–1.0
pref_flow_den	Fraction of the soil zone in which preferential flow occurs	soil-zone	0.0–0.1
sat_threshold	Water capacity between field capacity and total saturation	soil-zone	1.0–999.0
slowcoef_lin	Linear coefficient for interflow routing	soil-zone	0.001–0.5
slowcoef_sq	Non-linear coefficient for interflow routing	soil-zone	0.001–1.0
soil2gw_max	Maximum soil water excess that is routed directly to groundwater	soil-zone	0.0–0.5
soil_moist_max	Maximum available water holding capacity of soil zone	soil-zone	0.001–10.0
soil_rechr_max	Maximum available water holding capacity of recharge zone	soil-zone	0.001–5.0
ssr2gw_exp	Non-linear coefficient in equation used to route soil-zone water to groundwater	soil-zone	0.0–3.0
ssr2gw_rate	Linear coefficient in equation used to route soil-zone water to groundwater	soil-zone	0.05–0.8
transp_tmax	Temperature that determines start of the transpiration period	soil-zone	0.0–1000.0
gwflow_coef	Linear groundwater discharge coefficient	groundwater	0.001–0.5

tion lost from canopy interception, snow sublimation, and soil and plant losses from the root zone; (6) interflow (ssres_flow) – shallow lateral flow in the unsaturated zone to the connected stream segment; (7) baseflow (gwres_flow) – the component of flow from the saturated zone to the connected stream segment; and (8) runoff (hru_outflow) – the total flow from the HRU contributing to streamflow in the connected stream segment. It is assumed that these eight output variables are representative of the processes typically considered in hydrological studies with DPHMs. Details of how these processes are simulated by PRMS are described by Markstrom et al. (2015).

2.4 Performance statistics

For DPHMs, there are many different performance measures that have been developed for different purposes (Krause et

al., 2005; Gupta et al., 2008, 2009; Mendoza et al., 2015a, b). Because this study is an analysis of model sensitivity, the performance measures need only track changes in model output and do not necessarily need to include observed measurements. Consequently, performance statistics can be developed for processes that are not normally evaluated by performance measures. Archfield et al. (2014) demonstrated that seven fundamental daily streamflow statistics (FDSS) can be used to group streams by similar hydrologic response and tend to provide non-redundant information. In this study, all seven FDSS were computed for each of the eight PRMS time-series output variables corresponding to the processes. For the purpose of illustration, this article focuses on three of the FDSS: (1) mean; (2) coefficient of variation (CV); and (3) the autoregressive lag 1 correlation coefficient (AR-1). In an intuitive sense, these three statistics can be thought to represent changes in total volume, "spikiness" or "flashiness", and day-to-day timing, respectively. These performance statistics are computed on the daily time series of the process variables for the 10-year evaluation period.

2.5 FAST analysis

Parameter sensitivity analysis measures the variability of model output given variability of calibration parameter values. This is determined by partitioning the total variability in the model output or change in performance statistics to individual calibration parameters (Reusser et al., 2011). The Fourier amplitude sensitivity test (FAST) (Schaibly and Shuler, 1973; Cukier et al., 1973, 1975; Saltelli et al., 2006) was selected for this study because it has been demonstrated that it can efficiently estimate non-linear hydrologic model parameter sensitivity (Guse et al., 2014; Pfannerstill et al., 2015; Reusser et al., 2011). FAST is a variance-based global sensitivity algorithm that estimates the first-order partial variance of model output explained by each calibration parameter (hereafter referred to as *parameter sensitivity*). Specifically, this first-order variance is the variability in the output that is directly attributable to variations in any one parameter and is distinguishable from higher order variances associated with parameter interactions. An important caveat is that these higher order variances are not accounted for in the analysis. It is assumed that first-order partial variance is sufficient to identify sensitive parameters. This same assumption, as applied to process identification, may be more problematic. If there are sets of interactive sensitive parameters that have not been identified, then the associated process(es) will not be identified as such.

Selected parameters are varied within defined ranges at independent frequencies among different model runs. FAST identifies the variability of parameter sensitivities and their ranks, by means of their contribution to total power in the power spectrum. FAST has been implemented as the 'fast' library in the statistical software R (Reusser et al., 2011; Reusser, 2013; R Core Team, 2015) in two parts. In the first part, the user identifies the calibration parameters and respective value ranges for the test, then FAST generates sets of test calibration parameter values (hereafter referred to as *trials*). Calibration parameter values are varied across the trials according to non-harmonic fundamental frequencies. The user then runs the DPHM for each trial and computes corresponding performance statistics. Then the user runs the second part of the FAST package that performs a Fourier analysis of the performance statistics over the trial space looking for the frequency signatures associated with each calibration parameter.

The FAST methodology results in a simple procedure for computing parameter sensitivities on an HRU basis for all the CONUS. The steps in this process are as follows:

1. Assign appropriate ranges for the 35 calibration parameters (Markstrom et al., 2015; as in LaFontaine et al., 2013). These are shown in Table 1.

2. Run the first part of the FAST procedure (as described above) to develop over 9000 unique parameter sets, comprised of value combinations for the calibration parameters. The total number and content of these parameter sets, and the results from their simulation by PRMS, are completely determined by the first part of the FAST procedure in order to investigate the trial space. Each of the prescribed simulations are independent of each other so they can run in parallel on a computer cluster.

3. Compute the FDSS based performance statistics (mean, CV, and AR-1) for each process.

4. Run the second part of the FAST procedure (as described above) using output from step 3, resulting in PRMS parameter sensitivities, at each HRU, for the 56 combinations of seven performance statistics and eight processes (plus totals).

3 Results

3.1 Parameter sensitivity by process and performance statistic

Figure 2 shows parameter sensitivity as a set of maps ordered by process and performance statistic. This illustrates the spatial variability in parameter sensitivity and the importance that choice of performance statistic can make in terms of evaluation of hydrologic response. In these maps, the HRUs are colored according to the parameter sensitivity, which is computed by summing the first-order sensitivity for all 35 parameters separately for each of the 8 output variables, each corresponding to their respective process. (These sums do not necessarily add up to 1). Then each individual category of modeled process and performance statistic is scaled to account for total sensitivity. This summed sensitivity across the parameters, by each category, is hereafter referred to as *cumulative parameter sensitivity*. Parameter sensitivities associated with process (column labeled "Process average" in Fig. 2) are averaged across all of the parameter sensitivity values computed for the different performance statistics, while parameter sensitivities associated with the performance statistics (last row labeled "Performance statistic average" in Fig. 2) are averaged across all of the parameter sensitivity values computed for the different processes. These categories are indicated by their position in the rows and columns in Fig. 2. When looking at a single performance statistic for a single process, the cumulative parameter sensitivity can vary from near 0.0 (white colored HRUs) to near 1.0 (black colored HRUs). Low values in these maps indicate that there are no parameters that can be changed in any way to affect the performance statistic (this situation is hereafter referred to as an *inferior process*). Likewise, each HRU has a cumulative sensitivity value (i.e., the sum of all of the partial sensitivities for each process). The process with

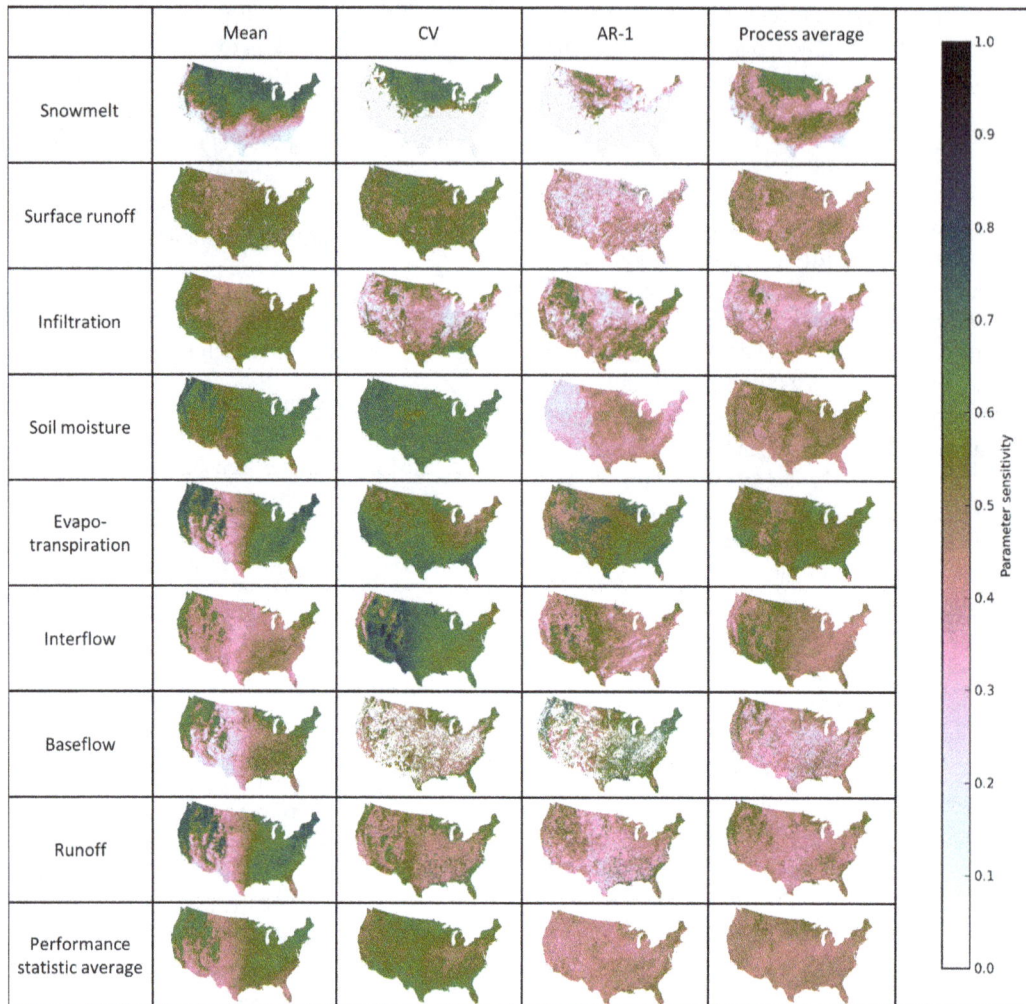

Figure 2. Maps of the conterminous US showing Precipitation–Runoff Modeling System parameter sensitivity by Hydrologic Response Unit (HRU) by process and performance statistic. The HRU parameter sensitivity is computed by summing the first-order sensitivity for all parameters. The process average maps are made by averaging the parameter sensitivity values computed for the different performance statistics. The performance statistic maps are made averaging the parameter sensitivity values computed for the different processes.

the largest sum on an HRU is referred to as the *dominant process* for that HRU.

An example of an inferior process is clearly seen in the case of the mean of the snowmelt process in the southern CONUS HRUs. This is because the occurrence of snow in these areas is very infrequent. Also, there were HRUs for which the value of some performance statistics were mathematically undefined for certain processes (e.g., AR-1 and CV for the baseflow and snowmelt processes). These cases occur when the output variable representing the process does not change at all through time, regardless of the parameter values, and are extreme examples of inferior processes. Likewise, a clear example of a dominant hydrologic process is the CV of interflow in the Intermountain West region of the CONUS (Figs. 1 and 2). This means that for these HRUs,

there exist some calibration parameters that can be varied, which affect this process to a very high degree.

Also apparent from Fig. 2 is that there are clear spatial patterns in the parameter sensitivity on the basis of the geographical features of the CONUS. Generally, many of the maps show a sharp break in parameter sensitivity between mountain ranges and comparatively lower elevations, northern contrasted with southern latitudes, and humid vs. arid climates. Specific contrasts can be seen in several maps such as when examining the humid Midwest as opposed to the Great Plains regions and the Pacific coastal areas and the Desert Southwest region of the CONUS (Fig. 1). Additionally, topographic features of the landscape are prominent (e.g., elevation for interflow), while in other maps, climate considerations seem to dominate (e.g., snowmelt). Another specific example is that the mean of each process, which indicates

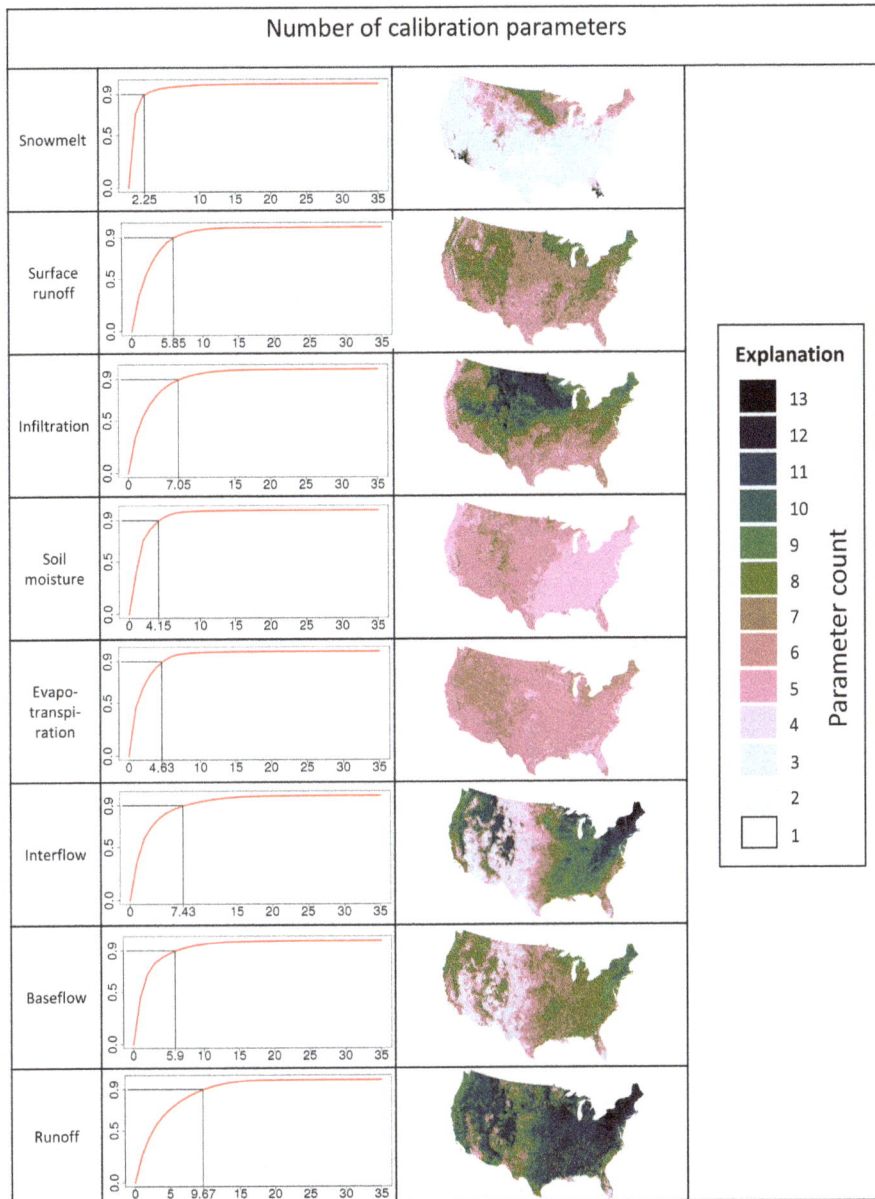

Figure 3. Cumulative parameter sensitivity across all Hydrologic Response Units (HRUs) in the CONUS Precipitation–Runoff Modeling System application are shown by process. **(a)–(h)** show the parameter count necessary to account for 90 % of the cumulative parameter sensitivity, summarized across all HRUs. For this count, the parameters are ranked and summed until the 90 % level is reached. The maps **(i)–(p)** show the count of ranked parameters required to reach the 90 % level on an HRU-by-HRU basis, by process.

the ability of any parameter(s) to change the total volume of water during a simulation, seems to have a low sensitivity band in the Great Plains region for all processes except for snowmelt (Fig. 1). This band of low sensitivity has been noted in other modeling studies (Newman et al., 2015; Bock et al., 2016).

3.2 Parameter count required to parameterize each process

To identify the expected count of parameters required to parameterize a particular process, cumulative parameter sensitivity across all HRUs of the CONUS has been computed and plotted (Fig. 3a–h). The sensitivity level accounted for by the most sensitive parameter, regardless of which parameter it is, for all HRUs across the CONUS is plotted in position 1 on the x axis of each of these plots (Fig. 3a–h). Then, cumulative sensitivity is plotted for the parameter in rank 2, and so on, until the cumulative sensitivity of all 35 calibration parameters is accounted for. The plots in Fig. 3a–h show that far fewer than the full 35 parameters are needed to account for most of the parameter sensitivity. In fact, to account for 90 % of the parameter sensitivity, this count varies from a low value of just over two for snowmelt to an average high value of over nine for runoff in selected HRUs.

The actual count of calibration parameters required to account for 90 % of the parameter sensitivity varies by process and region, as shown by the maps in Fig. 3i–p. These maps were generated by counting the number of parameters required to obtain the 90 % cumulative sensitivity level for each HRU. For example, Fig. 3o indicates that for the baseflow process, between three and nine parameters are needed to account for 90 % of the parameter sensitivity in the various HRUs across the CONUS, with the higher count needed in mountainous, Great Lakes, and New England regions. The maps also indicate that between 2 (Fig. 3i) and 13 parameters (Fig. 3k, n, and p) are required for parameterization of these processes. This analysis indicates that more parameters are needed to simulate the components of streamflow (e.g., baseflow, interflow, and surface runoff) than processes that do not result directly in flow (e.g., snowmelt, evapotranspiration, and soil moisture). In addition, simulated processes that are identified as being sensitive to parameters with which they are not normally associated, may indicate that these processes are a convolution of other processes, consequently making parameters sensitive that are not normally sensitive.

Visually, these maps (Fig. 3i–p) indicate that HRU calibration parameter counts vary regionally. For most processes, higher parameter counts are seen in the more mountainous regions of the Cascade, Sierra Nevada, Rocky, Ozark, and Appalachian mountains, although this is true to a much lesser extent for the evapotranspiration and soil moisture processes (Fig. 3m and l). Higher values also seem prevalent in the New England and Great Lake regions (Fig. 1). This result seems to indicate that, no matter which part of the hydrologic cycle is simulated, more parameters are required in these regions. In contrast, low parameter counts seem prevalent in the Great Plains and Desert Southwest regions.

Finally, Fig. 3 illustrates the extent to which it is possible to decompose the parameter estimation problem into a sub-set of independent problems, and hence reduce the dimensionality of the inference problem and avoid the troublesome nature of parameter interactions. By considering a single (or reduced set of) processes and performance statistic categories at a time, the sensitive parameter space can be substantially reduced. It also illustrates that there is a strong spatial component to this decomposition. In order to make the information presented in Fig. 3 more useful for DPHM application, the particular sensitive parameters have been determined for each HRU by ranking the calibration parameters by sensitivity for each category of process and performance statistic for each individual HRU and is summa-rized by counting the occurrence of each parameter across the HRUs and ranking them within their respective category of process and performance statistic (Table 2). To address the issue of the spatial variability of these parameters, the percentage of the total number of HRUs for which that parameter is sensitive is shown as the number in parentheses after the parameter name in Table 2. Higher percentage values would indicate that the corresponding parameter is sensitive across more of the CONUS. Refer to Table 1 for a complete description of these parameters.

When looking at the categorical parameter lists of Table 2, it is expected that different parameters would associate with different processes (i.e., along a column), but it is surprising to see how different the parameter lists are for different performance statistics (moving across a row) for the same process. An example of this is the baseflow process: the baseflow coefficient (PRMS parameter gwflow_coef) is the most sensitive parameter for performance statistics CV and AR-1, but is not even in the list of sensitive parameters for the performance statistics related to the mean of the process. This implies that this parameter is influential for affecting the timing of baseflow, while it does not have any effect on the total volume of baseflow.

Further inspection of Table 2 indicates that some calibration parameters occur in many of the 24 categories (8 processes times 3 performance statistics), while some parameters do not occur at all. A count of how many times each parameter occurs provides insight into how many process and/or performance statistic combinations that particular parameter influences. To investigate this for the CONUS application, another view of the information in Table 2 is shown in Fig. 4. The 25 sensitive calibration parameters from Table 2 are listed on the y axis of Fig. 4, ranked by order of the number of times that they appear in the process and/or performance statistic categories. Furthermore, each appearance is indicated by an adjacent circle. Independent of the number of times a parameter occurs within a category (number of circles), the color of the circle visually indicates the proportion of the CONUS HRUs that are affected by that parameter. Specifically, a red circle indicates that more HRUs are affected, while blue indicates that fewer HRUs are affected.

Figure 4 shows that 3 specific parameters affect 18 or more process and/or performance statistic categories; 7 parameters affect seven to 14 categories, and 15 specific parameters affect one to five categories. Finally, of the 35 parameters studied, 10 are never used for any combination of process and performance statistic (Table 2 and Fig. 4). It is apparent from Fig. 4, that for the CONUS application of PRMS, the parameters affecting the most process categories are soil_moist_max (maximum available water holding capacity), jh_coef (Jensen–Haise air temperature coefficient), and dday_intcp (intercept in degree–day solar radiation equation). Because these parameters affect so many categories, modelers would be wise to invest their resources in developing the best values possible for these parameters to avoid

Table 2. Ordered list of most sensitive Precipitation–Runoff Modeling System calibration parameters by process and performance statistic. The parameters listed in each cell of the table are those that are required to account for 90 % of the cumulative sensitivity across all hydrologic response units (HRUs). The number in parentheses following the parameter name is the proportion of the CONUS HRUs, in percent, in which that parameter is part of the set that accounts for 90 % of the cumulated sensitivity on an HRU-by-HRU basis. These parameters are described in Table 1.

Process	Performance statistic		
	Mean	CV	AR 1
Snowmelt	tmax_allsnow(96), tmax_allrain(92)	tmax_allsnow(39), tmax_allrain(38), rad_trncf(9), freeh2o_cap(8), dday_intcp(7)	tmax_allsnow(34), dday_intcp(29), rad_trncf(28), radmax(24), tmax_allrain(17), jh_coef(15), freeh2o_cap(14), cecn_coef(14), emis_noppt(13), jh_coef_hru(13), potet_sublim(10)
Surface runoff	smidx_exp(98), carea_max(98), soil_moist_max(98), smidx_coef(96), jh_coef(90), dday_intcp(33)	carea_max(93), smidx_exp(82), jh_coef(64), tmax_allsnow(55), smidx_coef(52), srain_intcp(33), soil_moist_max(23), tmax_allrain(22)	soil_moist_max(92), carea_max(83), jh_coef(65), smidx_exp(64), smidx_coef(42), tmax_allsnow(39), dday_intcp(25), srain_intcp(23), tmax_allrain(16), radmax(15)
Infiltration	smidx_exp(99), soil_moist_max(99), carea_max(99), smidx_coef(95), jh_coef(64), srain_intcp(50)	carea_max(80), tmax_allsnow(69), jh_coef(63), smidx_exp(62), srain_intcp(54), smidx_coef(54), tmax_allrain(48), radmax(37), freeh2o_cap(36), soil_moist_max(35), dday_intcp(31), rad_trncf(18)	carea_max(72), soil_moist_max(64), smidx_exp(61), tmax_allsnow(60), srain_intcp(60), tmax_allrain(42), jh_coef(35), smidx_coef(24), freeh2o_cap(16), dday_intcp(16)
Soil moisture	soil_moist_max(100), jh_coef(99), dday_intcp(94), radmax(82)	jh_coef(98), radmax(98), soil_moist_max(97), dday_intcp(94)	soil_moist_max(99), jh_coef(98), dday_intcp(89), radmax(35)
Evapo-transpiration	jh_coef(100), soil_moist_max(96), dday_intcp(96), radmax(92), jh_coef_hru(62), smidx_coef(37), dday_slope(25)	radmax(100), jh_coef(100), soil_moist_max(95), dday_intcp(73), dday_slope(67), soil_rechr_max(34)	jh_coef(100), radmax(100), dday_slope(75), soil_moist_max(74), dday_intcp(67), soil_rechr_max(49)

Table 2. Continued.

Process	Performance statistic		
	Mean	CV	AR 1
Interflow	soil_moist_max(99), soil2gw_max(94), pref_flow_den(90), jh_coef(84), carea_max(65), smidx_exp(45), dday_intcp(31), smidx_coef(19)	fastcoef_lin(100), soil_moist_max(87), pref_flow_den(71), jh_coef(61), carea_max(49), soil2gw_max(29), smidx_exp(25), tmax_allsnow(17), dday_intcp(16)	soil_moist_max(96), fastcoef_lin(89), slowcoef_sq(83), carea_max(72), jh_coef(61), pref_flow_den(47), smidx_exp(47), ssr2gw_exp(40), soil2gw_max(26), dday_intcp(18), tmax_allsnow(16)
Baseflow	jh_coef(100), soil_moist_max(91), dday_intcp(81), soil2gw_max(74), radmax(64), carea_max(37), jh_coef_hru(36)	gwflow_coef(48), soil_moist_max(40), jh_coef(28), soil2gw_max(28), smidx_coef(20), carea_max(16), tmax_allsnow(13), dday_intcp(12), smidx_exp(8)	gwflow_coef(48), soil_moist_max(44), soil2gw_max(22), carea_max(18)
Runoff	jh_coef(100), dday_intcp(96), soil_moist_max(96), radmax(93), jh_coef_hru(62), smidx_coef(37), dday_slope(26)	gwflow_coef(97), soil_moist_max(81), fastcoef_lin(76), pref_flow_den(71), carea_max(58), jh_coef(54), smidx_exp(49), smidx_coef(42), soil2gw_max(36), tmax_allsnow(15)	slowcoef_sq(90), soil2gw_max(90), gwflow_coef(82), carea_max(81), soil_moist_max(78), smidx_exp(72), smidx_coef(60), fastcoef_lin(36), pref_flow_den(35), jh_coef(30), slowcoef_lin(22)
Parameters not sensitive	adjmix_rain, fastcoef_sq, ppt_rad_adj, radj_sppt, radj_wppt, sat_threshold , ssr2gw_rate, tmax_index, transp_tmax, wrain_intcp		

unintended parameter interaction during calibration. Ideally, these parameters could be estimated from reliable external data, set for the model and not calibrated. The parameters that affect the least number of process categories (aside from the parameters that are never sensitive) are cecn_coef (convection condensation energy coefficient), ssr2gw_exp (coefficient in equation used to route water from the soil to the groundwater reservoir), emis_noppt (emissivity of air on days without precipitation), potet_sublim (fraction of potential evapotranspiration that is sublimated), and slowcoef_lin (slow interflow routing coefficient). Ideally, these parameters could be set to default values since there is limited value in calibrating them.

Also apparent from Fig. 4 is that there are many parameters between these two extreme groups. Parameters like smidx_coef (soil moisture index for contributing area calcu-

lation) can appear in several process categories, without any high rankings, while there are other parameters like slow-coef_sq (slow interflow routing coefficient) that appear in relatively few process categories, but have high rankings. This behavior may be due to the vertical routing order (i.e. processes that occur nearer to the surface happen before the deeper ones) of the associated processes (Yilmaz et al., 2008; Pfannerstill et al., 2015). In PRMS, the process of partitioning of precipitation into either direct surface runoff or infiltration (controlled directly by parameter smidx_coef) is "faster" and occurs in the vertical routing order before the process of interflow generation (controlled directly by parameter slow-coef_sq). These parameters may be the best candidates for calibration because they are sensitive, while at the same time interaction across processes is perhaps limited.

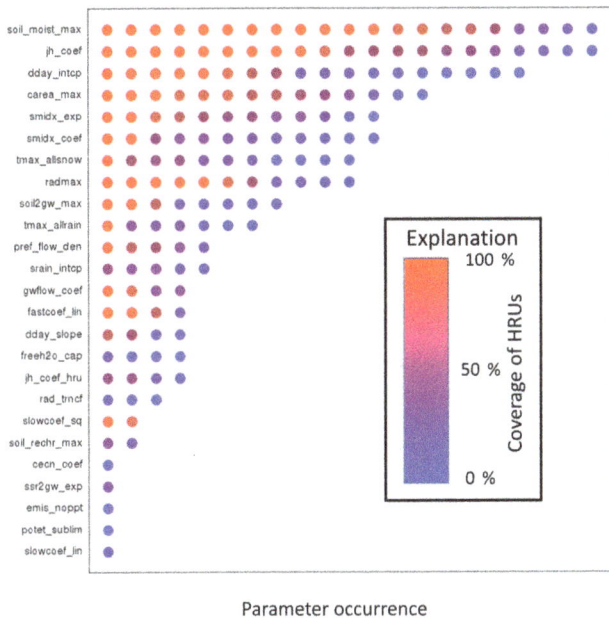

Figure 4. Summarizes the frequency of occurrence of the different calibration parameters in the process and/or performance statistic categories of Table 2. The circles in each row adjacent to a parameter name indicate how many times the respective parameter occurs in these different categories. Parameters with more circles are affecting more process categories. The color of each circle indicates the extent of the spatial coverage of that occurrence; specifically, red circles (as opposed to blue) indicate that more Hydrologic Response Units are affected by the respective parameter.

3.3 Identification of dominant and inferior processes by HRU

To identify the dominant and inferior process(es) by geographic area, the following procedure is done for each HRU:

1. the parameter sensitivity scores are summed for each parameter, resulting in a score for each parameter for each time-series output variable and performance statistic;

2. the parameter scores are averaged by performance statistics, resulting in a score for each process;

3. the process scores are ranked for each HRU;

4. the top (and bottom) ranked process determines the most dominant (and most inferior) single process for each HRU as shown in Fig. 5.

Generally, Fig. 5a shows that evapotranspiration is the most prevalent dominant process for the CONUS. This is probably because it is a major component of the hydrologic cycle and sensitive parameters are available to affect it in every HRU. However, this is not universal, and the dominant process varies by geographic region, with snowmelt being the

dominant process in the northern Great Planes and northern Rocky Mountains, total runoff being the most important in the Pacific Northwest, and with interflow important in bands across the Intermountain West (Fig. 1). Each process is dominant somewhere depending on local conditions. Equally informative are the locations of the most inferior processes (Fig. 5b). This clearly shows that PRMS snowmelt parameters are not sensitive across the Central Valley of California, and in the Deep South and the southwestern US (Fig. 1). Areas where runoff is more dominant than evapotranspiration, as in the Cascade Mountains and coastal areas of the Pacific Northwest, are locations where the runoff is a substantially greater part of the water budget. Interestingly, infiltration and baseflow appear to be equally inferior across most of CONUS, with pockets of HRUs that are insensitive to soil moisture, surface runoff, and interflow, depending on local conditions. There are no HRUs that rank evapotranspiration as the most inferior process.

Dominant and inferior processes can be identified for HRUs at the watershed scale as well. Figure 5c shows the most dominant process by HRU for the Apalachicola–Chattahoochee–Flint River watershed in the southeastern US. This watershed has been the subject of previous PRMS modeling studies (LaFontaine et al., 2013). When using this information at a finer resolution, it shows that evapotranspiration is the most dominant process watershed wide, but with pockets of HRUs in the northern part of the watershed where runoff is the most dominant and a pocket in the southern part of the watershed where infiltration is most dominant. Likewise, the most inferior process for each HRU is identified in Fig. 5d. This clearly indicates that parameters and performance statistics related to snowmelt, and to a lesser degree baseflow, do not need to be considered when modeling this watershed. Figure 5d also indicates that in the northern part of the watershed, infiltration and runoff are inferior processes as well, which could in part be due to impervious conditions around the Atlanta metropolitan area.

4 Discussion

4.1 Causes of parameter sensitivity

There are regions where parameter sensitivity is typically high for a particular performance statistic (e.g., New England region (Fig. 1) for performance statistic based on mean of processes) or typically low (e.g., Great Plains region (Fig. 1) for mean of processes) regardless of the process (Fig. 2). Why do the HRUs of some regions exhibit parameter sensitivity to almost all processes, while others exhibit parameter sensitivity to almost none? All other things being equal, there can only be two sources of these spatial patterns:

1. The physiography that is used to define the non-calibration parameters (e.g., elevation, vegetation type, soil type) renders all calibration parameters insensitive.

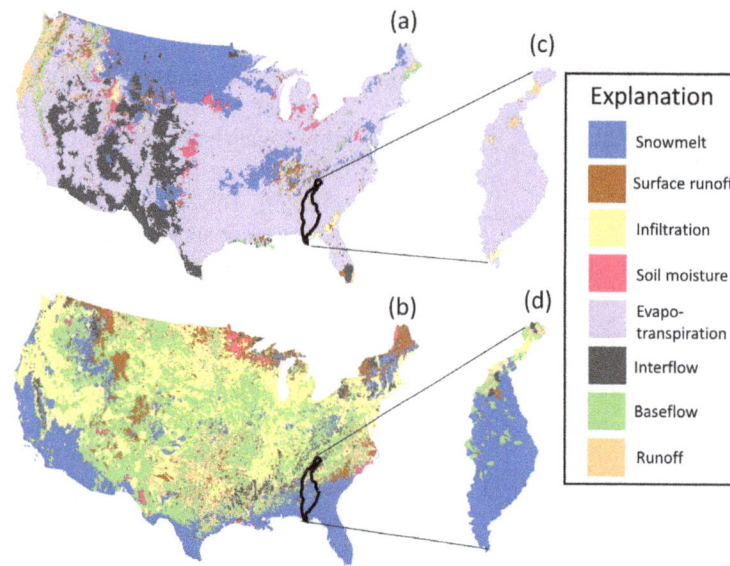

Figure 5. Precipitation–Runoff Modeling System parameter sensitivity organized by process ranked for each hydrologic response unit for the entire conterminous US – maps **(a)** and **(b)** – and for the Apalachicola–Chattahoochee–Flint River Basin – maps **(c)** and **(d)**. The maps on the top **(a, c)** show the most dominant process, while the maps on the bottom **(b, d)** show the most inferior process.

A theoretical example of this could be if an HRU is characterized as entirely impervious, resulting in the non-existence of any simulated soil water.

2. Patterns in the climate data used to drive the model (e.g., daily temperature and precipitation) could control model response. A theoretical example of this could be an HRU that receives no precipitation.

The hydrologic response of the HRUs in either case would always remain unchanged, regardless of changes in any parameter value. In either case, these sources of information are independent of the DPHM and could lead to the conclusion that the dominant processes identified by the methods outlined in this article could correspond to perceptible dominant processes in the physical world (i.e., how the "real world" works).

The number of unique calibration parameters for each process in Table 2 (i.e., counting the parameters across each row) may provide some insight into the complexity of each process as represented in the model structure of PRMS. In theory, more "complicated" hydrologic processes would require more parameters for parameterization than the "simpler" ones. According to this view, runoff (16 calibration parameters), infiltration (12 calibration parameters), and interflow (12 calibration parameters) are the most complex processes to simulate, with soil moisture (4) being the simplest. Baseflow (11 calibration parameters), snowmelt (11 calibration parameters), surface runoff (10 calibration parameters), and evapotranspiration (8 calibration parameters) are in between. This reflects the fact that in PRMS, runoff is a much more complicated calculation with many of the other processes di-

rectly contributing information. Also apparent is that more parameters are needed to simulate the components of streamflow (e.g., baseflow, interflow, and surface runoff) than processes that do not result directly in flow (e.g. snowmelt, evapotranspiration, and soil moisture). The only process that does not follow this pattern is infiltration. Storm-event-based infiltration is typically simulated with sub-daily time steps to account for the variability of time and intensity in this process. It is possible that PRMS must compensate for this shortcoming in structure with a more complex parameterization of the process.

Table 2 indicates that there are 10 calibration parameters that are never sensitive regardless of the process or performance statistic. This indicates that these parameters should always be set to the default value, with minimal resources used to estimate them, and never be calibrated. Additional modeling studies could reveal situations where these parameters actually do exhibit some sensitivity, perhaps in situations with smaller geographical domains or over different time periods. It is also possible that these parameters are never sensitive, indicating some structural problem or unwarranted complexity in the DPHM, and the removal of some algorithms from the source code of the DPHM is advised. Additional study is required of these 10 non-sensitive calibration parameters, and upon further review of the PRMS source code, a structural problem (e.g., unintended constraint, non-differentiable behavior, or software bug) might be revealed. Alternatively, the problem could be related to invalid parameter ranges in the FAST analysis or problems with the climate data used to drive the model. Finally, it could be that alternative or improved performance statistics could resolve this issue.

4.2 Choice of performance statistic

The maps of Fig. 2 clearly illustrate the importance that choice of performance statistic can make in terms of evaluation of hydrologic response. When the maps of performance statistics within a single hydrologic process are compared (i.e., the maps across a single row), the spatial patterns and magnitude of the parameter sensitivity can be very different. This could indicate that the performance statistics based on the FDSS truly are non-redundant and are accounting for different aspects of the processes.

Table 2 indicates that the baseflow coefficient (PRMS parameter gwflow_coef, Markstrom et al., 2015) is the most sensitive parameter for performance statistics CV and AR-1, but not sensitive to the mean of the baseflow process performance statistics. This points to the fact that despite having knowledge of a parameter being associated with the computation of a certain process, sensitivity analysis can reveal that the response of the simulation is completely different when the performance statistic changes. It also indicates that sensitivity analysis might be an important step in selection of an appropriate performance statistic and that uncritical application of performance statistics may be misleading.

4.3 Spatial aspects of dominant and inferior processes

When the dominant and inferior processes are determined for an HRU (Fig. 5), it is possible that certain parameters are included in both the most dominant and most inferior processes at the same time. This apparent contradiction is not necessarily a conflict but indicates that the calibration parameters must work in concert with the evaluation method. For example, there exist HRUs where the evapotranspiration process is dominant and at the same time the runoff or infiltration processes are inferior (Fig. 5a and b). The parameter soil_moist_max is indicated as being sensitive for all three of these processes (Table 2). This parameter would demonstrate equifinality if evaluated within the context of the inferior processes (i.e., those output variables and performance statistics associated with the inferior process) but would be a very effective calibration parameter resulting in optimal values when viewed within the context of parameters and variables of the dominant process.

This method of identification of inferior and dominant processes for a specific geographical location (i.e., HRU, watershed, or region), determined by sensitivity analysis, is defined within the context of the application of the DPHM and may not necessarily have the same meaning within a different context. However, this methodology does have the ability to spatially classify watersheds and identify dominant processes. This classification scheme depends not only on the physiographic nature of the watershed, but also on the scale,

resolution, and purpose that were considered by the modeler when the application was developed.

4.4 Further study

Providing modelers with reduced lists of calibration parameters on an HRU-by-HRU, watershed-by-watershed, or region-by-region basis is the first step in the path of this research. Subsequent steps to this approach could be developed into more sophisticated methods where orthogonal output variables and performance statistics could provide much more insight into methods of effective model calibration. Other advancements in this approach may identify groups of parameters that effectively behave together, thus reducing the number of parameters and making specific model output respond more directly to a single or a few parameters, reducing parameter interaction. This suggests that model parameterization and calibration might benefit from a step-by-step strategy, using as much information as possible to set non-interactive parameters and remove them from consideration before the more interactive parameters are calibrated, reducing the dimensionality of the problem (Hay et al., 2006; Hay and Umemoto, 2006).

Another question for future research is: Does the classification of dominant hydrologic processes, both geographical and categorical, as described in this study, apply in other contexts? Comparable findings from other modeling studies, such as those by Newman et al. (2015) and Bock et al. (2016), might indicate that there could be a connection. These other studies use the same input information (i.e., being driven with the same climate data and using the same sources of information for parameter estimation), and thus simulation results and model sensitivity to this information might be similar. Also, can real world watersheds be classified by sensitivity analysis using DPHMs? Based on the findings of the work presented so far, the answer is inconclusive. Clearly there are some results that indicate that it might be possible. For example, the methods described here effectively identify "snowmelt watersheds" in the mountainous and northern latitudes, but, is all of this necessary to accomplish this? Might simpler methods (e.g., an isohyetal snowfall map) identify snowmelt watersheds just as effectively?

Questions remain about using parameter sensitivity for identification of structural inadequacies within the CONUS application and specifically the PRMS model itself. A full analysis of these parameters and how they relate to their respective process(es) is beyond the scope of this article, but it could relate information about the structure of PRMS. In this study, certain hydrologic processes (e.g., depression storage, streamflow routing, flow-through lakes, and strong groundwater–surface-water interaction) were not considered because of additional data requirements and parameterization complexity. The PRMS model also allows for selection of alternative methods for many of the module types. Each of these modules uses different equations and calibration parameters. Future work might be to determine the effect of using different modules or maybe even to determine the selection of the PRMS modules through sensitivity analysis. Just as the spatial and temporal scope of any modeling project must be defined, the scope of the hydrologic processes, and the detail to which these processes are simulated, must be likewise defined. Also, alternative ways of defining HRUs (e.g., larger or smaller, or even based on dominant process instead of geographic location) could affect the analysis. Model development and application could perhaps proceed by first accounting for those factors that have the most effect.

5 Conclusion

Watersheds in the real world clearly exhibit hydrologic behavior determined by dominant processes based on geographic location (i.e., land surface conditions and climate forcings). A methodology has been developed to identify regions, watersheds, and HRUs according to dominant process(es) on the basis of parameter sensitivity response with respect to a distributed-parameter hydrology model. The parameters in this model were divided into two groups – those that are used for model calibration and those that were not. A global parameter sensitivity analysis was performed on the calibration parameters for all HRUs derived for the conterminous US. Categories of parameter sensitivity were developed in various ways, on the basis of geographic location, hydrologic process, and model response. Visualization of these categories provides insight into model performance, and useful information about how to structure the modeling application should take advantage of as much local information as possible.

By definition, an insensitive parameter is one that does not affect the output. Ideally, a distributed-parameter hydrology model would have just a few calibration parameters, all of them meaningful, each controlling the algorithms related to the corresponding process. This would result in low parameter interaction and a clear correspondence between input and output. However, this is not always the case, and despite the fact that parameter interaction is unavoidable in these types of models, this behavior is also seen in the real world. For instance, in watersheds where evaporation is very high, antecedent soil moisture is affected, which has a direct influence on infiltration. The real world process of evaporation has an effect on infiltration, just as evaporation parameters have an effect on simulation of infiltration in watershed hydrology models. Application of distributed-parameter hydrologic modeling application require that the uncertainty problem and the calibration problem be addressed at the same time. While the user of a DPHM can do nothing about the complexity of the model's internal structure, the apparent complexity can be reduced by limiting the parameters and the affected output under consideration.

Results of this study indicate that it is possible to identify the influence of different hydrologic processes when simulating with a distributed-parameter hydrology model on the basis of parameter sensitivity analysis. Factors influencing this analysis include geographic area, topography, land cover, soil, geology, climate, and other unidentified physical effects. Identification of these processes allows the modeler to focus on the more important aspects of the model input and output, which can simplify all facets of the hydrologic modeling application.

Edited by: E. Zehe

References

Ali, G., Tetzlaff, D., Soulsby, C., McDonnell, J. and Capell, R.: A comparison of similarity indices for catchment classification using a cross-regional dataset, Adv. Water Resour., 40, 11–22, doi:10.1016/j.advwatres.2012.01.008, 2012.

Amorocho, J. and Hart, W. E.: A critique of current methods in hydrologic systems investigation, Trans. Am. Geophys. Un., 45, 307–321, 1964.

Archfield, S. A., Kennen, J. G., Carlisle, D. M., and Wolock, D. M.: An objective and parsimonious approach for classifying nature flow regimes at a continental scale, River Res. Appl., 30, 1166–1183, 2014.

Battaglin, W. A., Hay, L. E., and Markstrom, S. L.: Simulating the potential effects of climate change in two Colorado Basins and at two Colorado ski areas, Earth Interact., 15, 1–23, 2011.

Beven, K., Changing ideas in hydrology – The case of physically-based models, J. Hydrol., 105, 157–172, 1989.

Beven, K: A manifesto for the equifinality thesis, J. Hydrol., 320, 18–36, doi:10.1016/j.jhydrol.2005.07.007, 2006.

Bock, A. R., Hay, L. E., McCabe, G. J., Markstrom, S. L., and Atkinson, R. D.: Parameter regionalization of a monthly water balance model for the conterminous United States, Hydrol. Earth Syst. Sci., 20, 2861–2876, doi:10.5194/hess-20-2861-2016, 2016.

Bower, D. E.: Evaluation of the Precipitation-Runoff Modeling System, Beaver Creek, Kentucky, US Geological Survey Scientific Investigations Report 84-4316, US Geological Survey, 1–39, 1985.

Boyle, D. P., Lamorey, G. W., and Huggins, A. W.: Application of a hydrologic model to assess the effects of cloud seeding in the Walker River basin, Nevada, J. Weather Modificat., 38, 66–67, 2006.

Brun, R., Reichert, P., and Kunsch, H. R.: Practical identifiability analysis of large environmental simulation models, Water Resour. Res., 37, 1015–1030, doi:10.1029/2000wr900350, 2001.

Cary, L. E.: Techniques for estimating selected parameters of the U.S. Geological Survey's Precipitation-Runoff Modeling System in eastern Montana and northeastern Wyoming, US Geological Survey Water-Resources Investigations Report 91-4068, US Geological Survey, Reston, Virgina, 1–39, 1991.

Chaney, N. W., Herman, J. D., Reed, P. M., and Wood, E. F.: Flood and drought hydrologic monitoring: the role of model parameter uncertainty, Hydrol. Earth Syst. Sci., 19, 3239–3251, doi:10.5194/hess-19-3239-2015, 2015.

Clark, M. P. and Vrugt, J. A.: Unraveling uncertainties in hydrologic model calibration: Addressing the problem of compensatory parameters, Geophys. Res. Lett., 33, L06406, doi:10.1029/2005gl025604, 2006.

Cukier, R. I., Fortuin, C. M., and Shuler, K. E.: Study of the sensitivity of coupled reaction systems to uncertainties in rate coefficients I, J. Chem. Phys., 59, 3873–3878, 1973.

Cukier, R. I., Schaibly, J. H., and Shuler, K. E.: Study of the sensitivity of coupled reaction systems to uncertainties in rate coefficients III, J. Chem. Phys., 63, 1140–1149, 1975.

Cuntz, M., Mai, J., Zink, M., Thober, S., Kumar, R., Schafer, D., Schron, M., Craven, J., Rakovec, O., Spieler, D., Prykhodko, V., Dalmasso, G., Musuuza, J., Langenberg, B., Attinger, S., and Samaniego, L.: Computationally inexpensive identification of noninformative model parameters by sequential screening, Water Resour. Res., 51, 6417–6441, doi:10.1002/2015WR016907, 2015.

Duan, Q., Schaake, J., Andréassian, V., Franks, S., Goteti, G., Gupta, H. V., Gusev, Y. M., Habets, F., Hall, A., Hay, L., Hogue, T., Huang, M., Leavesley, G., Liang, X., Nasonova, O. N., Noilhan, J., Oudin, L., Sorooshian, S., Wagener, T., and Wood, E. F.: Model Parameter Estimation Experiment (MOPEX): An overview of science strategy and major results from the second and third workshops, J. Hydrol., 320, 3–17, 2006.

Dudley, R. W.: Simulation of the quantity, variability, and timing of streamflow in the Dennys River Basin, Maine, by use of a precipitation-runoff watershed model, US Geological Survey Scientific Investigations Report 2008-5100, US Geological Survey, Reston, Virgina, 1–44, 2008.

Ewan, J.: Hydrograph matching method for measuring model performance, J. Hydrol., 408, 178–187, doi:10.1016/j.jhydrol.2011.07.038, 2011.

Freeze, R. A. and Harlan, R. L.: Blueprint for a physically-based, digitally-simulated hydrologic response model, J. Hydrol., 9, 237–258, 1969.

Grayson, R. B., Moore, I. D., and McMahon, T. A.: Physically based hydrologic modeling, 2, Is the concept realistic?, Water Resour. Res., 28, 2659–2666, 1992.

Gupta, H. V., Wagener, T., and Liu, Y. Q.: Reconciling theory with observations: elements of a diagnostic approach to model evaluation, Hydrol. Process., 22, 3802–3813, doi:10.1002/hyp.6989, 2008.

Gupta, H. V., Kling, H., Yilmaz, K. K., and Martinez, G. F.: Decomposition of the mean squared error and NSE performance criteria: Implications for improving hydrological modelling, J. Hydrol., 377, 80–91, doi:10.1016/j.jhydrol.2009.08.003. 2009.

Gupta, H. V., Clark, M. P., Vrugt, J. A., Abramowitz, G., and Ye, M.: Towards a comprehensive assessment of model structural adequacy, Water Resour. Res., 48, W08301, doi:10.1029/2011wr011044, 2012.

Guse, B., Reusser, D. E., and Fohrer, N.: How to improve the representation of hydrological processes in SWAT for a lowland catchment – Temporal analysis of parameter sensitivity and model performance, Hydrol. Process., 28, 2651–2670, doi:10.1002/hyp.9777, 2014.

Hay, L. E., Leavesley, G. H., Clark, M. P., Markstrom, S. L., Viger,

R. J., and Umemoto, M.: Step-wise, multiple-objective calibration of a hydrologic model for a snowmelt-dominated basin, J. Am. Water Resour., 42, 891–900, 2006.

Hay, L. E., Markstrom, S. L., and Ward-Garrison, C. D.: Watershed-scale response to climate change through the 21st century for selected basins across the United States, Earth Interact., 15, 1–37, 2011.

Hay, L. E. and Umemoto, M.: Multiple-objective stepwise calibration using Luca, US Geological Survey Open-File Report 2006-1323, US Geological Survey, Reston, Virgina, 1–25, 2006.

He, Y., Bardossy, A., and Zehe, E.: A catchment classification scheme using local variance reduction method, J. Hydrol., 411, 140–154, 2011.

Hrachowitz, M., Fovet, O., Ruiz, O., Euser, T., Gharari, S., Nijzink, R., Freer, J., Savenije, H. H. G., and Gascuel-Odoux, C.: Process consistency in models: The importance of system signatures, expert knowledge, and process complexity, Water Resour. Res., 50, 7445–7469, doi:10.1002/2014WR015484, 2014.

Jakeman, A. and Hornberger, G.: How much complexity is warranted in a rainfall-runoff model?, Water Resour. Res., 29, 2637–2649, 1993.

Kirchner, J. W., Hooper, R. P., Kendall, C., Neal, C., and Leavesley, G. H.: Testing and validating environmental models, Sci. Total Environ., 183, 33–47, 1996.

Koczot, K. M., Markstrom, S. L., and Hay, L. E.: Effects of baseline conditions on the simulated hydrologic response to projected climate change, Earth Interact., 15, 1–23, 2011.

Krause, P., Boyle, D. P., and Bäse, F.: Comparison of different efficiency criteria for hydrological model assessment, Adv. Geosci., 5, 89–97, doi:10.5194/adgeo-5-89-2005, 2005.

LaFontaine, J. H., Hay, L. E., Viger, R. J., Markstrom, S. L., Regan, R. S., Elliott, C. M., and Jones, J. W.: Application of the Precipitation-Runoff Modeling System (PRMS) in the Apalachicola–Chattahoochee–Flint River Basin in the southeastern United States, US Geological Survey Scientific Investigations Report 2013-5162, US Geological Survey, Reston, Virgina, 1–118, 2013.

Markstrom, S. L., Hay, L. E., Ward-Garrison, C. D., Risley, J. C., Battaglin, W. A., Bjerklie, D. M., Chase, K. J., Christiansen, D. E. Dudley, R. W., Hunt, R. J., Koczot, K. M., Mastin, M. C., Regan, R. S., Viger, R. J., Vining, K. C., and Walker, J. F.: Integrated watershed scale response to climate change for selected basins across the United States, US Geological Survey Scientific Investigations Report 2011-5077, US Geological Survey, Reston, Virginia, 1–143, 2012.

Markstrom, S. L., Regan, R. S., Hay, L. E., Viger, R. J., Webb, R. M. T., Payn, R. A., and LaFontaine, J. H.: PRMS-IV, the precipitation-runoff modeling system, version 4, book 6, chap. B7, US Geological Survey Techniques and Methods, US Geological Survey, Reston, Virgina, 1–158, doi:10.3133/tm6B7, 2015.

Mayer, D. G. and Butler, D. G.: Statistical validation, Ecol. Model., 68, 21–32, 1993.

McDonnell, J., Sivapalan, M., Vaché, K., Dunn, S., Grant, G., Haggerty, R., Hinz, C., Hooper, R., Kirchner, J., and Roderick, M.: Moving beyond heterogeneity and process complexity: a new vision for watershed hydrology, Water Resour. Res., 43, 6 pp., doi:10.1029/2006wr005467, 2007.

Mendoza, P. A., Clark, M. P., Barlage, M., Rajagopalan, B., Samaniego, L., Abramowitz, G. and Gupta H., Are we unnecessarily constraining the agility of complex process-based models?, Water Resour. Res., 51, 716–728, doi:10.1002/2014WR015820, 2015a.

Mendoza, P. A., Clark, M. P., Mizukami, N., Newman, A. J., Barlage, M. E., Gutmann, D. Rasmussen, R. M., Rajagopalan, B., Brekke, L. D., and Arnold, J. R.: Effects of hydrologic model choice and calibration on the portrayal of climate change impacts, J. Hydrometeorol., 16, 762–780, 2015b.

Merz, R. and Blöschl, G.: Regionalisation of catchment model parameters, J. Hydrol., 287, 95–123, 2004.

Newman, A. J., Clark, M. P., Sampson, K., Wood, A., Hay, L. E., Bock, A., Viger, R. J., Blodgett, D., Brekke, L., Arnold, J. R., Hopson, T., and Duan, Q.: Development of a large-sample watershed-scale hydrometeorological data set for the contiguous USA: data set characteristics and assessment of regional variability in hydrologic model performance, Hydrol. Earth Syst. Sci., 19, 209–223, doi:10.5194/hess-19-209-2015, 2015.

Perrin, C., Michel, C., and Andreassian, V.: Does a large number of parameters enhance model performance? Comparative assessment of common catchment model structures on 429 catchments, J. Hydrol., 242, 275–301, doi:10.1016/s0022-1694(00)00393-0, 2001.

Pfannerstill, M., Guse, B., Reusser, D., and Fohrer, N.: Process verification of a hydrological model using a temporal parameter sensitivity analysis, Hydrol. Earth Syst. Sci., 19, 4365–4376, doi:10.5194/hess-19-4365-2015, 2015.

R Core Team: R: A language and environment for statistical computing, R Foundation for Statistical Computing, Vienna, Austria, http://www.R-project.org/ (last access: August 2015), 2015.

Reusser, D. E.: Implementation of the Fourier amplitude sensitivity test (FAST), R package version 0.63, http://CRAN.R-project.org/package=fast (last access: August 2015), 2013.

Reusser, D. E. and Zehe, E.: Inferring model structural deficits by analyzing temporal dynamics of model performance and parameter sensitivity, Water Resour. Res., 47, W07550, doi:10.1029/2010WR009946, 2011.

Reusser, D. E., Buytaert, W., and Zehe, E.: Temporal dynamics of model parameter sensitivity for computationally expensive models with the Fourier amplitude sensitivity test, Water Resour. Res., 47, W07551, doi:10.1029/2010WR009947, 2011.

Saltelli, A., Ratto, M., and Tarantola, S.: Sensitivity analysis practices: strategies for model-based inference, Reliab. Eng. Syst. Saf., 10, 1109–1125, 2006.

Samaniego, L., Kumar, R., and Attinger, S.: Multiscale parameter regionalization of a grid-based hydrologic model at the mesoscale, Water Resour. Res., 46, W05523, doi:10.1029/2008WR007327, 2010.

Sanadhya, P., Giron, J., and Arabi, M.: Global sensitivity analysis of hydrologic processes in major snow-dominated mountainous river basins in Colorado, Hydrol. Process., 28, 3404–3418, doi:10.1002/hyp.9896, 2013.

Schaefli, B. and Gupta, H. V.: Do Nash values have value?, Hydrol. Process., 21, 2075–2080, 2007.

Schaibly, J. H. and Shuler, K. E.: Study of the sensitivity of coupled reaction systems to uncertainties in rate coefficients II, applications, J. Chem. Phys., 59, 3879–3888, 1973.

Seibert, J.: Regionalization of parameters for a conceptual rainfall-runoff model, Agr. Forest Meteorol., 98, 279–293, 1999.

Sivakumar, B.: Dominant processes concept in hydrology: moving forward, Hydrol. Process., 18, 2349–2353, 2004.

Sivakumar, B. and Singh, V. P.: Hydrologic system complexity and nonlinear dynamic concepts for a catchment classification framework, Hydrol. Earth Syst. Sci., 16, 4119–4131, doi:10.5194/hess-16-4119-2012, 2012.

Sivakumar, B., Jayawardena, A. W., and Li, W. K.: Hydrologic complexity and classification: a simple data reconstruction approach, Hydrol. Process., 21, 2713–2728, 2007.

Thornton, P. E., Thornton, M. M., Mayer, B. W., Wilhelmi, N., Wei, Y., Devarakonda, R., and Cook, R. B.: Daymet: daily surface weather data on a 1-km grid for North America, version 2, data set, available at: http://daac.ornl.gov from Oak Ridge National Laboratory Distributed Active Archive Center, Oak Ridge, Tennessee, USA, doi:10.3334/ORNLDAAC/1219, 2014.

Viger, R. J.: Preliminary spatial parameters for PRMS based on the geospatial fabric, NLCD2001 and SSURGO, US Geological Survey, Reston, Virgina, doi:10.5066/F7WM1BF7, 2014.

Viger, R. J. and Bock, A.: GIS features of the geospatial fabric for national hydrologic modeling, US Geological Survey, Reston, Virgina, doi:10.5066/F7542KMD, 2014.

Vogel, R. M.: Regional Calibration of Watershed Models, in: chap. 3 in Watershed Models, edited by: Singh, V. P. and Frevert, D. F., CRC Press, Boca Raton, Florida, 2005.

Wagener, T., McIntyre, N., Lees, M. J., Wheater, H. S., and Gupta, H. V.: Towards reduced uncertainty in conceptual rainfall–runoff modelling: dynamic identifiability analysis, Hydrol. Process., 17, 455–476, 2003.

Winter, T. C.: The concept of hydrologic landscapes, J. Am. Water Resour. Assoc., 37, 335–349, doi:10.1111/j.1752-1688.2001.tb00973.x, 2001.

Wolock, D. M., Winter, T. C., and McMahon, G.: Delineation and evaluation of hydrologic-landscape regions in the United States using geographic information system tools and multivariate statistical analyses, Environ. Manage., 34, S71–S88, doi:10.1007/s00267-003-5077-9, 2004.

Woods, R.: Seeing catchments with new eyes, Hydrol. Process., 16, 1111–1113, 2002.

Yilmaz, K. K., Gupta, H. V., and Wagener, T.: A process-based diagnostic approach to model evaluation: Application to the NWS distributed hydrologic model, Water Resour. Res., 44, W09417, doi:10.1029/2007WR006716, 2008.

7

Should seasonal rainfall forecasts be used for flood preparedness?

Erin Coughlan de Perez[1,3,4], Elisabeth Stephens[2], Konstantinos Bischiniotis[3], Maarten van Aalst[1,4],
Bart van den Hurk[5], Simon Mason[4], Hannah Nissan[4], and Florian Pappenberger[6]

[1]Red Cross Red Crescent Climate Centre, The Hague, 2521 CV, the Netherlands
[2]School of Archaeology, Geography and Environmental Science, University of Reading, Reading, RG6 6AH, UK
[3]Institute for Environmental Studies, VU University Amsterdam, 1081 HV, the Netherlands
[4]International Research Institute for Climate and Society, Columbia University, New York, 10964, USA
[5]Royal Netherlands Meteorological Institute (KNMI), De Bilt, 3731 GA, the Netherlands
[6]European Centre for Medium-Range Weather Forecasts, Reading, RG2 9AX, UK

Correspondence to: Erin Coughlan de Perez (coughlan.erin@gmail.com)

Abstract. In light of strong encouragement for disaster managers to use climate services for flood preparation, we question whether seasonal rainfall forecasts should indeed be used as indicators of the likelihood of flooding. Here, we investigate the primary indicators of flooding at the seasonal timescale across sub-Saharan Africa. Given the sparsity of hydrological observations, we input bias-corrected reanalysis rainfall into the Global Flood Awareness System to identify seasonal indicators of floodiness. Results demonstrate that in some regions of western, central, and eastern Africa with typically wet climates, even a perfect tercile forecast of seasonal total rainfall would provide little to no indication of the seasonal likelihood of flooding. The number of extreme events within a season shows the highest correlations with floodiness consistently across regions. Otherwise, results vary across climate regimes: floodiness in arid regions in southern and eastern Africa shows the strongest correlations with seasonal average soil moisture and seasonal total rainfall. Floodiness in wetter climates of western and central Africa and Madagascar shows the strongest relationship with measures of the intensity of seasonal rainfall. Measures of rainfall patterns, such as the length of dry spells, are least related to seasonal floodiness across the continent. Ultimately, identifying the drivers of seasonal flooding can be used to improve forecast information for flood preparedness and to avoid misleading decision-makers.

1 Introduction

Humanitarians have been investing significant attention and resources in the uptake and use of climate services to inform their work in disaster risk management. For example, disaster managers regularly participate in Regional Climate Outlook forums and climate service partnerships (Hewitt et al., 2012; ICPAC, 2016; Mwangi et al., 2014). While many early warning systems focus on short-term hydrological flood warnings, these climate service initiatives promote the use of forecasts of seasonal total rainfall. The use of such forecasts has yielded mixed results when used to prepare for heightened flood risk in Africa, such as prepositioning flood relief items (Braman et al., 2013) and evacuating vulnerable people (Anon, 2016). In this article we question whether seasonal rainfall forecasts have been overpromoted for their usefulness in flood preparation.

To clarify whether seasonal total rainfall forecasts indeed indicate increased risk of flooding, we identify the dominant indicators of seasonal flooding in different locations of sub-Saharan Africa. In many locations, it is likely that total rainfall is not the dominant driver, and other seasonal descriptors would give a better indication of the risk of flood hazards. Cumulative rainfall is not the dominant flood-generating process for floods in most river basins in the United States (Berghuijs et al., 2016), and monthly total rainfall has not been shown to be a good indicator of regional river "floodiness", or the percentage of regional rivers with extreme flooding (Stephens et al., 2015). We provide further discussion of "floodiness" in Sect. 2.2.

In the context of sub-Saharan Africa, we quantify the relationship between seasonal total rainfall and floodiness, and explore whether there might be alternative variables with a stronger relationship with floodiness at the seasonal level. In each river basin, the catchment size and the climate regime will affect the influence of hydraulic routing, soil dynamics, and precipitation patterns; we therefore identify which hydrometeorological variables are most related to seasonal flood risk in each location. We investigate the association between seasonal percentage floodiness and seasonal total rainfall, as well as the relationship with 14 other variables and their combinations.

2 Methods

Given the scarcity of hydrological data available for many parts of Africa, we offer an alternative methodology to that used by Berghuijs et al. (2016) for assessing the indicators of flood intensity and frequency in a region. Rainfall estimates from ERA-Interim Land (Balsamo et al., 2015) are used to force the Global Flood Awareness System, a global hydrological model (Alfieri et al., 2013). We calculate anomaly correlations between rainfall input and the predicted flooding, which is defined as the proportion of river cells that has extreme discharge in a region in a given time period (Stephens et al., 2015). We repeat this analysis with the 14 alternative variables, and develop a generalized linear model (glm) to identify which combinations of variables provided the greatest indication of flood hazard in each region.

Our methodology depends on the reanalysis for a climatology of rainfall and focuses on the hydrological model to estimate the consequences of this rainfall for river flows. This approach is not limited by a patchy observational network, and results can be compared across regions to inform regional policies. While the rainfall has been bias-corrected with observations, we would encourage the replication of this methodology using local rainfall observations for more detailed study of the local indicators of floodiness.

2.1 Rainfall

To calculate the rainfall indices, we use daily gridded reanalysis rainfall estimates from 1980 to 2010. The rainfall estimates are 24 h totals from the ERA-Interim Land reanalysis, which is adjusted from ERA-Interim calibrated using GPCP v2.1 data (Balsamo et al., 2015). Due to patchy observational networks, uncertainties in precipitation datasets over Africa are large (Sylla et al., 2013), and this bias correction was shown to improve the performance of river discharge simulations from ERA-Interim Land over Africa (Balsamo et al., 2015). The soil moisture estimates are also taken from the ERA-Interim Land dataset.

The area of study we have selected is sub-Saharan Africa, 16° N–35° S, 17° W–52° E. Because flooding primarily happens during the wet seasons, we applied a dry mask by eliminating all 3-month seasons that have an average of less than 15 % of the total annual rainfall and also less than 50 cm of rainfall in that season (Mason et al., 1999). To calculate seasonal total rainfall, we sum the daily rainfall estimates for each overlapping 3-month season (JFM, FMA, etc.) over a 2.5° grid box, as this is the resolution of many seasonal forecasting products from the Global Producing Centres for Long-Range Forecasts (Barnston et al., 2003; WMO, 2017).

2.2 Flooding

We use daily rainfall from ERA-Interim Land to drive a hydrological model to estimate river discharge. The system used here is the Global Flood Awareness System (GloFAS), which is comprised of a HTESSEL land surface model to generate surface and subsurface runoff and a Lisflood model to complete the routing and groundwater flows at a 0.1° resolution for the entire global land surface (Alfieri et al., 2013). In this study we focus on river flooding only; therefore, we only consider GloFAS river grid points which have a greater than 1000 km^2 upstream basin area. These river pixels are aggregated to the 2.5° resolution to match the rainfall scale.

There are several ways to define whether a location experienced "flooding", which is the variable of interest to the disaster manager. Here, we define flooding according to the return period of the discharge, such that extreme floods happen at approximately the same frequency throughout the study area. We focus on the 1 in 5 and 1 in 50-year events; these return periods are defined by fitting a Gumbel extreme value distribution to the daily flows (Alfieri et al., 2013).

To understand the magnitude of flooding in a 2.5° grid box, we calculate "floodiness" as defined in Stephens et al. (2015). Percentage floodiness is the percent of river pixels that have at least 1 day of flooding above the return period, and duration floodiness is the number of pixel days that have flooding during that season. Our results were very similar between percentage and duration floodiness; therefore, duration floodiness is not shown here.

2.3 Predictor variables

While seasonal total rainfall has demonstrated some predictability in this part of the world (Barnston et al., 2010b; Weisheimer and Palmer, 2014), there are other variables that might be predicted at the seasonal level: frequency of extreme events within a season, sub-seasonal rainfall patterns, soil moisture, and rainfall intensity. Here, we investigate whether variables in each of those categories could serve as a better indicator of flood risk in sub-Saharan Africa. In addition to seasonal total rainfall, we calculated 14 predictor variables at the seasonal level. These are defined as follows.

Extreme events within a season

- 1 day above 95th: number of days in the season during which daily precipitation is greater than the 95th percentile of daily precipitation of the entire time series.

- 1 day above 99th: number of days in the season during which daily precipitation is greater than the 99th percentile of daily precipitation of the entire time series.

- 3 days above 75th: number of 3-day events in the season during which 3-day precipitation is greater than the 75th percentile of 3-day precipitation of the entire time series.

- 3 days above 99th: number of 3-day events in the season during which 3-day precipitation is greater than the 99th percentile of 3-day precipitation of the entire time series.

- 5 days above 99th: number of 5-day events in the season during which 5-day precipitation is greater than the 99th percentile of 5-day precipitation of the entire time series.

Patterns of rainfall within a season

- Rainy days: seasonal count of the number of days in which daily precipitation is greater than 1 mm (Sillmann et al., 2013).

- Mean wet-spell length: average length of all wet spells in that season, where a wet spell is defined as the length of consecutive days in which daily precipitation is greater than 1 mm.

- Median dry-spell length: median length of all dry spells in that season, where a dry spell is defined as the length of consecutive days in which daily precipitation is less than 1 mm.

- Dry-spell autocorrelation: Spearman rank lag-1 autocorrelation of successive dry-spell lengths (Schleiss and Smith, 2016).

- 3-day autocorrelation: Spearman rank lag-3 autocorrelation of daily rainfall amounts.

Soil moisture and intensity

- Soil moisture: volumetric soil water layer 1: top soil layer 0–7 cm. Average daily soil moisture for the season in $\mathrm{kg\,m^{-3}}$.

- Intensity: total seasonal rainfall divided by the number of rainy days (see the definition above).

- Contribution of extremes: total rainfall falling in days of the 95th percentile or higher, divided by total seasonal rainfall (Alexander et al., 2013).

- Burstiness 15 day: burstiness as defined in Schleiss and Smith (2016): $\frac{\sigma_{\mu}-\mu}{\sigma_{\mu}+\mu}$, where μ is the average time between a specific amount of rainfall (interamount time), held at 15 days, and σ is the standard deviation of interamount times.

2.4 Comparison

We examine whether anomalously high values of these variables correlate with greater floodiness. Using seasonal anomalies for each variable, we calculate the Spearman rank correlation between the rainfall anomalies and floodiness at every grid point, as the data are not normally distributed. To assess our confidence in these results, we bootstrap the time series to generate 1000 replicates using a block bootstrap of five seasons. If less than 5 % of the rank correlations of these bootstrapped replicates have an opposite sign to the original result, we have confidence in our result. Only results with this level of confidence are plotted in the figures.

Basin hydrology can also lead to complex relationships between rainfall and flooding. We therefore explore the correlation between basin-level rainfall with basin-level floodiness. We average the rainfall variable and floodiness variable across food producing units (FPUs) (Cai and Rosegrant, 2002), which are defined by a combination of hydrological basins and geopolitical regions and are therefore relevant for decision-making purposes. We apply a dry mask for an entire FPU if more than half of the grid points in the FPU are in a dry season. With these aggregated results, we then apply the same correlation methods as for the grid points above.

Lastly, we fit a generalized linear model (glm) to three of the predictor variables from different categories that showed improvements in correlation relative to seasonal total rainfall. For the dependent variable, we use a binary dataset indicating the occurrence or not of floodiness above the 50-year return period. The model uses a binomial distribution with a logit link, and uses 10-fold cross-validation to fit the glm. We select the most parsimonious model within 1 standard error of the model with the minimum standard error, using the glmnet package for R (Friedman et al., 2010).

3 Results and discussion

Three-month seasonal total rainfall anomalies show significant correlation with floodiness in several regions (Fig. 1). The relationship is weakest in western and central Africa, and also weakens as flood severity increases.

When the rainfall and floodiness are aggregated by FPU and then correlated, the correlations improve in almost all locations, suggesting that seasonal total rainfall forecasts for FPUs (Fig. 1c and d) might be of greater use than grid-box forecasts (Fig. 1a and b) as a predictor of flood hazard. Different regional forecast aggregations could also be explored to determine whether this can be further optimized.

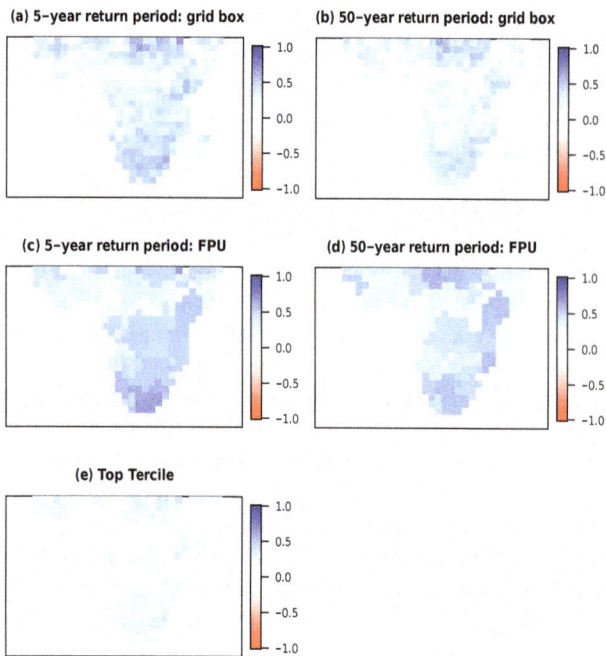

Figure 1. Anomaly rank correlations between seasonal total rainfall and percentage floodiness (Stephens et al., 2015) at the 5-year **(a)** and 50-year **(b)** return periods. Anomaly rank correlations between seasonal total rainfall for a 2.5° gridded food producing unit (FPU) and floodiness for that FPU at the 5-year **(c)** and 50-year **(d)** return periods. Correlations are only shown here if more than 95 % of all boostrapped replicates agreed on the sign of the result. The increase in probability of floodiness above the 5-year return period conditional on seasonal total rainfall falling in the top tercile **(e)**, expressed as the difference in probability relative to climatology.

While the correlations are significant in many regions, there is considerable variation in floodiness that remains unexplained by this variable. To demonstrate this, we calculate the probability of flooding (floodiness greater than 0) conditional on seasonal rainfall being in the top tercile of the distribution, which is the focus of many seasonal forecasts. Ultimately, even if a top-tercile rainfall forecast were given with 100 % certainty, it would represent only a small increase in the probability of flooding relative to climatology (Fig. 1e).

In Figs. 2–4 we display results from three different sets of possible predictor variables. In Fig. 2 we plot the anomaly rank correlations with floodiness for five different measures of extreme precipitation events within a season. None of these rainfall variables are a better predictor of floodiness in all locations (Fig. 2, second row); however, the number of rain events above the 99th percentile (1-, 3-, and 5-day events) tend to outperform seasonal total rainfall in the areas of western and central Africa (where seasonal total rainfall had the weakest correlations; see Fig. 1).

Next, we analyzed five different measures of rainfall patterns within a season, including the length of dry spells and wet spells. Apart from in isolated locations, these measures

do not have coherently stronger correlations with floodiness than seasonal total rainfall (Fig. 3).

The last set of variables we explored included soil moisture and several measures of seasonal rainfall intensity. Figure 4a shows that in most regions seasonal total rainfall is more strongly correlated with floodiness than soil moisture. In comparison, seasonal rainfall intensity shows a slightly higher correlation with floodiness across the continent (Fig. 4b), defined as the total precipitation divided by the number of rainy days. Similarly, the percent of seasonal rainfall occurring in the top 95th percentile days, here called the "contribution of extremes", shows higher correlations in the western and central Africa region (Fig. 4c). Both of these variables show less variation across Köppen climate regions, compared to seasonal total rainfall (Fig. 1). Burstiness (Schleiss and Smith, 2016) of a 15-day interamount time (Fig. 4d) does not show better correlations with floodiness than does seasonal total rainfall.

It is possible that a combination of these variables would outperform any of them in isolation, so we also test the combination of three different types of variables that each have strong correlations with floodiness: (1) 3 days above 99th, (2) soil moisture, and (3) contribution of extremes. To test whether a combination of these variables is better able to predict 50-year return period floodiness, we fit a logistic regression model for each grid point using these three variables. Because these variables are correlated with each other in several regions, we select the generalized linear model (glm) fit with the fewest variables that is still within 1 standard error of the optimal fitted model.

Results of the glm generally confirm the spatial patterns reflected in the correlation figures above, and indicate that a combination of these variables could be a useful indicator of floodiness in many regions. Figure 5 shows that the number of 3-day events above the 99th percentile was a meaningful contributor when added as a predictor independently, or in conjunction with another variable, in most of sub-Saharan Africa. Soil moisture is included as an additional predictor primarily in southern Africa, while the contribution of extremes was included primarily in central Africa. A combination of all three variables was recommended in eastern Africa and parts of southern Africa, while none of the predictors was selected as a meaningful contributor for much of western and central Africa.

4 Conclusions

In the analysis above, we have demonstrated that indicators of floodiness differ widely across the African continent, using a methodology that can be replicated for other data-scarce regions to assess the key indicators of flooding. Improvements to both the climatology of reanalysis rainfall and the skill of global hydrological models could further improve the understanding of predictability of these processes, and

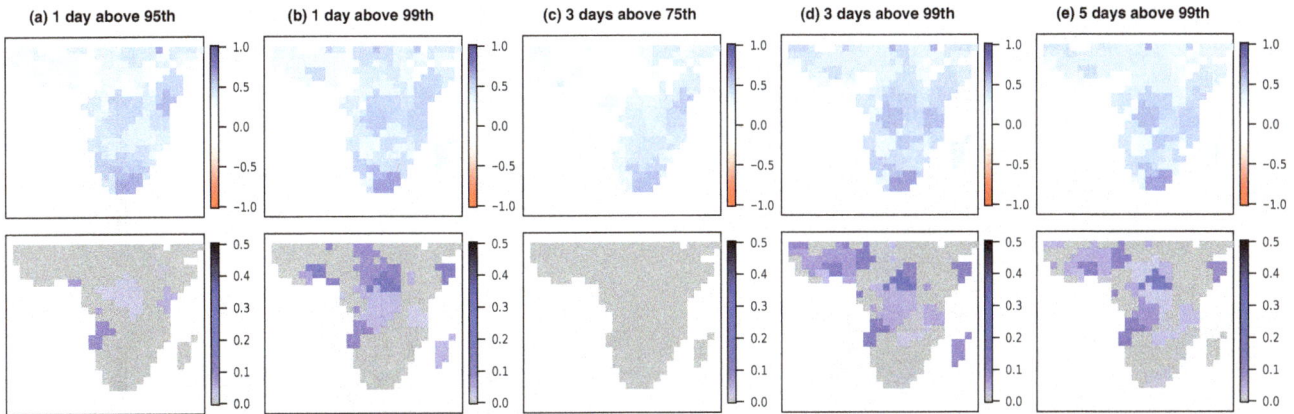

Figure 2. Correlation of the number of extreme events within a season and floodiness for FPUs in Africa. The top row shows the anomaly rank correlations between each variable and percentage floodiness at the 5-year return period at the FPU level. The bottom row is the improvement relative to seasonal total rainfall – locations in blue show a higher anomaly correlation for this variable than for seasonal total rainfall anomalies. Areas in which seasonal total rainfall has a higher or equal correlation are shown in grey. Note that results are only plotted for locations where more than 95 % of the boostrapped replicas agree on the sign of the change.

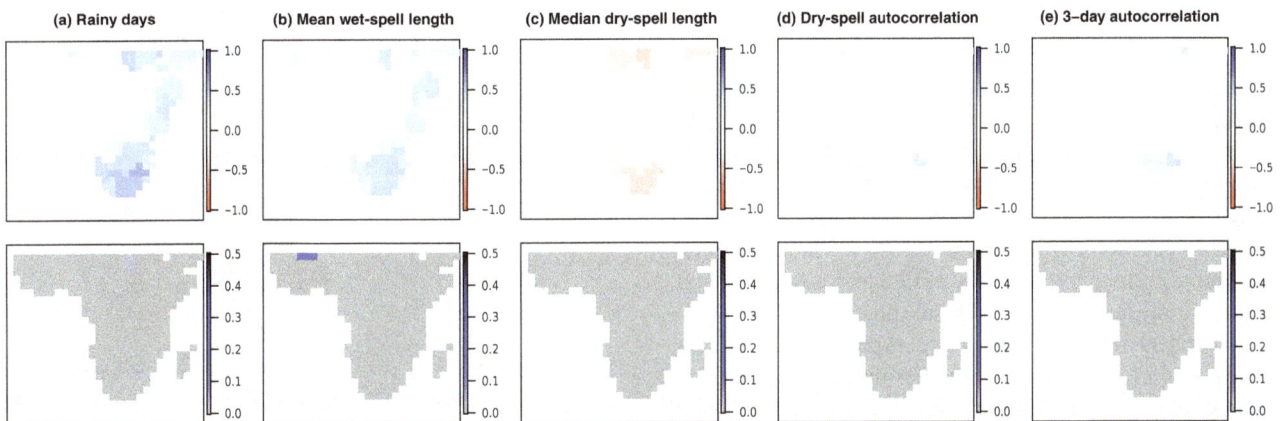

Figure 3. Same as Fig. 2 for the following variables. **(a)** Rainy days: number of days with more than 1 mm of rain. **(b)** Mean wet-spell length: mean length of consecutive days of rain greater than 1 mm. **(c)** Median dry-spell length: median length of consecutive dry days. **(d)** Dry-spell autocorrelation: Spearman rank lag-1 autocorrelation of successive dry spell lengths. **(e)** 3-day autocorrelation: Spearman rank lag-3 autocorrelation of daily rainfall amounts.

we encourage replication of this methodology using observations to further describe and validate the flood-generating processes in specific locations.

It is clear that seasonal total rainfall is not a reasonable proxy for floodiness in most of western Africa, central Africa, and Madagascar. Large portions of these regions fall into the "equatorial" Köppen classification, which includes tropical savannahs. Floodiness in these regions demonstrated a stronger relationship with measures of the intensity of rainfall during a season than in the rest of the continent. In these regions, the climate services community should reconsider their association of seasonal total rainfall with flood risk and flood preparation measures (Braman et al., 2013). When using forecasts in an operational context, imperfect forecast

skill of the rainfall proxy itself further reduces the usefulness of this information for flood preparedness.

On the other hand, much of eastern Africa, southern Africa, and the Sahel tends to show similar patterns in the dominant indicators of flooding. Seasonal total rainfall had some of the highest correlations in these regions, as well as the number of extreme events within a season. There are large "arid" areas in each of these regions, and these findings are consistent with studies done in other arid areas. Berghuijs et al. (2016) found that daily and multi-day rainfall events were the dominant flood-generating processes for river basins in arid regions of the United States, similar to the results in Fig. 2d.

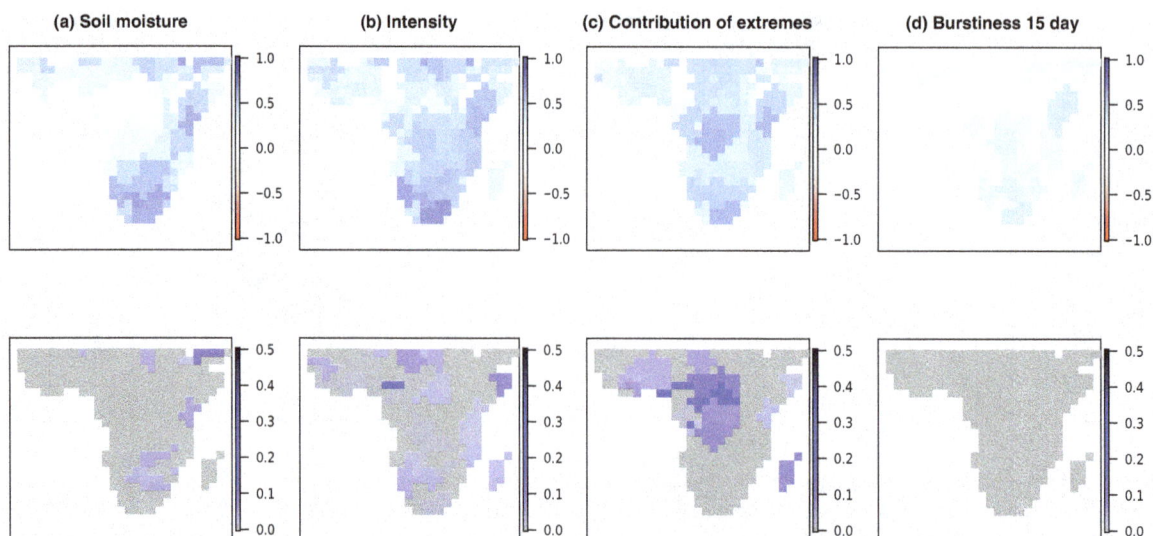

Figure 4. Same as Fig. 2 but for the following variables. **(a)** Soil moisture: seasonal average moisture in topsoil. **(b)** Intensity: total rainfall divided by the number of rainy days. **(c)** Contribution of extremes: total rainfall divided by the amount of rain contributed by the top 95th percentile days. **(d)** Burstiness 15 day: intermittency measure (Schleiss and Smith, 2016).

To maximize usefulness in these regions, forecasters could consider simple formatting alternatives to current forecasts that would provide a better indication of floodiness, such as replacing tercile forecasts with forecasts of the top percentiles of the distribution (Grieser, 2014), and offering aggregate forecasts for river basins or FPUs. The latter could also lend itself to greater forecast skill than for rainfall itself, and encourage regional-scale disaster preparedness.

Researchers developing new forecast products should consider several of the predictor variables discussed here. Forecasts of the frequency of extreme rainfall events would likely provide a better indication of floodiness, compared to seasonal total rainfall forecasts, for much of Sub-Saharan Africa. Studies have shown the potential predictability of this variable in several locations (Anderson et al., 2015; Higgins et al., 2000; Verbist et al., 2010). Seasonal forecasts of soil moisture could give a useful indication of flood risk in dry regions of Africa (Fig. 4), and these forecasts are also likely to have seasonal predictability in areas where they can be well initialized, notably due to the persistence of soil moisture (Kanamitsu et al., 2002; Koster et al., 2010; Poveda et al., 2001). This also takes evaporation into account.

Forecasts of rainfall intensity could give a better indication of flood risk in western and central Africa (Fig. 5). However, intensity is the least spatially coherent and therefore least likely to be predictable (Moron et al., 2007). Further research into the area is merited, as there are a few examples showing some potential predictability of rainfall intensity (Pineda and Willems, 2016).

Seasonal skill in forecasting total 3-month rainfall anomalies is varied around the world; the highest skill has been achieved during ENSO events in areas that have ENSO tele-

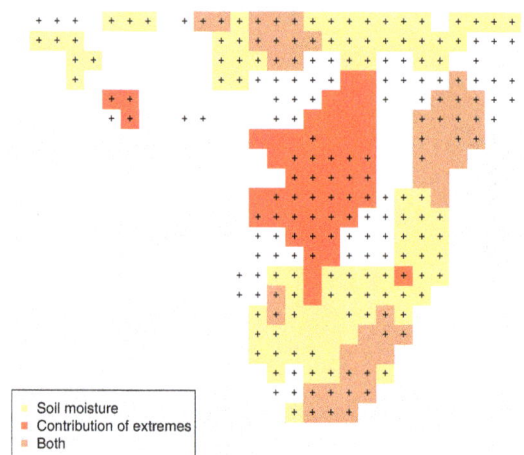

Figure 5. Results of optimizing a logistic regression model using a combination of the high-performing variables considered earlier. The model predicted whether there was any floodiness at the 50-year return period by using the following predictors: number of 3-day events in the 95th percentile (crosses), soil moisture (yellow), and the contribution of extremes (red). To optimize the model, we selected the most parsimonious combination of these three predictors that formed a glm that is within 1 standard error of the standard error that could be achieved by the maximum fit. FPUs that are plain white showed no value in using any of the predictors, while locations with colors/symbols show which predictors were retained in the optimized model, either alone or in combination with other predictors.

connections (Barnston et al., 2010a; Weisheimer and Palmer, 2014). Given the low correlations we have found here between floodiness and either seasonal total rainfall or other rainfall indicators, forecasts of any of these proxies are unlikely to provide strong signals of increased risk. However, there have been several studies using large-scale climate patterns and sea surface temperatures (SSTs) as predictors of flood risk, most focusing on the role of ENSO in changing global flood risk (Emerton et al., 2017; Ward et al., 2014, 2016). Further research on using SSTs and other climate patterns to directly forecast changes to flooding is merited, to explore whether such forecasts would give stronger indications of change in flood hazard than seasonal climate models of rainfall.

Ultimately, the most informative forecasts of flood hazard at the seasonal scale could be seasonal streamflow forecasts using hydrological models calibrated for individual river basins (Sahu et al., 2016). While this is more computationally and resource intensive, investments in better forecasts of seasonal flood risk could be of immense use to the disaster preparedness community.

In their work, disaster managers can support these forecasting efforts by better defining the meteorological and hydrological variables that relate to disaster. Sharing this information with forecasters can inform the development of forecast products that provide specific information about these "danger levels", thus better enabling stakeholders to take appropriate preparatory actions. Forecast-based finance initiatives are underway globally, with the aim of taking action and releasing financing proportional to the risk information in a forecast, before the potential disaster (Coughlan de Perez et al., 2016). Changes to forecast products to provide clearer and more targeted risk information can support this process, and enable humanitarians to better anticipate and prepare for disasters before they strike.

Competing interests. The authors declare that they have no conflict of interest.

Special issue statement. This article is part of the special issue "Sub-seasonal to seasonal hydrological forecasting". It is not associated with a conference.

Acknowledgements. We thank our colleagues for their insights and suggestions on indices to consider. We are grateful to the German Federal Foreign Office for their support of the development of forecast-based financing pilots around the world, which have inspired these research questions. This work was supported by the UK Natural Environment Research Council (NE/P000525/1). This work was also funded in part by grants/cooperative agreements from the National Oceanic and Atmospheric Administration (NA15OAR4310076 and NA13OAR4310184). The views expressed are those of the authors and do not necessarily reflect the views of NOAA or its subagencies. Elisabeth Stephens' time was funded by Leverhulme Early Career Fellowship ECF-2013-492.

Edited by: Quan J. Wang

References

Alexander, L., Yang, H., and Perkins, S.: ClimPACT: Indices and software, available at: https://www.google.com/url?sa=t (last access: 24 January 2017), 2013.

Alfieri, L., Burek, P., Dutra, E., Krzeminski, B., Muraro, D., Thielen, J., and Pappenberger, F.: GloFAS – global ensemble streamflow forecasting and flood early warning, Hydrol. Earth Syst. Sci., 17, 1161–1175, https://doi.org/10.5194/hess-17-1161-2013, 2013.

Anderson, B. T., Gianotti, D., and Salvucci, G. D.: Characterizing the Potential Predictability of Seasonal, Station-Based Heavy Precipitation Accumulations and Extreme Dry Spell Durations*, J. Hydrometeorol., 16, 843–856, https://doi.org/10.1175/JHM-D-14-0111.1, 2015.

Anon: Kenya: Slum Residents in Nyeri Refuse to Relocate Ahead of El Nino Rains, The Star, Kenya, 6 October 2015.

Balsamo, G., Albergel, C., Beljaars, A., Boussetta, S., Brun, E., Cloke, H., Dee, D., Dutra, E., Muñoz-Sabater, J., Pappenberger, F., de Rosnay, P., Stockdale, T., and Vitart, F.: ERA-Interim/Land: a global land surface reanalysis data set, Hydrol. Earth Syst. Sci., 19, 389–407, https://doi.org/10.5194/hess-19-389-2015, 2015.

Barnston, A., Mason, S. J., Goddard, L., Dewitt, D. G., and Zebiak, S. E.: Multimodel Ensembling in Seasonal Climate Forecasting at IRI, B. Am. Meteorol. Soc., 84, 1783–1796, https://doi.org/10.1175/BAMS-84-12-1783, 2003.

Barnston, A., Li, S., Mason, S. J., DeWitt, D. G., Goddard, L., and Gong, X.: Verification of the First 11 Years of IRI's Seasonal Climate Forecasts, J. Appl. Meteorol. Clim., 49, 493–520, https://doi.org/10.1175/2009JAMC2325.1, 2010a.

Barnston, A. G., Li, S., Mason, S. J., Dewitt, D. G., Goddard, L., and Gong, X.: Verification of the first 11 years of IRI's seasonal climate forecasts, J. Appl. Meteorol. Clim., 49, 493–520, https://doi.org/10.1175/2009JAMC2325.1, 2010b.

Berghuijs, W. R., Woods, R. A., Hutton, C. J., and Sivapalan, M.: Dominant flood generating mechanisms across the United States, Geophys. Res. Lett., 43, 4382–4390, https://doi.org/10.1002/2016GL068070, 2016.

Braman, L. M., van Aalst, M. K., Mason, S. J., Suarez, P., Ait-Chellouche, Y., and Tall, A.: Climate forecasts in disaster management?: Red Cross flood operations in West Africa, 2008, Disasters, 37, 144–164, https://doi.org/10.1111/j.1467-7717.2012.01297.x, 2013.

Cai, X. and Rosegrant, M. W.: Global Water Demand and Supply Projections, Water Int., 27, 159–169, https://doi.org/10.1080/02508060208686989, 2002.

Coughlan de Perez, E., van den Hurk, B., van Aalst, M. K., Amuron, I., Bamanya, D., Hauser, T., Jongma, B., Lopez, A., Mason, S., Mendler de Suarez, J., Pappenberger, F., Rueth, A., Stephens, E., Suarez, P., Wagemaker, J., and Zsoter, E.: Action-based flood forecasting for triggering humanitarian action, Hydrol. Earth Syst. Sci., 20, 3549–3560, https://doi.org/10.5194/hess-20-3549-2016, 2016.

Emerton, R., Cloke, H. L., Stephens, E. M., Zsoter, E., Woolnough, S. J., and Pappenberger, F.: Complex picture for likelihood of ENSO-driven flood hazard, Nature Communications, 8, 1–9, https://doi.org/10.1038/ncomms14796, 2017.

Friedman, J. H., Hastie, T., and Tibshirani, R.: Regularization Paths for Generalized Linear Models via Coordinate Descent, J. Stat.

Softw., 33, 22 pp., https://doi.org/10.18637/jss.v033.i01, 2010.

Grieser, J.: Flexible Forecasts?: Responding to User Needs, International Research Institute for Climate and Society, available at: http://iri.columbia.edu/news/flexible-forecasts-for-decision-makers/ (last access: 17 September 2016), 2014.

Hewitt, C., Mason, S., and Walland, D.: The Global Framework for Climate Services, Nature Climate Change, 2, 831–832, https://doi.org/10.1038/nclimate1745, 2012.

Higgins, R. W., Schemm, J. K. E., Shi, W., and Leetmaa, A.: Extreme precipitation events in the Western United States related to tropical forcing, J. Climate, 13, 793–820, https://doi.org/10.1175/1520-0442(2000)013<0793:EPEITW>2.0.CO;2, 2000.

ICPAC: The Forty-Fourth Greater Horn of Africa Climate Outlook Forum (GHACOF44) Announcement, available at: http://www.icpac.net/wp-content/uploads/GHACOF-44-Announcement.pdf, last access: 16 September 2016.

Kanamitsu, M., Lu, C.-H., Schemm, J., and Ebisuzaki, W.: The Predictability of Soil Moisture and Near-Surface Temperature in Hindcasts of the NCEP Seasonal Forecast Model, J. Climate, 16, 510–521, 2002.

Koster, R. D., Mahanama, S. P. P., Yamada, T. J., Balsamo, G., Berg, A. A., Boisserie, M., Dirmeyer, P. A., Drewitt, G., Gordon, C. T., Guo, Z., Jeong, J.-H., Lawrence, D. M., Lee, W.-S., Li, Z., Luo, L., Malyshev, S., Merryfield, W. J., Seneviratne, S. I., Stanelle, T., van den Hurk, B. J. J. M., Vitart, F., and Wood, E. F.: Contribution of land surface initialization to subseasonal forecast skill?: First results from a multi-model experiment, Geophys. Res. Lett., 37, L02402, https://doi.org/10.1029/2009GL041677, 2010.

Mason, S. J., Goddard, L., Graham, N. E., Yulaeva, E., Sun, L., and Arkin, P. A.: The IRI Seasonal Climate Prediction System and the 1997/98 El Niño Event, B. Am. Meteorol. Soc., 80, 1853–1873, 1999.

Moron, V., Robertson, A. W., Ward, M. N., and Camberlin, P.: Spatial Coherence of Tropical Rainfall at the Regional Scale, J. Climate, 20, 5244–5263, https://doi.org/10.1175/2007JCLI1623.1, 2007.

Mwangi, E., Wetterhall, F., Dutra, E., Di Giuseppe, F., and Pappenberger, F.: Forecasting droughts in East Africa, Hydrol. Earth Syst. Sci., 18, 611–620, https://doi.org/10.5194/hess-18-611-2014, 2014.

Pineda, L. E. and Willems, P.: Multisite Downscaling of Seasonal Predictions to Daily Rainfall Characteristics over Pacific – Andean River Basins in Ecuador and Peru Using a Nonhomogeneous Hidden Markov Model, J. Hydrometeorol., 17, 481–498, https://doi.org/10.1175/JHM-D-15-0040.1, 2016.

Poveda, G., Jaramillo, A., Gil, M. M., Quiceno, N., and Mantilla, R. I.: Seasonality in ENSO-related precipitation, river discharges, soil moisture, and vegetation index in Colombia, Water Resour. Res., 37, 2169–2178, 2001.

Sahu, N., Robertson, A. W., Boer, R., Behera, S., DeWitt, D. G., Takara, K., Kumar, M., and Singh, R. B.: Probabilistic seasonal streamflow forecasts of the Citarum River, Indonesia, based on general circulation models, Stoch. Env. Res. Risk A., 12 pp., https://doi.org/10.1007/s00477-016-1297-4, 2016.

Schleiss, M. and Smith, J. A.: Two Simple Metrics for Quantifying Rainfall Intermittency: The Burstiness and Memory of Interamount Times, J. Hydrometeorol., 17, 421–436, https://doi.org/10.1175/JHM-D-15-0078.1, 2016.

Sillmann, J., Kharin, V. V., Zwiers, F. W., Zhang, X., and Bronaugh, D.: Climate extremes indices in the CMIP5 multimodel ensemble: Part 2. Future climate projections, J. Geophys. Res.-Atmos., 118, 2473–2493, https://doi.org/10.1002/jgrd.50188, 2013.

Stephens, E., Day, J. J., Pappenberger, F., and Cloke, H.: Precipitation and Floodiness, Geophys. Res. Lett., 42, 10316–10323, https://doi.org/10.1002/2015GL066779, 2015.

Sylla, M. B., Giorgi, F., Coppola, E., and Mariotti, L.: Uncertainties in daily rainfall over Africa?: assessment of gridded observation products and evaluation of a regional climate model simulation, Int. J. Climatol., 33, 1805–1817, https://doi.org/10.1002/joc.3551, 2013.

Verbist, K., Robertson, A. W., Cornelis, W. M., and Gabriels, D.: Seasonal predictability of daily rainfall characteristics in central northern Chile for dry-land management, J. Appl. Meteorol. Clim., 49, 1938–1955, https://doi.org/10.1175/2010JAMC2372.1, 2010.

Ward, P. J., Eisner, S., Flörke, M., Dettinger, M. D., and Kummu, M.: Annual flood sensitivities to El Niño–Southern Oscillation at the global scale, Hydrol. Earth Syst. Sci., 18, 47–66, https://doi.org/10.5194/hess-18-47-2014, 2014.

Ward, P. J., Kummu, M., and Lall, U.: Flood frequencies and durations and their response to El Niño Southern Oscillation?: Global analysis, J. Hydrol., 539, 358–378, https://doi.org/10.1016/j.jhydrol.2016.05.045, 2016.

Weisheimer, A. and Palmer, T. N.: On the reliability of seasonal climate forecasts, J. R. Soc. Interface, 11, 20131162, https://doi.org/10.1098/rsif.2013.1162, 2014.

WMO: Global Producing Centres for Long-Range Forecasts, available at: http://www.wmo.int/pages/prog/wcp/wcasp/gpc/gpc.php, last access: 16 January 2017.

8

Identification of hydrological model parameter variation using ensemble Kalman filter

Chao Deng[1,2]**, Pan Liu**[1,2]**, Shenglian Guo**[1,2]**, Zejun Li**[1,2]**, and Dingbao Wang**[3]

[1]State Key Laboratory of Water Resources and Hydropower Engineering Science, Wuhan University, Wuhan, China
[2]Hubei Provincial Collaborative Innovation Center for Water Resources Security, Wuhan, China
[3]Department of Civil, Environmental & Construction Engineering, University of Central Florida, Orlando, FL, USA

Correspondence to: Pan Liu (liupan@whu.edu.cn)

Abstract. Hydrological model parameters play an important role in the ability of model prediction. In a stationary context, parameters of hydrological models are treated as constants; however, model parameters may vary with time under climate change and anthropogenic activities. The technique of ensemble Kalman filter (EnKF) is proposed to identify the temporal variation of parameters for a two-parameter monthly water balance model (TWBM) by assimilating the runoff observations. Through a synthetic experiment, the proposed method is evaluated with time-invariant (i.e., constant) parameters and different types of parameter variations, including trend, abrupt change and periodicity. Various levels of observation uncertainty are designed to examine the performance of the EnKF. The results show that the EnKF can successfully capture the temporal variations of the model parameters. The application to the Wudinghe basin shows that the water storage capacity (SC) of the TWBM model has an apparent increasing trend during the period from 1958 to 2000. The identified temporal variation of SC is explained by land use and land cover changes due to soil and water conservation measures. In contrast, the application to the Tongtianhe basin shows that the estimated SC has no significant variation during the simulation period of 1982–2013, corresponding to the relatively stationary catchment properties. The evapotranspiration parameter (C) has temporal variations while no obvious change patterns exist. The proposed method provides an effective tool for quantifying the temporal variations of the model parameters, thereby improving the accuracy and reliability of model simulations and forecasts.

1 Introduction

Hydrological model parameters are critically important for accurate simulation of runoff. Parameters of conceptual hydrological models can be considered as a simplified representation of the physical characteristics in hydrologic processes. Therefore, parameter values are closely related to the catchment conditions, such as climate change, afforestation and urbanization (Peel and Blöschl, 2011). In hydrological modeling, parameters are usually assumed to be stationary; i.e., the calibrated parameters are constants during the calibration period, and have extrapolative ability outside the range of the observations used for parameter estimation (Merz et al., 2011). The estimated parameters usually depend on the calibration period since the calibration period may contain different climatic conditions and hydrological regimes compared to the simulation period (Merz et al., 2011; Zhang et al., 2011; Coron et al., 2012; Seiller et al., 2012; Westra et al., 2014; Patil and Stieglitz, 2015). The model parameters may change as a response to the variations in climatic conditions and catchment properties. For example, land use and land cover changes contribute to temporal changes of model parameters (Andréassian et al., 2003; Brown et al., 2005; Merz et al., 2011). Therefore, it is no longer appropriate to treat parameters as time invariant.

Time-variant hydrological model parameters have been reported in a few recent publications (Merz et al., 2011; Brigode et al., 2013; Jeremiah et al., 2013; Thirel et al., 2015; Westra et al., 2014; Patil and Stieglitz, 2015). For example, Ye et al. (1997) and Paik et al. (2005) mentioned the seasonal variations of hydrological model parameters. Merz et al. (2011) analyzed the temporal changes of model parameters, which were calibrated by using six consecutive 5-year periods between 1976 and 2006 for 273 catchments in Austria. Recently, Westra et al. (2014) proposed a strategy to cope with nonstationarity of hydrological model parameters, which were represented as a function of a time-varying covariate set before using an optimization algorithm for calibration. Previous studies provided two main methods to estimate the time-variant model parameters: (1) available historical records are divided into consecutive subsets, and parameters are calibrated separately for each subset using an optimization algorithm (Merz et al., 2011; Thirel et al., 2015); (2) a functional form of selected time-variant model parameters is constructed, and the parameters for the function are estimated using an optimization algorithm based on the entire historical record (Jeremiah et al., 2013; Westra et al., 2014).

The data assimilation (DA) actually provides another method to identify the potential temporal variations of model parameters by updating them in real time when observations are available (Liu and Gupta, 2007; Xie and Zhang, 2013). The DA method has been widely applied in hydrology for soil moisture estimation (Han et al., 2012; Kumar et al., 2012; Yan et al., 2015) and flood forecasting (Y. Li et al., 2013; Liu et al., 2012; Abaza et al., 2014). It has also been successfully used to estimate model parameters (Moradkhani et al., 2005; Kurtz et al., 2012; Montzka et al., 2013; Panzeri et al., 2013; Vrugt et al., 2013; Xie and Zhang, 2013; Shi et al., 2014; Xie et al., 2014). For example, Vrugt et al. (2013) proposed two Particle-DREAM (DiffeRential Evolution Adaptive Metropolis) methods, i.e., Particle-DREAM for time-variant and time-invariant parameters, to track the evolving target distribution of HyMOD parameters, while both results were approximately similar and statistically coherent since only 3 years of data were used. Xie and Zhang (2013) used a partitioned forecast-update scheme based on the ensemble Kalman filter (EnKF) to retrieve optimal parameters in a distributed hydrological model. Although the DA method has been used to estimate model parameters, these studies are focused on the estimation of constant parameters. Little attention has been paid to the identification of time-variant model parameters by using the DA method.

The aim of this study is to assess the capability of the EnKF to identify the temporal variations of the model parameters for a monthly water balance model. Thus, a synthetic experiment, including four scenarios with different parameter variations and one scenario with time-invariant parameters, is designed for parameter estimation at different uncertainty levels. Furthermore, two case studies are implemented to estimate the model parameter series and to interpret the parameter variations in response to the changes in catchment characteristics, i.e., land use and land cover. The remainder of this paper is organized as follows. Section 2 presents a brief review of the monthly water balance model and the EnKF method. Following the methodology, Sect. 3 describes the synthetic experiment and the application to two case studies. Results and discussion are presented in Sect. 4, followed by conclusions in Sect. 5.

2 Methodology

2.1 Monthly water balance model

The two-parameter monthly water balance model (TWBM), developed by Xiong and Guo (1999), has been widely applied for monthly runoff simulation and forecast (Guo et al., 2002, 2005; Xiong and Guo, 2012; S. Li et al., 2013; Zhang et al., 2013; Xiong et al., 2014). The inputs of the model include monthly areal precipitation and potential evapotranspiration. The actual monthly evapotranspiration is calculated as follows:

$$E_i = C \times \mathrm{EP}_i \times \tanh\left(\frac{P_i}{\mathrm{EP}_i}\right), \tag{1}$$

where E_i represents the actual monthly evapotranspiration; EP_i and P_i are the monthly potential evapotranspiration and precipitation, respectively; C is the first model parameter; and i is the time step.

The monthly runoff is dependent on the soil water content and is calculated by the following equation:

$$Q_i = S_i \times \tanh\left(\frac{S_i}{\mathrm{SC}}\right), \tag{2}$$

where Q_i is the monthly runoff and S_i is the soil water content. As the second model parameter, SC represents the water storage capacity of the catchment in millimeters. The available water for runoff at the ith month is computed by $S_{i-1} + P_i - E_i$. Then, the monthly runoff is calculated as

$$Q_i = (S_{i-1} + P_i - E_i) \times \tanh\left(\frac{S_{i-1} + P_i - E_i}{\mathrm{SC}}\right). \tag{3}$$

Finally, the soil water content at the end of each time step is updated based on the water conservation law:

$$S_i = S_{i-1} + P_i - E_i - Q_i. \tag{4}$$

2.2 Ensemble Kalman filter

As a sequential data assimilation technique, EnKF is essentially the Monte Carlo implementation of the Kalman filter, producing an ensemble of state simulations for updating the state variables and their covariance matrices (Evensen, 1994; Burgers et al., 1998; Moradkhani et al., 2005; Shi et al., 2014). It is applicable to a variety of nonlinear problems (Evensen, 2003; Weerts and El Serafy, 2006) and has been widely applied to hydrological models (Abaza et al., 2014; DeChant and Moradkhani, 2014; Delijani et al., 2014; Samuel et al., 2014; Tamura et al., 2014; Xue and Zhang,

Table 1. States and parameters of the two-parameter monthly water balance model.

Parameters and state variables		Description	Ranges and unit
Parameter	C	Evapotranspiration parameter	0.2–2.0 (–)
	SC	Catchment water storage capacity	100–4000 (mm)
State variable	S	Soil water content	mm

2014; Deng et al., 2015). Furthermore, the EnKF has been successfully used in time-invariant parameter estimations for hydrological models (Moradkhani et al., 2005; Wang et al., 2009; Xie and Zhang, 2010, 2013).

In this paper, the EnKF is applied to simultaneously estimate state variables and parameters (Table 1) in the TWBM model. The augmented state vector includes both states and model parameters (Wang et al., 2009), i.e., $Z = (\theta, x)^T$, where θ includes the evapotranspiration parameter (C) and the catchment water storage capacity (SC), and x is the soil water content (S). The model forecast is conducted for each ensemble member as follows:

$$\begin{pmatrix} \theta_{i+1|i}^k \\ x_{i+1|i}^k \end{pmatrix} = \begin{pmatrix} \theta_{i|i}^k \\ f\left(x_{i|i}^k, \theta_{i+1|i}^k, u_{i+1}\right) \end{pmatrix} + \begin{pmatrix} \delta_i^k \\ \varepsilon_i^k \end{pmatrix},$$

where $\delta_i^k \sim N(0, U_i), \varepsilon_i^k \sim N(0, G_i)$, (5)

$\theta_{i+1|i}^k$ is the kth ensemble member forecast of model parameters at time $i + 1$; $\theta_{i|i}^k$ is the kth updated ensemble member of model parameters at time i; $x_{i+1|i}^k$ is the kth ensemble member forecast of model state at time $i + 1$; $x_{i|i}^k$ is the kth updated ensemble member of model state at time i; f is the forecasting model operator, i.e., the TWBM model; u_{i+1} is the forcing data for the hydrological model, including precipitation and potential evapotranspiration; ε_i^k and δ_i^k are the independent white noise for the forecasting model, following a Gaussian distribution with zero mean and specified covariance G_i and U_i, respectively. Note that the parameters in Eq. (5) are propagated by adding random disturbances to the parameter member between time steps (Wang et al., 2009).

The observation ensemble member can be written as

$$y_{i+1}^k = h\left(x_{i+1|i}^k, \theta_{i+1|i}^k\right) + \xi_{i+1}^k, \quad \xi_{i+1}^k \sim N(0, W_{i+1}), \quad (6)$$

where y_{i+1}^k is the kth ensemble member of the model simulated runoff at time $i + 1$; h is the observation operator which represents the relationship between the observation and the state variables; ξ_{i+1}^k is the noise term, which follows a Gaussian distribution with zero mean and specified covariance W_{i+1}.

Based on the available state and observation equations, the model parameters and state are updated according to the following equation:

$$Z_{i+1|i+1}^k = Z_{i+1|i}^k + K_{i+1}\left(y_{i+1}^k - h\left(Z_{i|i}^k\right)\right),\quad (7)$$

where Z is the augmented state vector that includes both states and parameters; y_{i+1}^k is the kth observation ensemble member generated by adding the observation error ξ_{i+1}^k to the observed runoff:

$$y_{i+1}^k = y_{i+1} + \xi_{i+1}^k, \quad (8)$$

K_{i+1} is the Kalman gain matrix that represents the weight between the forecasts and observations. It can be calculated as (Evensen, 1994, 2003; Evensen and van Leeuwen, 1996; Moradkhani et al., 2005)

$$K_{i+1} = \sum_{i+1|i}^{zy}\left(\sum_{i+1|i}^{yy} + W_{i+1}\right)^{-1},\quad (9)$$

where $\sum_{i+1|i}^{zy}$ is the cross-covariance of the forecasted state and parameters and $\sum_{i+1|i}^{yy}$ is the error covariance of the forecasted output. The error covariance matrix is calculated based on the forecasted ensemble members:

$$\sum_{i+1|i} = \frac{1}{N-1} Z_{i+1|i}\, Z_{i+1|i}^T,\quad (10)$$

where $Z_{i+1|i} = \left(z_{i+1|i}^1 - \overline{z}_{i+1|i}, \cdots, z_{i+1|i}^N - \overline{z}_{i+1|i}\right)$; $\overline{z}_{i+1|i}$ is the ensemble mean of the forecasted members, and N is the ensemble size.

Since the parameters are limited within a range, the constrained EnKF (Wang et al., 2009) is used in this study. The ensemble size, uncertainties in input and output have significant impacts on the assimilation performance of the EnKF, and they are specified following the previous studies (Moradkhani et al., 2005; Wang et al., 2009; Xie and Zhang, 2010; Nie et al., 2011; Lü et al., 2013; Samuel et al., 2014). The ensemble size is set to 1000 for the synthetic experiment and the two case studies. In the present study, the uncertainties, including state variable and parameter errors (ε and δ in Eq. 5, respectively) and runoff observation error (ξ in Eq. 6), are assumed to follow a Gaussian distribution with zero mean and specified covariance. Note that the model parameter errors should vary depending on the hydrological model used and the study basin (Clark et al., 2008). Larger standard deviation can generate greater perturbations to model parameters, and it can improve the coverage of updated parameters but also may cause fluctuations in the estimates. In this study, the parameter errors are determined empirically; i.e., the standard deviation of C is set to 0.01 for all the cases, while that of SC is set to 5.0, 1.0 and 0.5 in the synthetic experiment, Wudinghe basin and Tongtianhe basin, respectively. The standard deviations of both model state and observation errors are assumed to be proportional to the magnitude of true values (Wang et al., 2009; Lü et al., 2013). The proportional factors of the model state are set to 0.05 for all the cases. Different proportional factors of runoff observation and precipitation (Table 3) are evaluated to examine the capability of the EnKF in the synthetic experiment, whereas the proportional factors of runoff observation are set to 0.1 and zero precipitation errors are assumed in the two case studies.

2.3 Evaluation index

Two evaluation criteria, including the Nash–Sutcliffe efficiency (NSE) (Nash and Sutcliffe, 1970) and the volume error (VE) are used to evaluate the runoff assimilation results for the synthetic experiment and the application to real catchments (Deng et al., 2015; Li et al., 2015).

$$NSE = 1 - \frac{\sum_{i=1}^{n}\left(Q_{\text{sim},i} - Q_{\text{obs},i}\right)^2}{\sum_{i=1}^{n}\left(Q_{\text{obs},i} - \overline{Q}_{\text{obs}}\right)^2}, \quad (11)$$

$$VE = \frac{\sum_{i=1}^{n} Q_{\text{sim},i} - \sum_{i=1}^{n} Q_{\text{obs},i}}{\sum_{i=1}^{n} Q_{\text{obs},i}}, \quad (12)$$

where $Q_{\text{sim},i}$ and $Q_{\text{obs},i}$ are the simulated and observed runoff for the ith month, $\overline{Q}_{\text{obs}}$ is the mean value of the observed runoff and n is the total number of data points. The NSE ranges from $-\infty$ to 1 and has been widely used to assess the goodness of fit for hydrological modeling. A NSE value of 1 stands for a perfect match of simulated runoff to the observations, whereas a value of 0 indicates that the model simulations are equivalent to the mean value of the runoff observations; negative NSE values indicate that the mean observed runoff is better than the model simulations. The VE is a measure of bias between the simulated and observed runoff. For example, VE with the value of 0 denotes no bias, and a negative value means an underestimation of the total runoff volume.

The assimilated parameter results are evaluated using the following criteria, including the Pearson correlation coefficient (R), the root mean square error (RMSE) and mean absolute relative error (MARE):

$$R = \frac{\sum_{i=1}^{n}\left(\theta_{\text{sim},i} - \overline{\theta}_{\text{sim}}\right)\left(\theta_{\text{obs},i} - \overline{\theta}_{\text{obs}}\right)}{\sqrt{\sum_{i=1}^{n}\left(\theta_{\text{sim},i} - \overline{\theta}_{\text{sim}}\right)^2 \left(\theta_{\text{obs},i} - \overline{\theta}_{\text{obs}}\right)^2}}, \quad (13)$$

$$RMSE = \sqrt{\frac{1}{n}\sum_{i=1}^{n}\left(\theta_{\text{sim},i} - \theta_{\text{obs},i}\right)^2}, \quad (14)$$

$$MARE = \frac{1}{n}\sum_{i=1}^{n}\frac{\left|\theta_{\text{sim},i} - \theta_{\text{obs},i}\right|}{\theta_{\text{obs},i}}, \quad (15)$$

where $\theta_{\text{sim},i}$ and $\theta_{\text{obs},i}$ are the assimilated and true model parameters for the ith month, $\overline{\theta}_{\text{sim}}$ and $\overline{\theta}_{\text{obs}}$ are the mean of the assimilated and true model parameters for the ith month and n is the total number of data points.

3 Data and study area

3.1 Synthetic experiment

A synthetic experiment is designed to evaluate the capability of the assimilation procedure to identify the temporal variation of model parameters. Five scenarios of different parameter variations are developed, as shown in Table 2. The model parameters in the first four scenarios are time variant, and those in the last scenario are constant. Parameter C, the evapotranspiration parameter, is considered to be sinusoidal reflecting potential seasonal variations in hydrological model parameters (Paik et al., 2005; Ye et al., 1997). An increasing trend is also considered to account for the potential annual or long-term variability. The change of parameter SC is considered to be gradual and abrupt, since the catchment water storage capacity can be affected by land use and land cover changes, such as afforestation and dam construction. The parameters in scenario 5 are treated as constants like in conventional hydrological modeling. Observations for precipitation and potential evapotranspiration are generated by adding a Gaussian disturbance to the corresponding data from a real catchment, and runoff is then produced using the TWBM model. The data set used in this experiment is 672 months long. The first 24-month period is set for model warm-up to reduce the impact of the initial soil moisture conditions. The steps toward identifying temporal variation of model parameters are as follows:

1. Time series of model parameters are synthetically generated, including the time-variant parameters and the constant parameters. Model parameter sets are produced using a sinusoidal function and/or a linear trend function within the specified ranges shown in Table 1. The runoff observations for each scenario are computed from the TWBM model taking monthly potential evapotranspiration, monthly precipitation and the parameters as inputs.

2. The initial ensembles of model parameters and state variables are generated using uniform distributions within the specified ranges in Table 1. The ensemble size and the total number of assimilation time steps are specified.

3. After the initialization of parameters and state variables, the hydrological model parameters and states are updated by assimilating the runoff observations obtained in step (1). The additive errors for generating the ensemble members of model parameters, state variables and runoff observations are obtained from Gaussian distributions with zero mean and specified variance.

To evaluate the effects of errors on identifying parameter variation, different levels of observation uncertainty are considered in the synthetic experiment, as detailed in Table 3. The uncertainties from the observed precipitation and runoff are characterized by adding Gaussian noises, where standard

Table 2. Different variations of model parameters in the synthetic experiment.

Scenario	Description
Scenario 1	C has a periodic variation, and SC has an increasing trend
Scenario 2	C has a periodic variation, and SC has an abrupt change
Scenario 3	C has a periodic variation with an increasing trend, and SC has an increasing trend
Scenario 4	C has a periodic variation with an increasing trend, and SC has an abrupt change
Scenario 5	Both C and SC are constant

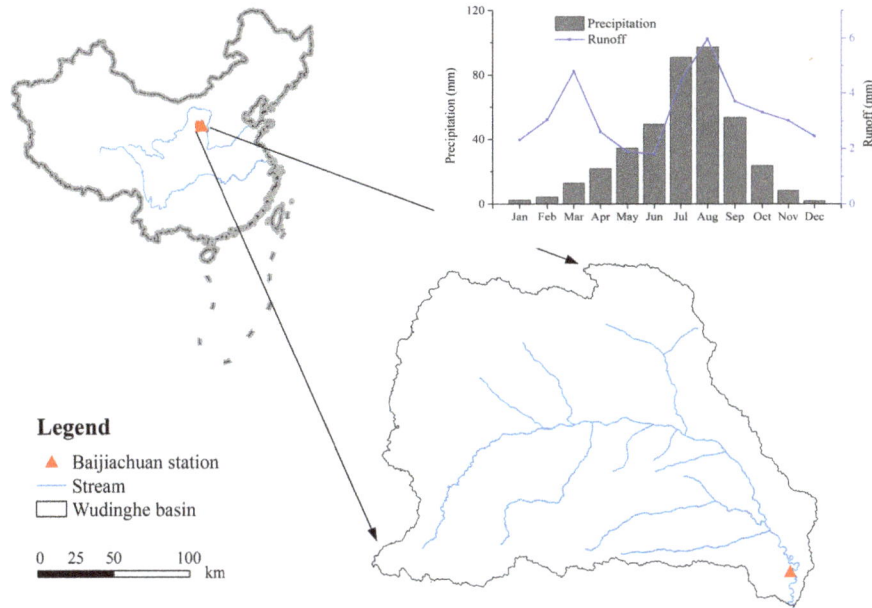

Figure 1. Location and mean monthly precipitation and runoff from 1956 to 2000 of the Wudinghe basin.

Table 3. Proportional factors of the standard deviations for precipitation (γ_P) and runoff (γ_Q) uncertainties.

Type	Low level	Medium level	High level
γ_P	0	0.05	0.10
γ_Q	0.05	0.10	0.20

deviations are assumed to be proportional to the magnitude of the true values, and the corresponding proportional factors are denoted as γ_P and γ_Q. The proportional factors are set to account for the practical measurement error (Wang et al., 2009; Xie and Zhang, 2010).

3.2 Study area

3.2.1 Case 1: Wudinghe basin

The method is applied to the Wudinghe basin (Fig. 1), which is a sub-basin of the Yellow River basin and located in the southern fringe of the Maowusu Desert and the northern part of the Loess Plateau in China, where the climate is semiarid climate. It has a drainage area of approximately 30 261 km^2 and a total length of 491 km. The Wudinghe basin has an average slope of 0.2 %, and its elevation ranges from 600 to 1800 m above the sea level. The Baijiachuan gauge station, which is the most downstream station of the Wudinghe basin, drains 98 % of the total basin area. The mean annual precipitation over the basin is 401 mm, of which 72.5 % occurs in the rainy season from June to September (Fig. 2). The mean annual potential evapotranspiration is 1077 mm, and the mean annual runoff is about 39 mm with a runoff coefficient of 0.1.

The soil erosion is severe in the Wudinghe basin, owing to the highly erodible loess and sparse vegetation. Since the 1960s, the soil and water conservation measures have been undertaken. Several engineering measures, including tree and grass plantation, check dam and reservoir construction, and land terracing, were effectively implemented during several decades. The land use changes caused by the soil and water conservation measures had a significant effect on increasing water storage capacity (Xu, 2011).

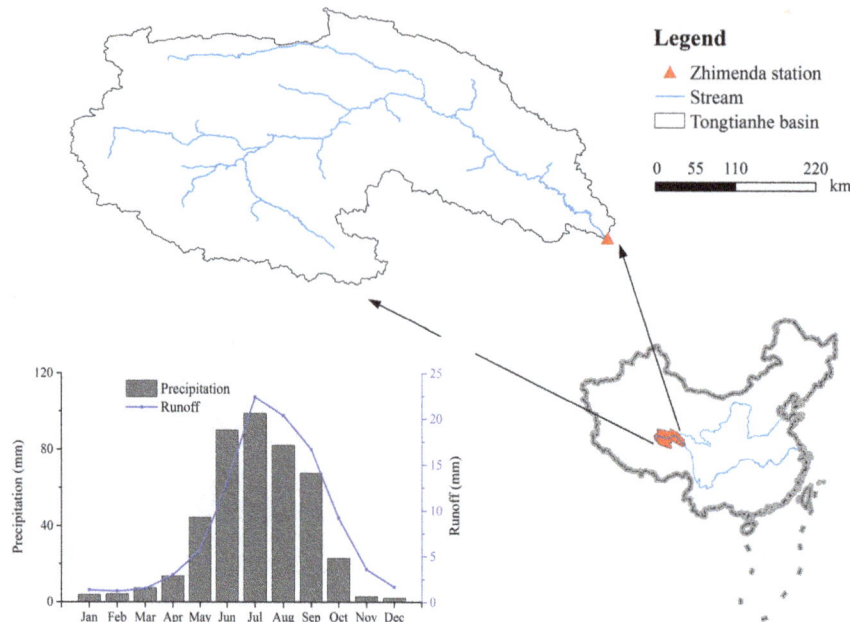

Figure 2. Location and mean monthly precipitation and runoff from 1980 to 2013 of the Tongtianhe basin.

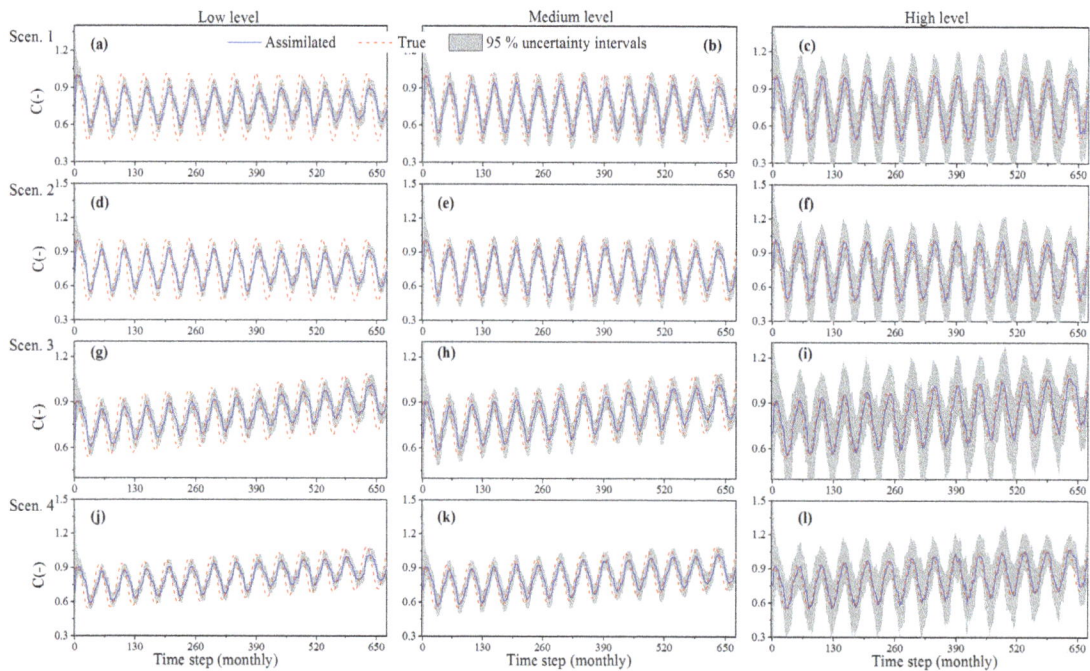

Figure 3. Comparison between estimated C and its true values for various parameter changes under different uncertainty levels. The gray areas represent the 95 % prediction uncertainty intervals.

3.2.2   Case 2: Tongtianhe basin

The Tongtianhe basin (Fig. 3) is located in the southwestern Qinghai Province, China, with a continental climate. It belongs to the source area of the Yangtze River basin with a drainage area of about $140\,000\,km^2$ and a total main stream length of 1206 km. The elevation of the Tongtianhe basin approximately ranges from 3500 to 6500 m a.s.l. Zhimenda is the basin outlet. The mean annual precipitation over the basin is 440 mm, of which 76.9 % occurs in the period from June to September (Fig. 4). The mean annual potential evapotranspiration is 796 mm, and the mean annual runoff is about

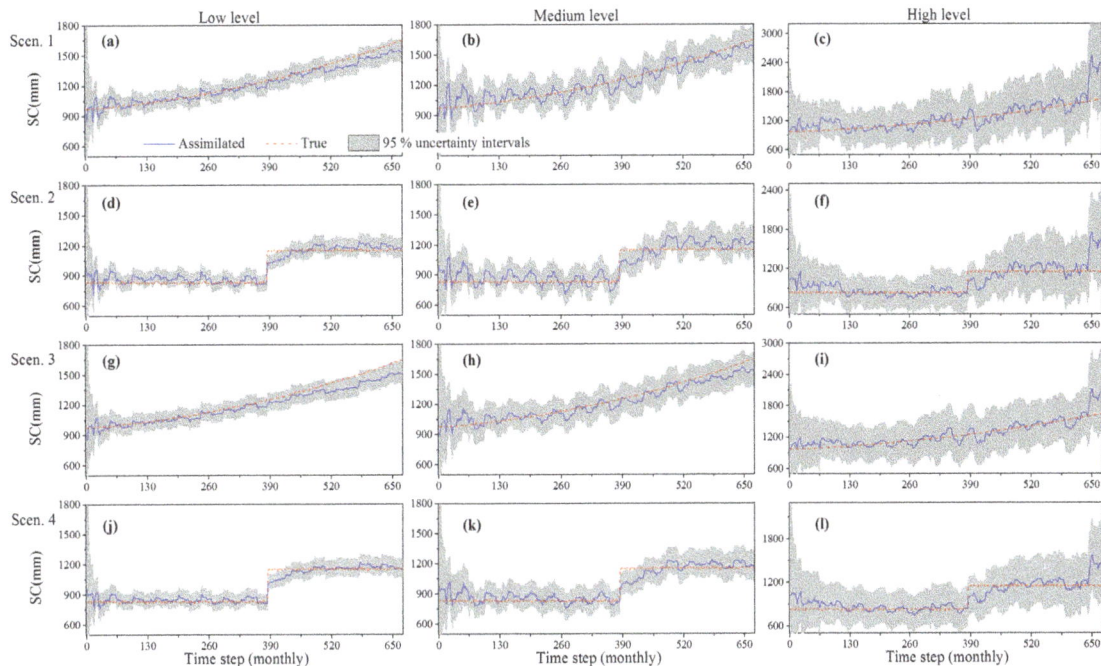

Figure 4. Comparison between estimated SC and its true values for various parameter changes under different uncertainty levels. The gray areas represent the 95 % prediction uncertainty intervals.

99 mm with a runoff coefficient of 0.23. The Tongtianhe basin is rarely affected by human activities, owing to the water source protection guidelines conducted by the government. The Tongtianhe basin is used for comparison on model parameter identification.

3.2.3 Data

The data sets used in this study include monthly precipitation, potential evapotranspiration and runoff in the Wudinghe basin (from 1956 to 2000) and the Tongtianhe basin (from 1980 to 2013). The potential evapotranspiration is estimated using the Penman–Monteith equation (Allen et al., 1998) based on the meteorological data from the China Meteorological Data Sharing Service System (http://data.cma.cn). To reduce the impact of the initial conditions, a 2-year data set, i.e., from 1956 to 1957 for Wudinghe basin and from 1980 to 1981 for Tongtianhe basin, is reserved as the warm-up period.

4 Results and discussion

4.1 Synthetic experiment

The comparisons of the estimated and true model parameters under different scenarios are presented in Figs. 3, 4 and 5. Tables 4 and 5 show the evaluated statistics for the parameters and runoff estimations. The assimilated parameter values are obtained from the ensemble mean at each time step. The

estimation of parameters C and SC have the similar trends to the true parameter series. The temporal variations of the estimated C agree well with the true series, although it has biases on the peaks of the periodic changes. For SC, the temporal estimates can capture the different changes in Table 2, especially for the abrupt change where the estimated values respond immediately. Different uncertainty levels are considered to examine the capability of the EnKF method. The results in Fig. 3 show that the estimated C has more accurate peaks with smaller RMSE and higher R values under the high-level uncertainty (Table 4); whereas, the SC estimates in Fig. 4 have some fluctuations when the uncertainty level increases. This is due to the estimated values vary with increasing uncertainty levels in the assimilation process. In the synthetic experiment, the true C is assumed to be periodic with a higher degree of variation, whereas the true SC series have less variation.

It should be noted that there are time lags between the assimilated and true C. The observation at the current time step is used to adjust the state variables and parameters in EnKF, and the updates of parameters depend on the Kalman gain for parameters. A runoff observation at the current time is determined by states at the current and previous time steps (Pauwels and Lannoy, 2006). The Kalman gain is dependent on the relative value of observation error to model error. The updated states are closer to the observation with a higher Kalman gain (Tamura et al., 2014). The synthetic C series were assumed to be periodic when many peak values exist, whereas the variation of SC series is less. The time lag be-

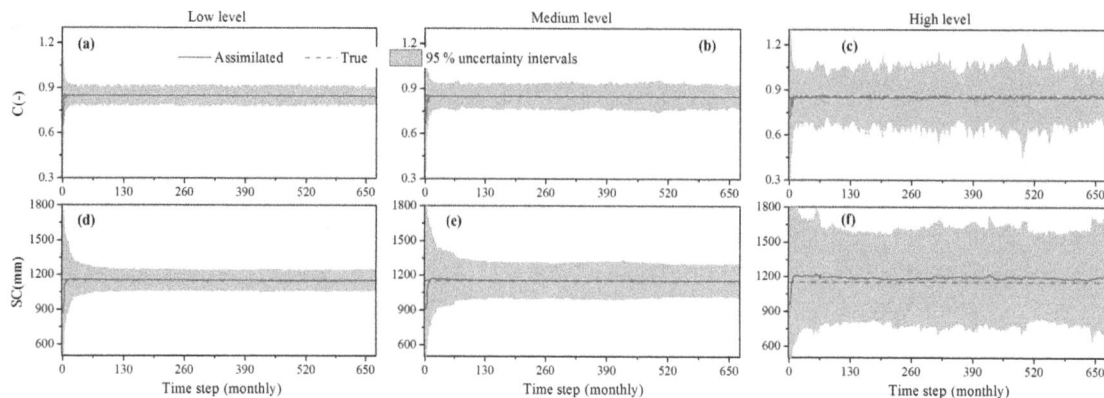

Figure 5. Estimations of time-invariant C and SC under different uncertainty levels. The gray areas represent the 95 % prediction uncertainty intervals.

Table 4. Performance statistics for various changes of **(a)** parameter C and **(b)** SC estimations under different levels of uncertainty in the synthetic experiment.

Scenario	Low level			Medium level			High level		
	RMSE	MARE	R	RMSE	MARE	R	RMSE	MARE	R
(a) Parameter C									
Scenario 1	0.15	0.21	0.55	0.16	0.18	0.68	0.18	0.11	0.89
Scenario 2	0.16	0.19	0.63	0.17	0.16	0.75	0.18	0.09	0.91
Scenario 3	0.12	0.13	0.64	0.13	0.11	0.72	0.14	0.07	0.91
Scenario 4	0.13	0.12	0.70	0.13	0.10	0.77	0.14	0.06	0.93
Scenario 5	0	–	–	0	–	–	0	–	–
(b) Parameter SC									
Scenario 1	182.87	0.03	0.99	187.76	0.05	0.94	253.35	0.83	0.83
Scenario 2	158.30	0.04	0.96	167.47	0.07	0.91	189.59	0.80	0.80
Scenario 3	180.20	0.03	0.99	183.06	0.04	0.97	215.04	0.88	0.88
Scenario 4	156.42	0.03	0.97	158.50	0.05	0.93	170.90	0.86	0.86
Scenario 5	1.54	–	–	3.67	–	–	20.54	–	–

tween assimilated and true values exists particularly when peak values occur (Clark et al., 2008; Samuel et al., 2014).

The results for the scenario of constant parameters are shown in Fig. 5, demonstrating that the estimated parameters can approach their true values after the initial 24 assimilation steps. The gray areas represent the 95 % prediction uncertainty intervals, which reduce quickly and approach a stable spread. The performance of the estimated parameters is correlated with the uncertainty level. Higher precipitation and runoff observation errors correspond to the greater RMSE values (Table 4) of estimated parameters and uncertainty ranges. The performance of runoff estimations for various parameter changes under different levels of uncertainty is shown in Table 5, suggesting that the EnKF perfectly matches the observations with NSEs higher than 0.95 and absolute VEs smaller than 0.02. The EnKF can successfully capture the temporal variations of the true parameters, although the uncertainty levels of the observations can affect

its performance to a certain degree. The above results demonstrate that the EnKF is able to identify the temporal variation of the model parameters by updating the state variables and parameters based on the runoff observations.

4.2 Case studies

Figure 6 shows the double mass curve between monthly runoff and precipitation for the Wudinghe and Tongtianhe basins, respectively. Figure 6a shows the linear relationship between cumulative runoff and precipitation pre- and post-1972 in the Wudinghe basin, which is similar to the result presented by Xu (2011) and Li et al. (2014). The results show two straight lines with different slopes for the relationships between precipitation and runoff, indicating that an abrupt change occurred in 1972; i.e., the runoff generation had been changed from this year due to the soil and water conservation measures. On the other hand, Fig. 6b demonstrates that

Table 5. Performance of runoff estimations for various parameter changes under different levels of uncertainty in the synthetic experiment.

Scenario	Low level		Medium level		High level	
	NSE	VE	NSE	VE	NSE	VE
Scenario 1	0.999	−0.0003	0.988	−0.0046	0.967	−0.0230
Scenario 2	0.999	0.0001	0.990	−0.0028	0.967	−0.0141
Scenario 3	0.999	−0.0011	0.990	−0.0013	0.974	−0.0264
Scenario 4	0.999	−0.0009	0.992	0.0002	0.959	−0.0147
Scenario 5	0.999	−0.0022	0.992	−0.0077	0.961	−0.0187

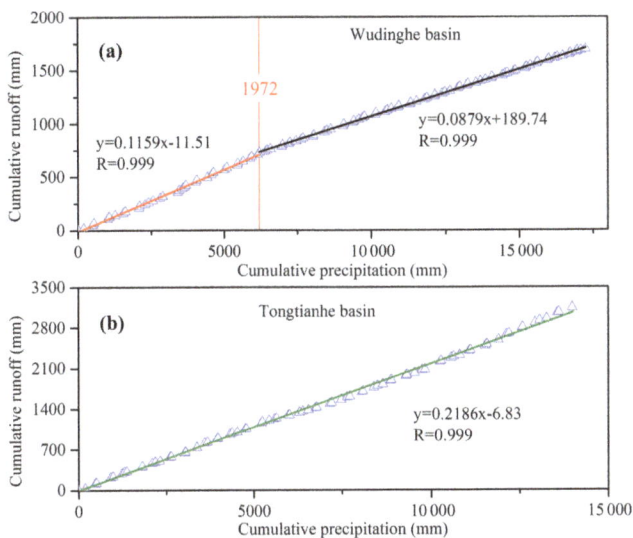

Figure 6. Double mass curve between monthly runoff and precipitation for Wudinghe basin within the period of 1958–2000 (**a**) and Tongtianhe basin within the period of 1982–2013 (**b**).

a single linear relationship fits all the data for the Tongtianhe basin, suggesting a stable precipitation–runoff relationship during the 1982–2013 period.

The estimated parameters and the associated 95 % prediction uncertainty intervals are shown in Fig. 7. The time series of estimated SC shows an apparent increasing trend, with two different trends for pre- and post-turning points in Fig. 6a. The temporal variation of the water storage capacity is correlated with the changes of land use and land cover. Both the trends in Fig. 7c show an increase of SC because the implementation of the large-scale engineering measures significantly improved the water holding capacity of the Wudinghe basin, especially for the reservoir and check dam construction. The trend slopes of the two periods, one from 1956 to 1971 and the other from 1972 to 2000, are different because the degree of implementing engineering measures varied during the period of 1958–2000. Moreover, the increase of the water holding capacity slowed down during the 1980s due to the sedimentation in reservoirs and check dams after periods of operation (Wang and Fan, 2003). Figure 8a shows the

long-term time series of precipitation and potential evaporation in the Wudinghe basin. The result shows that the runoff decreases significantly while precipitation changes slightly and potential evaporation has no trend, indicating that the actual evaporation increases significantly due to impacts of human activities, i.e., soil and water conservation measures. Figure 8b presents the runoff reduction caused by all the soil and water conservation measures, i.e., land terracing, tree and grass plantation and check dam and reservoir construction. The runoff reduction positively relates to the water holding capacity, namely the SC value. The slope for the period of 1958–1971 is higher than that for the period of 1972–1996, suggesting that the SC in the former period has a higher increasing trend. On the other hand, results of Tongtianhe basin show that the estimated SC has no detectable trend with a small R value. Moreover, the ranges and standard deviation of the estimated SC values are much smaller than those in the Wudinghe basin (Fig. 7), suggesting that the estimated SC has no obvious temporal variations.

For parameter C, the results show that the estimates have no significant temporal patterns because the trend line slopes are almost zero and the standard deviations are relatively small for the two basins (Fig. 7a and b); however, it can be treated as a time-variant parameter since temporal variations exist in the estimated C series. The temporal variations of the estimated C are related to the variation of monthly actual evaporation, which is affected by multiple climatic factors, such as air temperature, soil moisture and solar irradiance (Su et al., 2015). The gray regions represent the 95 % prediction uncertainty intervals obtained from the parameter ensembles. The stable and narrow uncertainty bounds shown in Fig. 7 indicate that the EnKF can provide superior performance of parameter estimation. The runoff simulations for both basins match well with the runoff observations. Specifically, the NSE and VE for the Wudinghe basin are 0.93 and 0.07, respectively. While the corresponding index values for the Tongtianhe basin are 0.99 and 0.04.

In summary, the above results demonstrate that the EnKF can identify the temporal variation of model parameters well by updating both state variables and parameters based on the runoff observations. The trends of parameter SC can be explained by the changes of catchment characteristics (i.e., land use and land cover) in the Wudinghe basin. However,

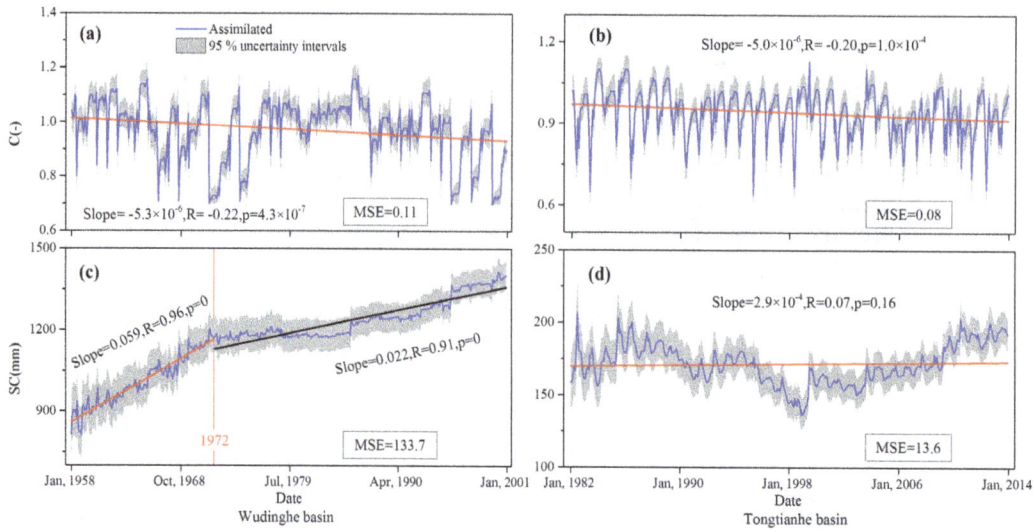

Figure 7. Estimated parameter values of C and SC for (1) Wudinghe basin within the period of 1958–2000, and (2) Tongtianhe basin within the period of 1982–2013. The gray areas represent the 95 % prediction uncertainty intervals. Note that the MSE denotes the standard deviation of the estimated parameter values.

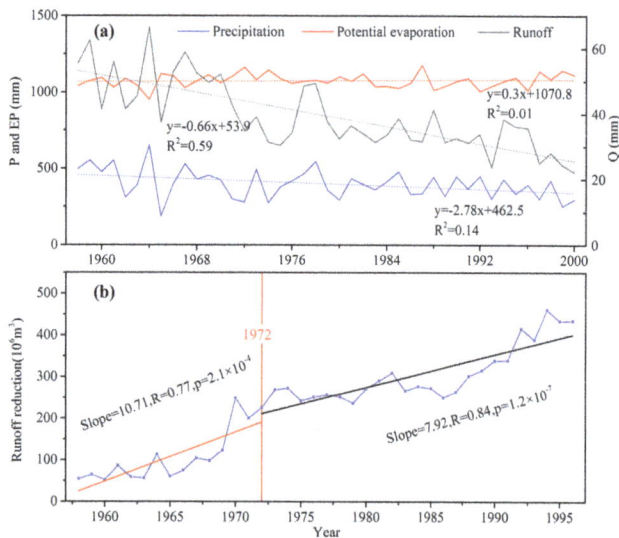

Figure 8. (a) Yearly precipitation, potential evaporation and runoff in Wudinghe basin during the period of 1958–2000; **(b)** Runoff reduction in Wudinghe basin caused by all the soil and water conservation measures, i.e., land terracing, tree and grass plantation and check dam and reservoir construction for the period of 1958–1996. Note that the data are from Wang and Fan (2003) and are only available from 1956 to 1996.

the estimated SC for the Tongtianhe basin is approximately stable with a small standard deviation because the basin is located in a water protection zone and has no significant changes on water storage capacity caused by human activities. The parameter C has temporal variations and can be treated as a time-variant parameter for both basins, although the estimates have no obvious temporal patterns. Therefore,

the EnKF is capable of identifying the temporal variations of model parameters.

5   Conclusions

This study proposes an ensemble Kalman filter (EnKF) to identify the temporal variation of model parameters of the two-parameter monthly water balance model (TWBM) by assimilating runoff observations. A synthetic experiment, which contains four scenarios with different changes of model parameters and one scenario with constant parameters, is designed to examine the capability of the proposed approach. Furthermore, three different levels of observation uncertainty are taken to assess the performance of the EnKF. The main conclusions are as follows. For the time-variant parameters, the EnKF provides superior performance even though slight time lags exist for parameters with periodic variations. The true values of the constant parameters can be approached quickly after 24 time steps of the assimilation process. The temporal variations of the parameters can be successfully captured even under a high level of observation uncertainties, which would have an influence on the performance of the EnKF.

The EnKF method is applied to the Wudinghe basin in China, aiming to detect the temporal variations of the model parameters and to provide an explanation for the parameter variation from the perspective of catchment characteristic changes. Meanwhile, a comparison is implemented to investigate the variation of model parameters in the Tongtianhe basin, which is barely affected by human activities. The parameter of water storage capacity (SC) for the monthly water balance model shows a significant increasing trend for the period of 1958–2000 in the Wudinghe basin. The soil and water conservation measures, including land terracing, tree and grass plantation and check dam and reservoir construction, were implemented from 1958 to 2000, resulting in the increase of the water holding capacity of the basin, which explains the increasing trend of SC. Moreover, the magnitudes of the engineering measures in different time periods play an important role in the degree of increasing trend for SC. In the Tongtianhe basin, the parameter SC has no significant trend for the period of 1982–2013, which is consistent with the relatively stationary catchment characteristics. The evapotranspiration parameter (C) has temporal variations and can be treated as a time-variant parameter, but no obvious trends exist.

The method proposed in this paper provides an effective tool for the time-variant model parameter identification. Future work will be focused on the influence of the correlations between/among model parameters and performance comparison of multiple data assimilation methods.

Acknowledgements. This study was supported by the Excellent Young Scientist Foundation of NSFC (51422907) and the Open Foundation of State Key Laboratory of Water Resources and Hydropower Engineering Science in Wuhan University (2015SWG01). The authors thank the China Meteorological Data Sharing Service System for providing part of the data used in this study. The authors would like to thank the editor and the anonymous reviewers for their comments that helped to improve the quality of the paper.

Edited by: A. Guadagnini

References

Abaza, M., Anctil, F., Fortin, V., and Turcotte, R.: Sequential streamflow assimilation for short-term hydrological ensemble forecasting, J. Hydrol., 519, 2692–2706, doi:10.1016/j.jhydrol.2014.08.038, 2014.

Allen, R. G., Pereira, L. S., Raes, D., and Smith, M.: Crop Evapotranspiration-Guidelines for Computing Crop Water Requirements-FAO Irrigation and Drainage Paper 56, Food and Agriculture Organization of the United Nations, Rome, Italy, 1998.

Andréassian, V., Parent, E., and Michel, C.: A distribution-free test to detect gradual changes in watershed behavior, Water Resour. Res., 39, 1252, doi:10.1029/2003WR002081, 2003.

Brigode, P., Oudin, L., and Perrin, C.: Hydrological model parameter instability: A source of additional uncertainty in estimating the hydrological impacts of climate change?, J. Hydrol., 476, 410–425, doi:10.1016/j.jhydrol.2012.11.012, 2013.

Brown, A. E., Zhang, L., McMahon, T. A., Western, A. W., and Vertessy, R. A.: A review of paired catchment studies for determining changes in water yield resulting from alterations in vegetation, J. Hydrol., 310, 28–61, doi:10.1016/j.jhydrol.2004.12.010, 2005.

Burgers, G., van Leeuwen, P. J., and Evensen, G.: Analysis scheme in the ensemble Kalman filter, Mon. Weather Rev., 126, 1719–1724, doi:10.1175/1520-0493(1998)126<1719:ASITEK>2.0.CO;2, 1998.

Clark, M. P., Rupp, D. E., Woods, R. A., Zheng, X., Ibbitt, R. P., Slater, A. G., Schmidt, J., and Uddstrom, M. J.: Hydrological data assimilation with the ensemble Kalman filter: Use of streamflow observations to update states in a distributed hydrological model, Adv. Water Resour., 31, 1309–1324, doi:10.1016/j.advwatres.2008.06.005, 2008.

Coron, L., Andréassian, V., Perrin, C., Lerat, J., Vaze, J., Bourqui, M., and Hendrickx, F.: Crash testing hydrological models in contrasted climate conditions: An experiment on 216 Australian catchments, Water Resour. Res., 48, W05552, doi:10.1029/2011WR011721, 2012.

DeChant, C. M. and Moradkhani, H.: Toward a reliable prediction of seasonal forecast uncertainty: Addressing model and initial condition uncertainty with ensemble data assimilation and sequential Bayesian combination, J. Hydrol., 519, 2967–2977, doi:10.1016/j.jhydrol.2014.05.045, 2014.

Delijani, E. B., Pishvaie, M. R., and Boozarjomehry, R. B.: Subsurface characterization with localized ensemble Kalman filter employing adaptive thresholding, Adv. Water Resour., 69, 181–196, doi:10.1016/j.advwatres.2014.04.011, 2014.

Deng, C., Liu, P., Guo, S., Wang, H., and Wang, D.: Estimation of nonfluctuating reservoir inflow from water level observations using methods based on flow continuity, J. Hydrol., 529, 1198–1210, doi:10.1016/j.jhydrol.2015.09.037, 2015a.

Deng, C., Liu, P., Liu, Y., Wu, Z. H., and Wang, D.: Integrated hydrologic and reservoir routing model for real-time water level forecasts, J. Hydrol. Eng., 20, 05014032, doi:10.1061/(ASCE)HE.1943-5584.0001138, 2015b.

Evensen, G.: Sequential data assimilation with a nonlinear quasi-geostrophic model using Monte Carlo methods to forecast error statistics, J. Geophys. Res., 99, 10143–10162, doi:10.1029/94JC00572, 1994.

Evensen, G.: The Ensemble Kalman filter: theoretical formulation and practical implementation, Ocean Dynam., 53, 343–367, doi:10.1007/s10236-003-0036-9, 2003.

Evensen, G. and van Leeuwen, P. J.: Assimilation of Geosat altimeter data for the Agulhas Current using the ensemble Kalman filter with a quasigeostrophic model, Mon. Weather Rev., 124, 85–96, doi:10.1175/1520-0493(1996)124<0085:AOGADF>2.0.CO;2, 1996.

Guo, S., Wang, J., Xiong, L., Ying, A., and Li, D.: A macroscale and semi-distributed monthly water balance model to predict climate change impacts in China, J. Hydrol., 268, 1–15, doi:10.1016/S0022-1694(02)00075-6, 2002.

Guo, S., Chen, H., Zhang, H., Xiong, L., Liu, P., Pang, B., Wang, G., and Wang, Y.: A semi-distributed monthly water balance model and its application in a climate change impact study in the middle and lower Yellow River basin, Water Int., 30, 250–260, doi:10.1080/02508060508691864, 2005.

Han, E., Merwade, V., and Heathman, G. C.: Implementation of surface soil moisture data assimilation with watershed scale distributed hydrological model, J. Hydrol., 416–417, 98–117, doi:10.1016/j.jhydrol.2011.11.039, 2012.

Jeremiah, E., Marshall, L., Sisson, S. A., and Sharma, A.: Specifying a hierarchical mixture of experts for hydrologic modeling:

Gating function variable selection, Water Resour. Res., 49, 2926–2939, doi:10.1002/wrcr.20150, 2013.

Kumar, S. V., Reichle, R. H., Harrison, K. W., Peters-Lidard, C. D., Yatheendradas, S., and Santanello, J. A.: A comparison of methods for a priori bias correction in soil moisture data assimilation, Water Resour. Res., 48, W03515, doi:10.1029/2010WR010261, 2012.

Kurtz, W., Hendricks Franssen, H.-J., and Vereecken, H.: Identification of time-variant river bed properties with the ensemble Kalman filter, Water Resour. Res., 48, W10534, doi:10.1029/2011WR011743, 2012.

Li, S., Xiong, L., Dong, L., and Zhang, J.: Effects of the Three Gorges Reservoir on the hydrological droughts at the downstream Yichang station during 2003–2011, Hydrol. Process., 27, 3981–3993, doi:10.1002/hyp.9541, 2013.

Li, X.-N., Xie, P., Li, B.-B., and Zhang, B.: A probability calculation method for different grade drought event under changing environment-Taking Wuding River basin as an example, Shuili Xuebao, J. Hydraul. Eng., 45, 585–594, doi:10.13243/j.cnki.slxb.2014.05.010, 2014 (in Chinese).

Li, Y., Ryu, D., Western, A. W., and Wang, Q. J.: Assimilation of stream discharge for flood forecasting: The benefits of accounting for routing time lags, Water Resour. Res., 49, 1887–1900, doi:10.1002/wrcr.20169, 2013.

Li, Z., Liu, P., Deng, C., Guo, S., He, P., and Wang, C.: Evaluation of the estimation of distribution algorithm to calibrate a computationally intensive hydrologic model, J. Hydrol. Eng., 21, 04016012, doi:10.1061/(ASCE)HE.1943-5584.0001350, 2015.

Liu, Y. and Gupta, H. V.: Uncertainty in hydrologic modeling: Toward an integrated data assimilation framework, Water Resour. Res., 43, 1–18, doi:10.1029/2006WR005756, 2007.

Liu, Y., Weerts, A. H., Clark, M., Hendricks Franssen, H.-J., Kumar, S., Moradkhani, H., Seo, D.-J., Schwanenberg, D., Smith, P., van Dijk, A. I. J. M., van Velzen, N., He, M., Lee, H., Noh, S. J., Rakovec, O., and Restrepo, P.: Advancing data assimilation in operational hydrologic forecasting: progresses, challenges, and emerging opportunities, Hydrol. Earth Syst. Sci., 16, 3863–3887, doi:10.5194/hess-16-3863-2012, 2012.

Lü, H. S., Hou, T., Horton, R., Zhu, Y. H., Chen, X., Jia, Y. W., Wang, W., and Fu, X. L.: The streamflow estimation using the Xinanjiang rainfall runoff model and dual state-parameter estimation method, J. Hydrol., 480, 102–114, doi:10.1016/j.jhydrol.2012.12.011, 2013.

Merz, R., Parajka, J., and Blöschl, G.: Time stability of catchment model parameters: Implications for climate impact analyses, Water Resour. Res., 47, W02531, doi:10.1029/2010WR009505, 2011.

Montzka, C., Grant, J. P., Moradkhani, H., Franssen, H.-J. H., Weihermüller, L., Drusch, M., and Vereecken, H.: Estimation of radiative transfer parameters from L-band passive microwave brightness temperatures using advanced data assimilation, Vadose Zone J., 12, 1–17, doi:10.2136/vzj2012.0040, 2013.

Moradkhani, H., Sorooshian, S., Gupta, H. V., and Houser, P. R.: Dual state–parameter estimation of hydrological models using ensemble Kalman filter, Adv. Water Resour., 28, 135–147, doi:10.1016/j.advwatres.2004.09.002, 2005.

Nash, J. E. and Sutcliffe, J. V.: River flow forecasting through conceptual models part I: A discussion of principles, J. Hydrol., 10, 282–290, doi:10.1016/0022-1694(70)90255-6, 1970.

Nie, S., Zhu, J., and Luo, Y.: Simultaneous estimation of land surface scheme states and parameters using the ensemble Kalman filter: identical twin experiments, Hydrol. Earth Syst. Sci., 15, 2437–2457, doi:10.5194/hess-15-2437-2011, 2011.

Paik, K., Kim, J. H., Kim, H. S., and Lee, D. R.: A conceptual rainfall-runoff model considering seasonal variation, Hydrol. Process., 19, 3837–3850, doi:10.1002/hyp.5984, 2005.

Panzeri, M., Riva, M., Guadagnini, A., and Neuman, S. P.: Data assimilation and parameter estimation via ensemble Kalman filter coupled with stochastic moment equations of transient groundwater flow, Water Resour. Res., 49, 1334–1344, doi:10.1002/wrcr.20113, 2013.

Patil, S. D. and Stieglitz, M.: Comparing spatial and temporal transferability of hydrological model parameters, J. Hydrol., 525, 409–417, doi:10.1016/j.jhydrol.2015.04.003, 2015.

Pauwels, V. R. N. and Lannoy, G. J. M. D.: Improvement of Modeled Soil Wetness Conditions and Turbulent Fluxes through the Assimilation of Observed Discharge, J. Hydrometeorol., 7, 458–477, doi:10.1175/JHM490.1, 2006.

Peel, M. C. and Blöschl, G.: Hydrological modelling in a changing world, Prog. Phys. Geog., 35, 249–261, doi:10.1177/0309133311402550, 2011.

Samuel, J., Coulibaly, P., Dumedah, G., and Moradkhani, H.: Assessing model state and forecasts variation in hydrologic data assimilation, J. Hydrol., 513, 127–141, doi:10.1016/j.jhydrol.2014.03.048, 2014.

Seiller, G., Anctil, F., and Perrin, C.: Multimodel evaluation of twenty lumped hydrological models under contrasted climate conditions, Hydrol. Earth Syst. Sci., 16, 1171–1189, doi:10.5194/hess-16-1171-2012, 2012.

Shi, Y., Davis, K. J., Zhang, F., Duffy, C. J., and Yu, X.: Parameter estimation of a physically based land surface hydrologic model using the ensemble Kalman filter: A synthetic experiment, Water Resour. Res., 50, 706–724, doi:10.1002/2013WR014070, 2014.

Su, T., Feng, T., and Feng, G.: Evaporation variability under climate warming in five reanalyses and its association with pan evaporation over China, J. Geophys. Res.-Atmos., 120, 8080–8098, doi:10.1002/2014JD023040, 2015.

Tamura, H., Bacopoulos, P., Wang, D., Hagen, S. C., and Kubatko, E. J.: State estimation of tidal hydrodynamics using ensemble Kalman filter, Adv. Water Resour., 63, 45–56, doi:10.1016/j.advwatres.2013.11.002, 2014.

Thirel, G., Andréassian, V., Perrin, C., Audouy, J. N., Berthet, L., Edwards, P., Folton, N., Furusho, C., Kuentz, A., Lerat, J., Lindström, G., Martin, E., Mathevet, T., Merz, R., Parajka, J., Ruelland, D., and Vaze, J.: Hydrology under change: an evaluation protocol to investigate how hydrological models deal with changing catchments, Hydrolog. Sci. J., 60, 1184–1199, doi:10.1080/02626667.2014.967248, 2015.

Vrugt, J. A., ter Braak, C. J. F., Diks, C. G. H., and Schoups, G.: Hydrologic data assimilation using particle Markov chain Monte Carlo simulation: Theory, concepts and applications, Adv. Water Resour., 51, 457–478, doi:10.1016/j.advwatres.2012.04.002, 2013.

Wang, D., Chen, Y., and Cai, X.: State and parameter estimation of hydrologic models using the constrained ensemble Kalman filter, Water Resour. Res., 45, W11416, doi:10.1029/2008WR007401, 2009.

Wang, G. and Fan, Z.: A study of water and sediment changes in the Yellow River, Publishing House of Yellow River Water Conservancy, Zhengzhou, China, 2003 (in Chinese).

Weerts, A. H. and El Serafy, G. Y. H.: Particle filtering and ensemble Kalman filtering for state updating with hydrological conceptual rainfall-runoff models, Water Resour. Res., 42, 1–17, doi:10.1029/2005WR004093, 2006.

Westra, S., Thyer, M., Leonard, M., Kavetski, D., and Lambert, M.: A strategy for diagnosing and interpreting hydrological model nonstationarity, Water Resour. Res., 50, 5090–5113, doi:10.1002/2013WR014719, 2014.

Xie, X. and Zhang, D.: Data assimilation for distributed hydrological catchment modeling via ensemble Kalman filter, Adv. Water Resour., 33, 678–690, doi:10.1016/j.advwatres.2010.03.012, 2010.

Xie, X. and Zhang, D.: A partitioned update scheme for state-parameter estimation of distributed hydrologic models based on the ensemble Kalman filter, Water Resour. Res., 49, 7350–7365, doi:10.1002/2012WR012853, 2013.

Xie, X., Meng, S., Liang, S., and Yao, Y.: Improving streamflow predictions at ungauged locations with real-time updating: application of an EnKF-based state-parameter estimation strategy, Hydrol. Earth Syst. Sci., 18, 3923–3936, doi:10.5194/hess-18-3923-2014, 2014.

Xiong, L. and Guo, S.: A two-parameter monthly water balance model and its application, J. Hydrol., 216, 111–123, doi:10.1016/S0022-1694(98)00297-2, 1999.

Xiong, L. and Guo, S.: Appraisal of Budyko formula in calculating long-term water balance in humid watersheds of southern China, Hydrol. Process., 26, 1370–1378, doi:10.1002/hyp.8273, 2012.

Xiong, L., Yu, K.-X., and Gottschalk, L.: Estimation of the distribution of annual runoff from climatic variables using copulas, Water Resour. Res., 50, 7134–7152, doi:10.1002/2013WR015159, 2014.

Xu, J.: Variation in annual runoff of the Wudinghe River as influenced by climate change and human activity, Quatern. Int., 244, 230–237, doi:10.1016/j.quaint.2010.09.014, 2011.

Xue, L. and Zhang, D.: A multimodel data assimilation framework via the ensemble Kalman filter, Water Resour. Res., 50, 4197–4219, doi:10.1002/2013WR014525, 2014.

Yan, H., DeChant, C. M., and Moradkhani, H.: Improving soil moisture profile prediction with the particle filter-Markov chain Monte Marlo method, IEEE T. Geosci. Remote, 53, 6134–6147, doi:10.1109/tgrs.2015.2432067, 2015.

Ye, W., Bates, B. C., Viney, N. R., Sivapalan, M., and Jakeman, A. J.: Performance of conceptual rainfall-runoff models in low-yielding ephemeral catchments, Water Resour. Res., 33, 153–166, doi:10.1029/96WR02840, 1997.

Zhang, D., Liu, X. M., Liu, C. M., and Bai, P.: Responses of runoff to climatic variation and human activities in the Fenhe River, China, Stoch. Env. Res. Risk A., 27, 1293–1301, doi:10.1007/s00477-012-0665-y, 2013.

Zhang, H., Huang, G. H., Wang, D., and Zhang, X.: Multi-period calibration of a semi-distributed hydrological model based on hydroclimatic clustering, Adv. Water Resour., 34, 1292–1303, doi:10.1016/j.advwatres.2011.06.005, 2011.

9

Snow cover dynamics in Andean watersheds of Chile (32.0–39.5° S) during the years 2000–2016

Alejandra Stehr[1,2] **and Mauricio Aguayo**[1,2]

[1]Centre for Environmental Sciences EULA-CHILE, University of Concepción, Concepción, Chile
[2]Faculty of Environmental Sciences, University of Concepción, Concepción, Chile

Correspondence to: Alejandra Stehr (astehr@udec.cl)

Abstract. Andean watersheds present important snowfall accumulation mainly during the winter, which melts during the spring and part of the summer. The effect of snowmelt on the water balance can be critical to sustain agriculture activities, hydropower generation, urban water supplies and wildlife. In Chile, 25 % of the territory between the region of Valparaiso and Araucanía comprises areas where snow precipitation occurs. As in many other difficult-to-access regions of the world, there is a lack of hydrological data of the Chilean Andes related to discharge, snow courses, and snow depths, which complicates the analysis of important hydrological processes (e.g. water availability). Remote sensing provides a promising opportunity to enhance the assessment and monitoring of the spatial and temporal variability of snow characteristics, such as the snow cover area (SCA) and snow cover dynamic (SCD). With regards to the foregoing questions, the objective of the study is to evaluate the spatiotemporal dynamics of the SCA at five watersheds (Aconcagua, Rapel, Maule, Biobío and Toltén) located in the Chilean Andes, between latitude 32.0 and 39.5° S, and to analyse its relationship with the precipitation regime/pattern and El Niño–Southern Oscillation (ENSO) events. Those watersheds were chosen because of their importance in terms of their number of inhabitants, and economic activities depending on water resources. The SCA area was obtained from MOD10A2 for the period 2000–2016, and the SCD was analysed through a number of statistical tests to explore observed trends. In order to verify the SCA for trend analysis, a validation of the MOD10A2 product was done, consisting of the comparison of snow presence predicted by MODIS with ground observations. Results indicate that there is an overall agreement of 81 to 98 % between SCA determined from ground observations and MOD10A2, showing that the MODIS snow product can be taken as a feasible remote sensing tool for SCA estimation in southern–central Chile. Regarding SCD, no significant reduction in SCA for the period 2000–2016 was detected, with the exception of the Aconcagua and Rapel watersheds. In addition to that, an important decline in SCA in the five watersheds for the period of 2012 and 2016 was also evident, which is coincidental with the rainfall deficit for the same years. Findings were compared against ENSO episodes that occurred during 2010–2016, detecting that Niña years are coincident with maximum SCA during winter in all watersheds.

1 Introduction

Snowmelt-driven watershed systems are highly sensitive to climate change, because their hydrologic cycle depends on both precipitation and temperature, and because water is already a scarce resource subject to ever-increasing pressure for its use (Barnett et al., 2005; Vicuña et al., 2011; Meza et al., 2012; Valdés-Pineda et al., 2014). Snowmelt controls the shape of the annual hydrograph, and affects the water balance at monthly and shorter timescales (Verbunt et al., 2003; Cortés et al., 2011). The effect of snowmelt on the water balance can be critical to sustain agriculture activities, hydropower generation, urban water supply and wildlife habitat quality (e.g. Vicuña et al., 2012, 2013).

The Andean watersheds present an important snowfall accumulation mainly during the austral winter; snow melts during spring and usually also during part of the summer, depending on relative altitude and ambient temperature. At higher elevations, a snowpack stores significant volumes of water, which are released to the surface runoff and groundwater when solar radiation increases.

In particular, 25 % of the Chilean territory between the Valparaiso and Araucanía regions is contained in areas where snow precipitation occurs (DGA, 1995). As in many other difficult-to-access regions of the world, the Chilean Andes – unlike western North America or the European Alps – have limited availability in temporal and spatial extent of hydrological data like discharge data, snow courses, and snow depths (Ragettli et al., 2013), which complicates the analysis of important hydrological processes and the validation of water quantity prediction models.

In this regard, remote sensing provides a promising opportunity to enhance the assessment and monitoring of the spatial and temporal variability of different variables involved in the precipitation–runoff processes in areas where data availability for hydrological modelling is scarce (Simic et al., 2004; Boegh et al., 2004; Melesse et al., 2007; Montzka et al., 2008; Milzowa et al., 2009, Er-Raki et al., 2010; Stehr et al., 2009, 2010).

Satellite-derived SCA from products like NOAA-AVHRR or MODIS can be used to enhance the assessment and monitoring of the spatial and temporal variability of snow characteristics (Lee et al., 2006; Li and Wang, 2010; Marchane et al., 2015; Wang et al., 2015), especially when it is combined with field data and snowpack models (Kuchment et al., 2010).

Specific spectral reflectance of snow (higher reflectance in the visible spectrum, compared to the mid-infrared electromagnetic spectrum) allows SCA to be accurately discriminated from snow-free areas using optical remote sensing methods (in the absence of clouds or vegetation canopies). Compared with other remote sensing techniques such as microwaves – which can be used to map snow water equivalent (SWE) – optical remote sensing, which is used to map snow areal extent (SAE), has a much higher spatial resolution (Zhou et al., 2005; Zeinivand and De Smedt, 2009).

Previous a Previous studies have compared MODIS snow maps with ground observations and snow maps produced by the National Operational Hydrologic Remote Sensing Center, USA (NOHRSC) (Hall et al., 2002; Klein and Barnett, 2003; Tekeli et al., 2005; Aulta et al., 2006). Klein and Barnett (2003) compared MODIS and NOHRSC products with ground observations (SNOTEL measurements), obtaining an overall accuracy of 94 and 76 %, respectively. The MOD10A2 snow product is capable of predicting the presence of snow with good precision (over 90 %) when the sky is clear (Zhou et al., 2005; Liang et al., 2008a; Wang et al., 2008; Huang et al., 2011; Zheng et al., 2017). All the previous studies were done in watersheds with a surface topography not as complex as in the Chilean Andes, where we have very steep slopes and a great oceanic effect due to the short distance between the coastline and the mountains.

Many of the studies on SCA changes and variability were done in the Northern Hemisphere, where topographic and orographic conditions are different from the ones we have in Chile. Research studies generally show negative trends in snow cover extent and snow water equivalent across both North America and Eurasia.

In Europe, Krajcí et al. (2016) analysed the main Slovak watersheds; for the 2001–2006 period, they obtained an increase in mean SCA; however, a significantly lower SCA is observed in the next 2007–2012 period. Their results indicate that there is no significant change in the mean watershed SCA in the period 2001–2014. Dietz et al. (2012), in their study of snow cover characteristics in Europe, found some abnormal events when comparing the mean conditions with single snow cover seasons. For the season 2005/2006 in particular, an increased snow cover with a later snow cover melt was detected. Marchane et al. (2015) studied the SCD over the Moroccan Atlas mountain range from 2000 until 2013, concluding that SCA has a strong inter-annual variability and that there is no statistical evidence of a trend in that period.

In Asia, two important regions have been studied: the Tibetan Plateau and the Himalayan region. In the case of the Tibetan Plateau, Zhang et al. (2012) have studied the SCD of four lake watersheds for the period 2001–2010; results indicate that spatial distribution and patterns of snow-covered days are very stable from year to year, and that there is no trend of snow cover change for each watershed. For the same period, Tang et al. (2013) found a high inter-annual variability of SCA, with parts of the studied area showing a declining trend in SCA, and other parts showing increasing trends in SCA. Wang et al. (2015) evaluated trends in SCA, showing a decrease in snow-covered areas from 2003 until 2010. For the Himalayan region, Maskey et al. (2011) studied the trends and variability of snow cover changes above 3000 m during the 2000–2008 period, showing a decreasing trend of snow cover during January and increasing trends during March. Snehmani et al. (2016) studied the 2001–2012 winter period (November–April) in order to obtain clear negative snow cover trends in the basin. Azmat et al. (2017) indicated that the watershed shows a consistent or slightly decreasing trend of snow cover, particularly over the high-altitude parts of the watershed during 2000–2009, and in the 14-year analysis (2000–2013), a slight expansion in the snow-covered area was observed in the whole basin. Gurung et al. (2017), based on the analysis SCA data between 2003 and 2012, also obtained a decline in SCA, with a statistically significant negative correlation between SCA and temperature, which indicates that this trend is partly a result of increasing temperatures.

In North America, Pederson et al. (2013) found that, since 1980, in the northern and southern Rocky Mountains the 1 April snow water equivalent (SWE) has been changing synchronously, and generally declining, associated with springtime warming. Fassnacht and Hultstrand (2015) examined trends in snow depth and SWE from three long-term snow course stations in northern Colorado from 1936 to 2014. They found negative trends at two of the stations for all the period and, at the third station, a positive trend was found for the first half of the record, and a decrease over the second half. Harpold et al. (2012) analysed SNOTEL SWE for the central and southern Rocky Mountains for the period 1984–2009; they found widespread decreases in maximum SWE and duration of snow cover. After a review of various studies in North America, Kunkel et al. (2016) concluded that all of them are consistent in indicating a decrease in snow on the ground, with some of the most extreme low values occurring in the last 10–15 years.

Studies of snowpack variation done in the central Andes of Chile and Argentina show that the average regional maximum value of the SWE series displays a positive (though non-significant) trend with marked interannual variability ranging from 6 to 257 % of the 1966–2004 mean (Masiokas et al., 2006). Results from Cornwell et al. (2016), who made a SWE reconstruction between 2001 and 2014, indicate that the years 2002 and 2005 stand out for displaying large positive anomalies throughout the entire model domain. The northern part shows above-average accumulation in only 3 out of the 14 simulated years (2002, 2005 and 2007), whereas the other part of the study area shows above-average accumulation for 6 years (2001, 2002, 2005, 2006, 2008 and 2009). In particular, the years 2007 and 2009 show a bimodal spatial structure, with excess accumulation (deficit) in the northern (southern) area and the inverse pattern in the latter year.

From the literature review, it is evident that SCD is site-specific and exhibits high variability among the different sites around the world. The Andes mountain range (on the Chilean side) has a great oceanic effect due to the short distance between the coastal line and the mountains, unlike the Northern Hemisphere, which has a big continental influence. This makes the isotherm 0 higher than in the Northern Hemisphere; in addition, the topographic features characterized by mean heights above 3000 m, with very steep slopes that produce a huge orographic effect that forces the rise of the western winds and condensation of moisture, strongly affect the regime of precipitation and temperatures. This causes small-scale spatial variations in weather, which are sometimes difficult to identify in satellite imagery. All the previously mentioned aspects make it necessary to do a validation of MOD10A2 before using it for other analyses. Indeed, one of the main sources of error in the classification of satellite images is the interpretation of topographic features, especially in areas with rugged slopes. The irregular topography produces shading and lighting effects that change the radiometric response of the surface, which depends on the local slope and its orientation (Riaño et al., 2003). As the climate-controlling hydrologic processes in the Andes are influenced by El Niño events (Escobar and Aceituno, 1993; Ayala et al., 2014), as well as by warm winter storms (Garreaud, 2013), assessment of SCD is of special interest in this region. Pioneer studies have been presented in recent years for the Mataquito River watershed (Demaría et al., 2013; Vicuña et al., 2013). The aim of the study is to evaluate the spatiotem-

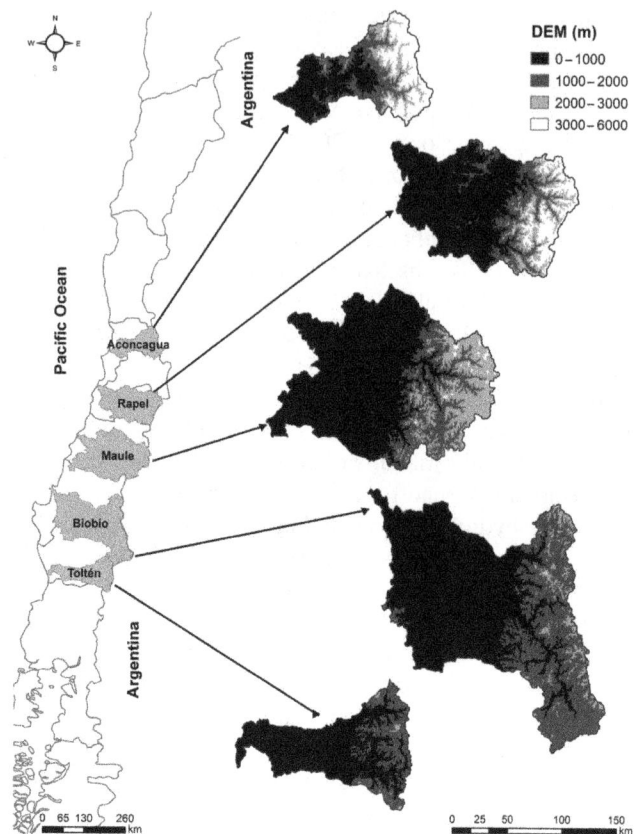

Figure 1. Study sites.

poral dynamics of the SCA in the southern–central Chilean Andes, and to analyse its relationship with the precipitation regime/pattern and El Niño–Southern Oscillation (ENSO) events. To prove this, we selected five watersheds located in the Chilean Andes, between latitude 32.0 and 39.5° S. We investigated trends and variability of snow cover changes at different temporal scales (seasonal and annual) and we used Moderate Resolution Imaging Spectroradiometer (MODIS) snow cover products (Hall et al., 2007) from 2000 to 2017. In situ precipitation measurements from the National General Water Directorate (DGA) were used.

2 Materials and methods

2.1 Study sites

The study site includes five watersheds located in central and southern Chile: Aconcagua, Rapel, Maule, Biobío and Toltén (Fig. 1). The watersheds were chosen considering their population and dependence on water-resourced economic activities.

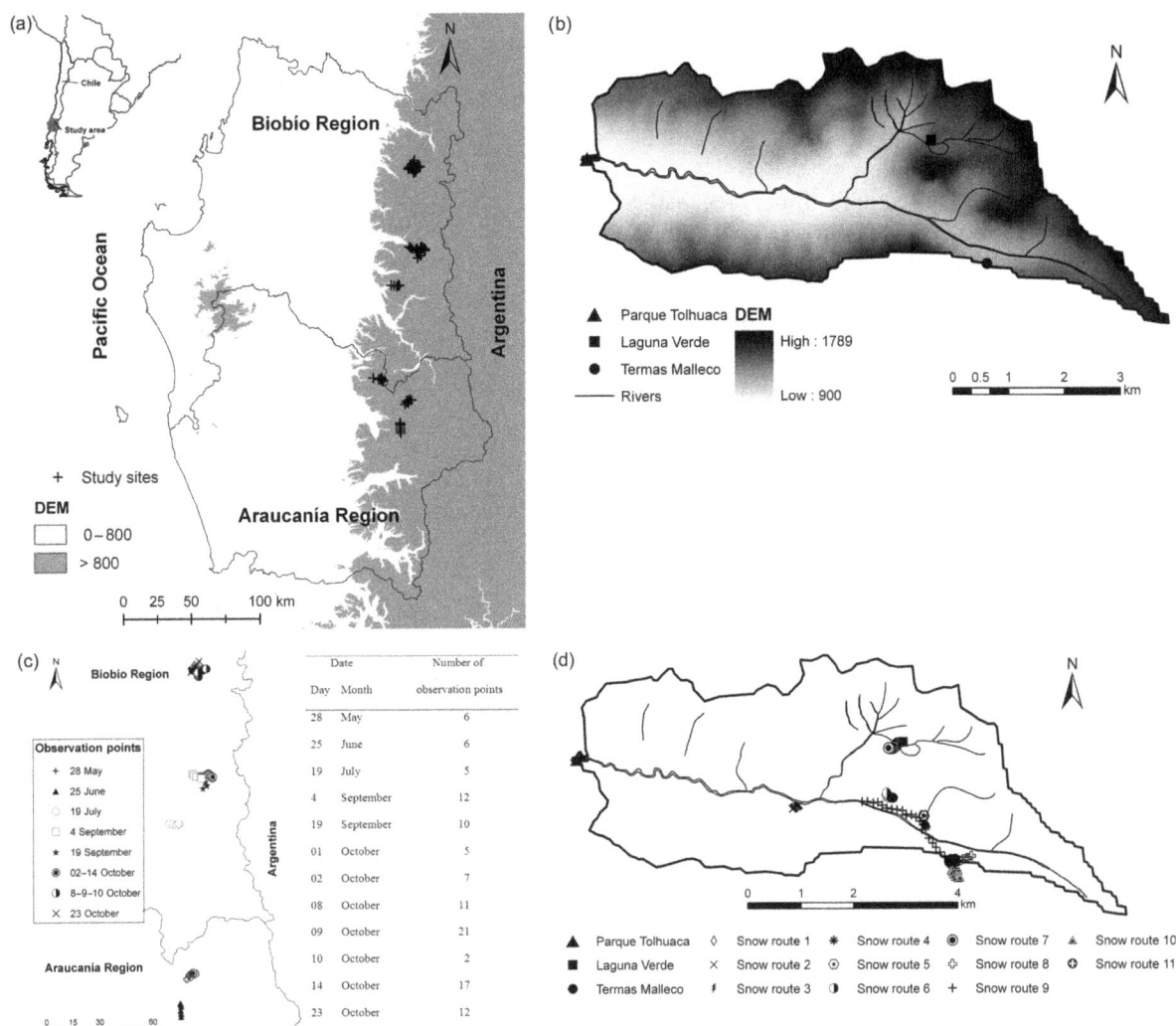

Figure 2. (a) Study sites for MOD10A2 validation. **(b)** Location of the meteorological stations for continuous monitoring of snow depth. **(c)** Location and date of 1-day observation points. **(d)** Location of the snow courses.

The Aconcagua watershed is located in the Valparaiso region, between the parallels 32°14′–33°09′ S and 69°59′–71°33′ W, with an area of 7340 km² and a maximum elevation of 5843 m a.s.l. Approximately 40 % of the watershed lies above the snowline (average altitude above which snow can be found in winter), which is located at 2100 m (Garreaud, 1992). The climate in the watershed is temperate Mediterranean with a long dry season of 7 to 8 months and a wet season of approximately 4 months (May–August) during which more than 80 % of the precipitation occurs. The average annual precipitation is 529 mm. Principal economic activities at the watershed are agriculture, mining and industries. It has a population of around 600 000 inhabitants (4 % of the Chilean population).

The Rapel watershed is located in the General Libertador Bernardo O'Higgins region between the parallels 33°52′–35°00′ S and 70°00′–71°53′ W, with an area of 13 695 km²

and a maximum elevation of 5138 m a.m.s.l. Approximately 30 % of the watershed lies above the snowline which is located at 1500 m (Peña and Vidal, 1993). The climate in the watershed is temperate Mediterranean, with a dry season of 4 to 5 months (November–March) and a wet season of approximately 4 months (May–August) during which more than 75 % of the precipitation occurs. The average annual precipitation is 960 mm. The main economic activities at the watershed are agriculture and mining. It has a population of around 570 000 inhabitants (3.8 % of the Chilean population).

The Maule watershed is located in the Maule region, between the parallels 35°05′–36°35′ S and 70°18′–72°42′ W, with an area of 20 295 km² and a maximum elevation of 3931 m a.m.s.l. Approximately 32 % of the watershed lies above the snowline, which is located at 1150 m (Peña and Vidal, 1993). The climate in the watershed is temperate Mediterranean, with a 6-month dry season (November–

April) and a wet season of approximately 4 months (May–August) during which more than 75 % of the precipitation occurs. The average annual precipitation is 1471 mm. The main economic activity at the watershed is agriculture. It has a population of around 410 000 inhabitants (2.7 % of the Chilean population).

The Biobío watershed is located in the Biobío region, between the parallels 36°45′–38°49′ S and 71°00′–73°20′ W, with an area of 24 264 km^2 and a maximum elevation of 3487 m a.m.s.l. Approximately 41 % of the watershed lies above the snowline, which is located at 850 m (Peña and Vidal, 1993). The climate in the watershed is Mediterranean, with a 5-month dry season (November–March) and a wet season of approximately 4 months (May–August), during which more than 55 % of the precipitation occurs. The average annual precipitation is 1891 mm. Principal economic activities related to water resources at the watershed are agriculture, forestry and industries. It has a population of around 630 000 inhabitants (4.2 % of the Chilean population).

The Toltén watershed is located in the Araucanía region, between the parallels 38°32′–39°38′ S and 71°21′–73°16′ W, with an area of 8398 km^2 and a maximum elevation of 3710 m a.m.s.l. Approximately 37 % of the watershed lies above the snowline, which is located at 750 m (Peña and Vidal, 1993). The climate in the watershed is temperate rainforest with Mediterranean influence, characterized by precipitation throughout the year but having less rain during the summer months than in the winter ones. The average annual precipitation is 2870 mm. Main economic activities related to water resources at the watershed are tourism, agriculture and forestry. It has a population of around 170 000 inhabitants (1.1 % of the Chilean population).

2.2 Ground observations of SCA

For validation of MOD10A2, ground observations were performed, including continuous monitoring of snow depth with meteorological stations, snow courses, and 1-day observations of snow presence and depth at some mountain trails. Figure 2a shows the locations of measuring sites.

2.2.1 Continuous monitoring of snow depth

In order to perform a continuous measurement of snow depth, three meteorological stations were installed in the upper part of the Biobío watershed. The stations are Parque Tolhuaca, located at 900 m a.s.l., Termas Malleco at 1190 m a.s.l., and Laguna Verde at 1410 m a.s.l. Also, snow depth data from three DGA meteorological stations (Portillo, 3005 m a.s.l., Laguna Negra, 2709 m a.s.l., and Volcan Chillan, 1964 m a.s.l.) were used. Figure 2b shows the location of the meteorological stations.

Snow depth was measured using acoustic snow depth sensors (Campbell Scientific, SR50A) with a frequency of 15 min. Corrections for variation of the speed of sound in air

Table 1. Dates and repetition times for each snow route during winter 2011.

Snow route	Dates	No. of times route was repeated
1	30 Jun	1
2	30 Jun	1
3	30 Jun, 19 Jul, 31 Aug, 14 Sep	4
4	1 Jul, 19 Jul, 31 Aug, 15 Sep	4
5	1 Jul	1
6	1 Jul	1
7	1 Jul	1
8	31 Aug, 15 Sep	2
9	1 Sep, 15 Sep	2
10	31 Aug, 14 Sep	2
11	31 Aug, 14 Sep	2

were made considering the air temperature measured at the same time intervals as the snow depth, using the temperature and relative humidity probe (Vaisala, HMP60). The data were collected from April 2010 to December 2011 at the Termas Malleco station, and from July 2011 to December 2011 at the Parque Tolhuaca and Laguna Verde stations. In the case of the DGA station, available data from April 2013 to December 2015 were used. Snow data were grouped considering the average snow depth over 8 days (same 8-day period of MOD10A2) and then reclassified as snow (1) if the average snow depth was > 0 cm and as no-snow (0) for snow depth = 0 cm.

2.2.2 One-day observation points

To cover a large spatial domain and to achieve a better spatial and temporal representation of SCA, a total of 124 different single measurements of snow depth were conducted during field campaigns at different mountain trails. Field measurements were taken from the end of June until the beginning of October 2011. The location of each observation point was recorded with a GPS and snow depth was measured with a Black Diamond QuickDraw Tour Probe 190. Figure 2c shows the location of the observation points and the date and number of observations for each day.

2.2.3 Snow courses

During 6 days of the 2011 winter season, 11 snow courses were conducted in the upper Malleco watershed. Figure 2d shows the location of snow courses. Each route was recorded with a GPS. Snow depth and density were measured with a Black Diamond QuickDraw Tour Probe 190 and Snow Sampling Tubes (3600 Federal Snow Tubes, Standard-Metric), respectively. Table 1 shows the dates and repetitions for each snow route performed.

2.3 MODIS snow cover products

The Moderate Resolution Imaging Spectroradiometer (MODIS) is on the Earth Observing System, which employs a cross-track scan mirror and a set of individual detector elements to provide imagery of the Earth's surface and clouds in 36 discrete and narrow spectral bands ranging in wavelength from 0.405 to 14.385 µm. It provides medium-to-coarse resolution imagery with a high temporal repeat cycle (1–2 days). The main purpose of the MODIS is to facilitate the study of global vegetation and land cover, vegetation properties, global land surface changes, surface albedo, surface temperature as well as snow and ice cover, on a daily or nearly daily basis. The MODIS snow cover products are one of the many geophysical standard products derived from MODIS data. The MODIS snow cover products are provided on a daily basis (MOD10A1) and as 8-day composites (MOD10A2), both at 500 m resolution over the Earth's land surfaces. MOD10A1 consists of 1200 km by 1200 km tiles of 500 m resolution data gridded in a sinusoidal map projection and MOD10A2 is a composite of MOD10A1 especially produced to show maximum snow extent. For this study, MOD10A2v005 (Hall et al., 2016) was used. Classification of SCA using MODIS collected data was done based on the Normalized Difference Snow Index (NDSI) (Hall et al., 1995; Klein et al., 1998; Riggs et al., 2006). Images were reprojected to WGS84 UTM 19S using the MODIS Reprojection Tool (MRT).

MODIS estimates of SCA were validated for the period between April 2010 and December 2011 by comparing them with ground observations. Eighty snow maps from MOD10A2 were analysed. Only cloud-free observations for each cell of the map grid were used.

2.4 Validation of MOD10A2

To validate the correspondence between the image classification and ground observations, a confusion matrix was used (Congalton and Mead, 1983; Story and Congalton, 1986, 1991; Foody, 2002). The confusion matrix is a simple cross tabulation of classified data versus observed ones, providing a basis for accuracy assessment (Campbell, 1996; Canters, 1997). Figure 3 presents the confusion matrix and how the different indexes were calculated.

2.5 Assessment of SCD

SCA variation in the five selected watersheds was evaluated using MOD10A2 for the period covering the years 2000–2016 using a total of 774 images. All images were processed in ArcGis, reprojecting them to WGS84 19S and cut according to the desired area, in this case for the Aconcagua, Rapel, Maule, Biobío and Toltén watersheds. SCA and clouds were quantified for each available image, and annual and seasonal averages of snow cover were calculated for each watershed.

SCD was analysed through a number of statistical tests to explore observed trends in SCA during the study period. "Non-parametric" statistical tests were chosen because they are more robust than the "parametric" test. A Mann–Kendall test was applied to determine the existence of monotonic trends. Sen's method (Gilbert, 1987) was applied to determine the rate of observed changes.

An analysis of the correlation between SCA and mean annual precipitation for the upper part of the watershed was done. Considering the spatial location of SCA, only precipitation stations located over the 700 m a.s.l. were considered. Availability of daily precipitation was evaluated for each year at each station; only years with more than 80 % of data were used in order to obtain the mean annual precipitation. The precipitation for the watershed was obtained considering the arithmetic mean between available stations. Table 2 shows the stations that were used and the availability of data.

3 Results

The MOD10A2 product was validated for determination of SCA in the watersheds under study and the SCD was analysed.

3.1 Validation of MOD10A2 through ground observations

The composite images MOD10A2 were compared with ground observations. For the study period 23, 75, 26, 119, 123 and 32 images were available for comparison with observations at Parque Tolhuaca, Termas Malleco, Laguna Verde, Portillo, Laguna Negra and Volcan Chillan stations, respectively. Clouds were present in 2 (9 %), 1 (1 %), 1 (4 %), 3 (3 %), 3 (2 %) and 2 (6 %) images for each aforementioned station, respectively. Table 3 presents the confusion matrix and the indexes of agreement. Overall accuracies of 86, 81, 88, 92, 83 and 97 % were observed at Parque Tolhuaca, Termas Malleco, Laguna Verde, Portillo, Laguna Negra and Volcan Chillan stations, respectively, which reached the target of 85 % (Thomlinson et al., 1999).

A total of 117 images corresponding to the study period were available for comparison with 1-day observations, with none of them classified as covered by "clouds". All ground observations were done over areas with snow presence. When comparing it to MOD10A2 images, the agreement was 97 %. MODIS that did not coincide with ground observations occurred during the beginning of the snowfall season or at the end of the melting period. In both cases, snow patches were observed in the field, covering areas at the subpixel scale of the MOD10A2 image.

For comparison with snow courses, a total of 282 images were available during the study period, all of them without clouds. Ground observations were done on areas covered by

k,k	A	B	C	...	q	$\sum$
A	n_{AA}	n_{AB}	n_{AC}	...	n_{Aq}	n_{A+}
B	n_{BA}	n_{BB}	n_{BC}	...	n_{Bq}	n_{B+}
C	n_{CA}	n_{CB}	n_{CC}	...	n_{Cq}	n_{C+}
⋮	⋮	⋮	⋮	...	⋮	⋮
q	n_{qA}	n_{qB}	n_{qC}	...	n_{qq}	n_{q+}
$\sum$	n_{+A}	n_{+B}	n_{+C}	...	n_{+q}	n

$$\text{Percentage correct} = \frac{\sum_{k=1}^{q} n_{kk}}{n} \times 100$$

$$\text{User's accuracy} = \frac{n_{kk}}{n_{k+}} \times 100$$

$$\text{Producer's accuracy} = \frac{n_{kk}}{n_{+k}} \times 100$$

Figure 3. Confusion matrix and percentage correct (overall accuracy), user accuracy and producer accuracy.

Table 2. Precipitation stations and availability of data.

Watershed	Station	Elevation m a.s.l.	Number of years with more than 80 % of data	% of availability
Aconcagua	Resguardo Los Patos	1220	16	98 %
	Rio Putaendo En Resguardo Los Patos	1218	11	81 %
	Jahuel	1020	15	97 %
	Los Andes	820	16	100 %
	Rio Aconcagua En Chacabuquito	950	16	100 %
	Vilcuya	1100	16	100 %
	Riecillos	1290	16	100 %
	Las Chilcas	850	16	98 %
Rapel	Rio Pangal En Pangal	1500	13	86 %
	Rio Cachapoal 5 km. Aguas Abajo Junta Cortaderal	1127	12	88 %
	Central Las Nieves	700	5	35 %
	La Rufina	743	16	99 %
Maule	Fundo El Radal	685	16	99 %
	Vilches Alto	1058	16	100 %
	Hornillo	810	16	99 %
	Rio Melado En El Salto	730	12	78 %
Biobío	Embalse Ralco	742	5	35 %
	Rio Biobío En Llanquen	767	11	74 %
	Laguna Malleco	894	14	96 %
	Liucura	1043	16	98 %
Toltén	Lago Tinquilco	850	16	100 %
	Puesco (Aduana)	620	14	94 %

snow only. The overall accuracy of MOD10A2 for predicting SCA was 98 %.

Results indicate that MOD10A2 has a satisfactory agreement with ground observations, and therefore the 8-day composite images are suitable for analysis of SCD in the Andean watersheds.

3.2 Assessment of SCD

Figure 4 shows the SCD in Aconcagua, Rapel, Maule, Biobío and Toltén watersheds for the period 2000–2016. All water-sheds show the same SCD with more SCA in winter than the other seasons. It can be appreciated that in the Maule, Biobío and Toltén watersheds the maximum SCA in 2016 is considerably lower than in the 5 previous years, and is one of the lowest of the whole study period.

Figure 5 shows mean SCA for each season and annual precipitation in the Aconcagua, Rapel, Maule, Biobío and Toltén watersheds for the period 2000–2016. Northern watersheds, i.e. Aconcagua and Rapel, present a higher SCA percentage. It can be appreciated that years with more precipitation do not necessarily have more SCA; instead, years with lower

Table 3. Confusion matrix and indexes of agreement for Parque Tolhuaca, Termas Malleco and Laguna Verde stations.

		Ground observation								
		Parque Tolhuaca Station			Termas Malleco Station			Laguna Verde Station		
		snow	no-snow	User's accuracy	snow	no-snow	User's accuracy	snow	no-snow	User's accuracy
MOD10A2	Snow	6	2	0.75	22	13	0.63	16	1	0.94
	no-snow	1	12	0.92	1	38	0.97	2	6	0.75
Producer's accuracy		0.86	0.86		0.96	0.75		0.89	0.86	
Overall accuracy		0.86			0.81			0.88		

		Ground observation								
		Portillo Station			Laguna Negra Station			Volcan Chillan Station		
		snow	no-snow	User's accuracy	snow	no-snow	User's accuracy	snow	no-snow	User's accuracy
MOD10A2	Snow	43	7	0.86	45	0	1.00	11	1	0.92
	no-snow	2	64	0.97	21	54	0.72	0	18	1.00
Producer's accuracy		0.96	0.90		0.68	1.00		1.00	0.95	
Overall accuracy		0.92			0.93			0.97		

precipitation have a higher SCA. Figure 6 displays the relation between SCA and annual precipitation for the different watersheds; it is clear that only at Aconcagua is there a good adjustment to the regression line ($R^2 = 0.7$) with a positive slope. In Rapel and Maule we can see a positive relation but with R^2 smaller than 0.5, which indicates a deficient adjustment. At Biobío we can graphically see a negative slope, i.e. when there is more precipitation there is less SCA, but statistically we have no adequate adjustment.

Figure 7 shows the variability of spatial distribution of maximum yearly SCA for the five watersheds. Years with maximum and minimum SCA of yearly maximum SCA are not coincident between watersheds. A considerable spatial variability of maximum SCA can be appreciated. There are no trends in the data.

Figure 8 shows annual and seasonal trends of SCA at the Aconcagua, Rapel, Maule, Biobío and Toltén watersheds. Trend analysis of the SCA series was performed with the non-parametric Mann–Kendall test. Decreasing trends in annual mean SCA are observed (p-value < 0.01) for the Aconcagua and Rapel watersheds, with a decreasing slope of $30 \, \mathrm{km^2 \, yr^{-1}}$ (Fig. 8a, b). No significant annual trend was observed for the other three watersheds. In autumn (Fig. 8a), only the Aconcagua watershed shows a decreasing trend of mean SCA variation at a level of significance ≤ 0.05, with a decreasing slope of $54 \, \mathrm{km^2 \, yr^{-1}}$. There is no significant variation in SCA in the winter (Fig. 8). In spring (Fig. 8b, c) the Rapel and Maule watersheds have a decreasing trend at a level of significance ≤ 0.05, with a decreasing slope of 38 and $79 \, \mathrm{km^2 \, yr^{-1}}$, respectively. In summer (Fig. 8c), the Rapel watershed has a decreasing trend at a level of significance ≤ 0.05, with a slope of $24 \, \mathrm{km^2 \, yr^{-1}}$; the Aconcagua and Maule watersheds show some decreasing trends with

level of significance ≤ 0.1 and slopes of 10 and $11 \, \mathrm{km^2 \, yr^{-1}}$ (Fig. 8a, c). It is important to remark that the Rapel watershed has large glacier areas. All the above-mentioned results are coincident with the outcomes given by the Pettitt homogeneity test, which shows that time series are not homogenous between two given times (p-value < 0.01). The change in the average in most of the cases is observed between 2008 and 2011.

Considering snow accumulation and melt seasons, data were grouped in two periods, i.e. autumn–winter (accumulation) and spring–summer (melt). For these two periods, a trend analysis was also done. Results indicated that there is a negative trend in average SCA during both seasons; the Aconcagua watershed has a decreasing slope (p-value < 0.01) of $29 \, \mathrm{km^2 \, yr^{-1}}$ during the snow accumulation season and the Maule watershed has a decreasing slope (p-value < 0.05) of $26 \, \mathrm{km^2 \, yr^{-1}}$ during the melt season. In the case of yearly maximum SCA, we found a significant positive trend for the Biobío watershed (p-value < 0.01); the other watersheds show a non-significant linear trend.

4   Discussion

In general terms, we can say that MOD10A2 has good agreement with ground observations, with always over 80 % precision when the sky is clear. Our results are a bit lower than the 90 % obtained in studies of the Northern Hemisphere (Zhou et al., 2005; Liang et al., 2008a; Wang et al., 2008; Huang et al., 2011; Zheng et al., 2017); this difference can be attributed to the complex topography of the Chilean Andes. Greater disagreements between MOD10A2 and ground observations were found at Termas Malleco and Laguna Negra stations. These stations are located close to the snowline where most

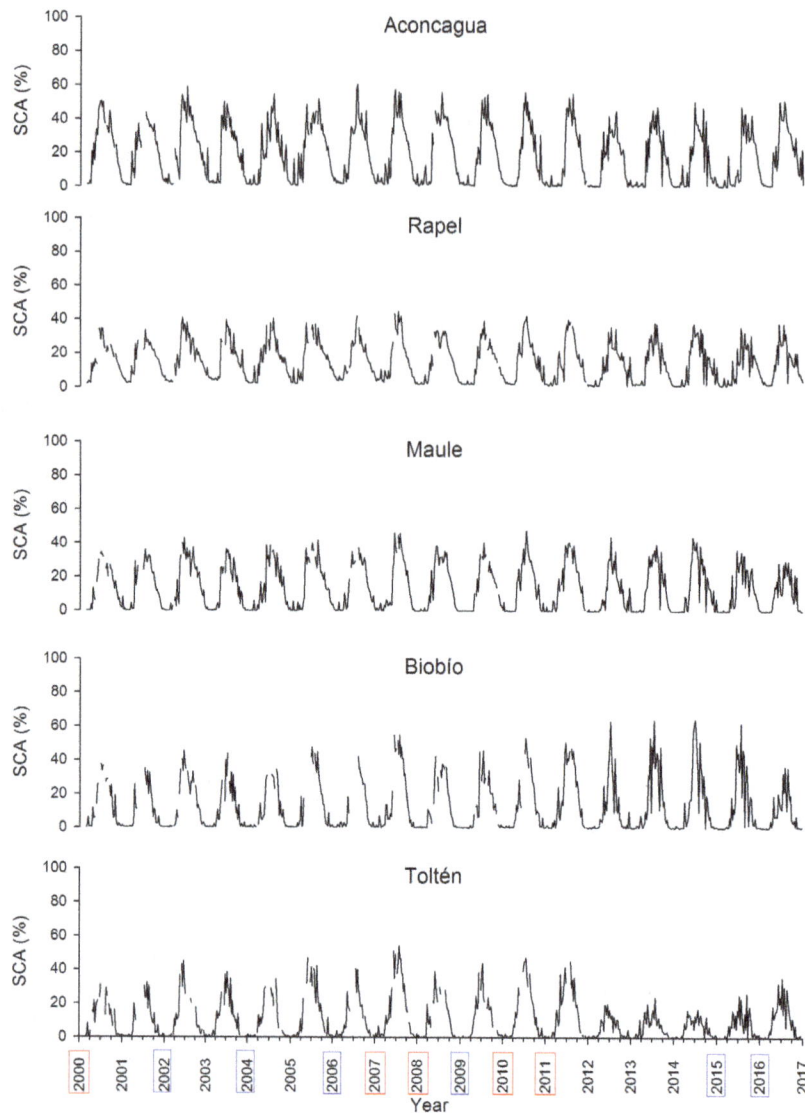

Figure 4. SCD for the 2000–2016 period at Aconcagua, Rapel, Maule, Biobío and Toltén watersheds. Red boxes indicate Niña years, blue ones Niño years.

of the time the soil is not covered by snow or is covered by snow for small periods of time; in this case, the presence of small snow patches is expected to be highly variable in time; it can even change at the subdaily scale and therefore it could reduce the capacity of images to capture snow coverage in the proximity of the snowline. Regarding SCA, we can see in the results presented in Figs. 4 and 5 that SCA has a similar behaviour in the different watersheds during the study period, i.e. all watersheds follow the same snow accumulation and melt dynamics and the maximum and minimum % of SCA are also coincident in time. SCA has a strong inter-annual variability, which is in agreement with the results obtained by other studies (Masiokas et al., 2006; Marchane et al., 2015; Tang et al., 2013; Zhang et al., 2012). Comparing the result with those obtained by Karici et al. (2016) in Slovak water-

sheds, we can appreciate similar results for the Aconcagua, Rapel and Maule watersheds and opposite ones for Biobío and Toltén, i.e. for the 2001–2006 period a decrease in mean SCA with respect to the mean of the whole period, and more SCA for the period 2007–2012.

Comparing our results of SCA with those of SWE obtained by Cornwell et al. (2016), we can appreciate that they are in agreement only with respect to the years 2002 and 2005 displaying the largest SCA of the entire study period. The other results are in dissension, as we have more years with above mean SCA during the beginning of the study period, and below throughout the end of the period (2012–2016).

In the case of trend analysis only the watersheds located in the northern part of the study area have a significant de-

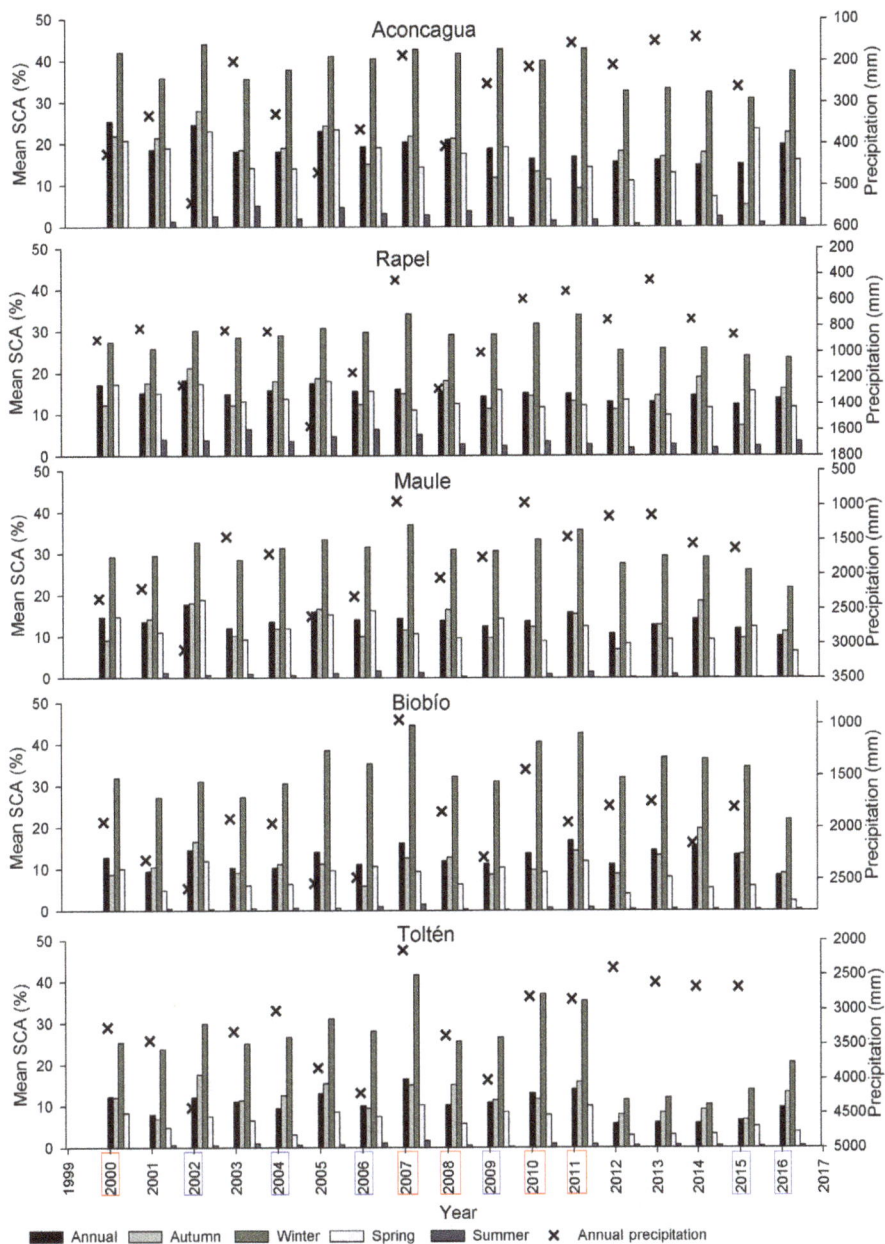

Figure 5. Mean seasonal and annual SCA (%) in Aconcagua, Rapel, Maule, Biobío and Toltén watersheds during 2000–2016. Red boxes indicate Niña years, blue ones Niño years.

creasing trend during accumulation and melt season, which is in agreement with results from different studies in Asia and North America (Tang et al., 2013; Maskey et al., 2011; Gurung et al., 2017; Fassnacht and Hultstrand, 2015; Kunkel et al., 2016). During wintertime we could not see any significant trend, which is consistent with results from Dietz et al. (2012).

The SCA magnitude for each watershed was contrasted with historical data from El Niño/La Niña–Southern Oscilla-

tion (ENSO) episodes from 2000 to 2016[1]. Normal or neutral years correspond to historical conditions, as Niño and Niña years correspond to wet and dry years, respectively, in the case of central and southern Chile. During 2000 and 2016, five episodes of El Niño (2002, 2004–2005, 2006–2007, 2009 and 2015–2016) and three episodes of La Niña (1999–2000, 2007–2008 and 2010–2011) occurred. Niño and Niña episodes were defined using the Oceanic Niño Index (ONI).

---

[1] http://www.cpc.noaa.gov/products/analysis_monitoring/ensostuff/ensoyears.shtml

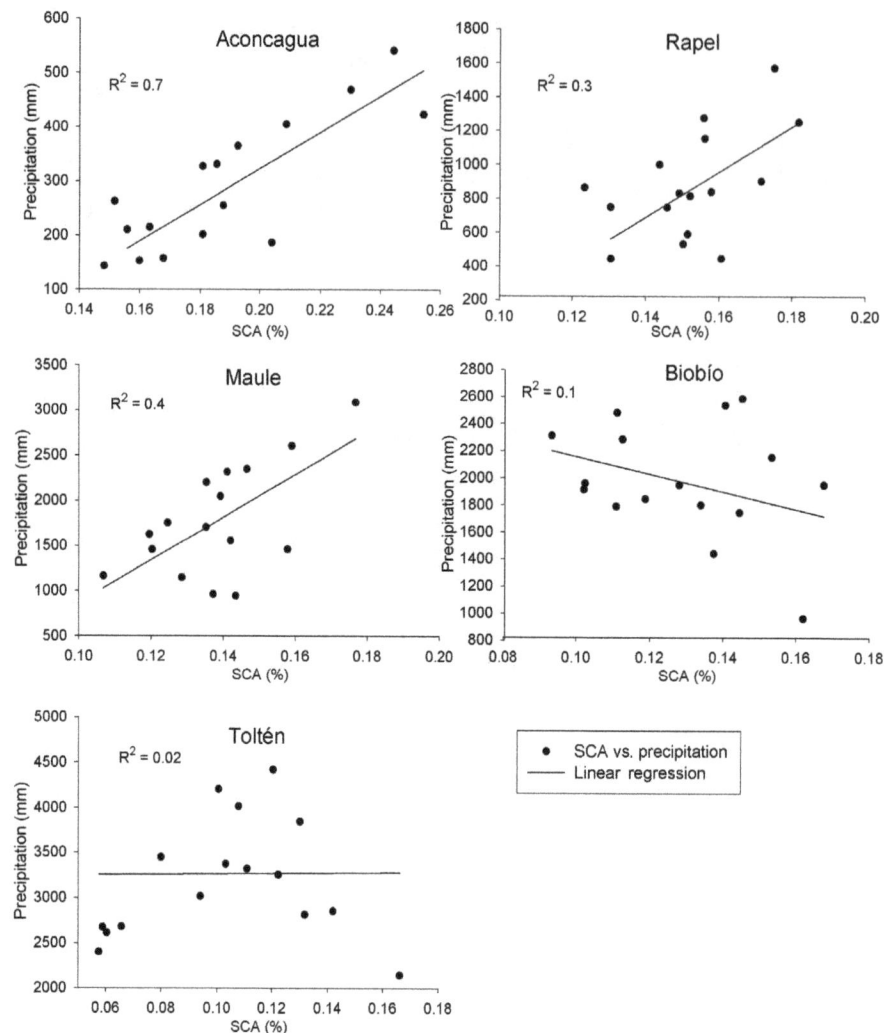

Figure 6. Relation between % of mean annual SCA and annual precipitation at Aconcagua, Rapel, Maule, Biobío and Toltén watersheds.

Niña years are coincident with maximum SCA in wintertime in all watersheds, with similar amounts of SCA between different Niña episodes. The 2007 Niña event is coincident with the highest % of SCA during winter for the study period (2000–2016) in all watersheds. Winter SCAs for Niña and Niño episodes are all above or just at the mean SCA for the 14 years under study. Years under the mean of the SCA area are all not coincident with Niña and Niño episodes, i.e. normal years are all under the mean SCA value. There is also no linear relation between the amount of precipitation and SCA, with the exception of the Aconcagua watershed (Fig. 6). This study only analysed snow coverage and not snow depth, i.e. volume of snow, which implies that a bigger SCA is not necessarily related to more water stored as snow. Considering the existing data availability in time and space of the DGA regarding snow depth, it was not possible to carry out an analysis of changes in snow depth in this study. During the course of the last 5 years, an important decline in SCA

was perceived in the entire watershed, with the exception of the Biobío watershed, where no difference can be seen until the last year (2016). This decline is bigger at the Toltén watershed, where the average for the last 5 years is 50 % of the mean for the entire period. In the analysis of the precipitation data we can observe that in all watersheds there was a rainfall deficit during 2012–2016: Aconcagua 26 %, Rapel 29 %, Maule 24 %, Biobío 19 %, and Toltén 27 % (DGA, 2012, 2013, 2014, 2015, 2016; DMC, 2013, 2014).

Masiokas et al. (2006) studied snowpack variation between 1951 and 2005 in Chile between latitude 30 and 37° S. They found that snow accumulation is positively related to El Niño, but they could not find a clear relationship with La Niña. The correlations that they found between snowfall and annual amount of rainfall in central Chile fit well with the known positive correlation of precipitation with El Niño during the Southern Hemisphere winter (June–August) (Waylen and Caviedes, 1990; Montecinos and Aceituno, 2003). It

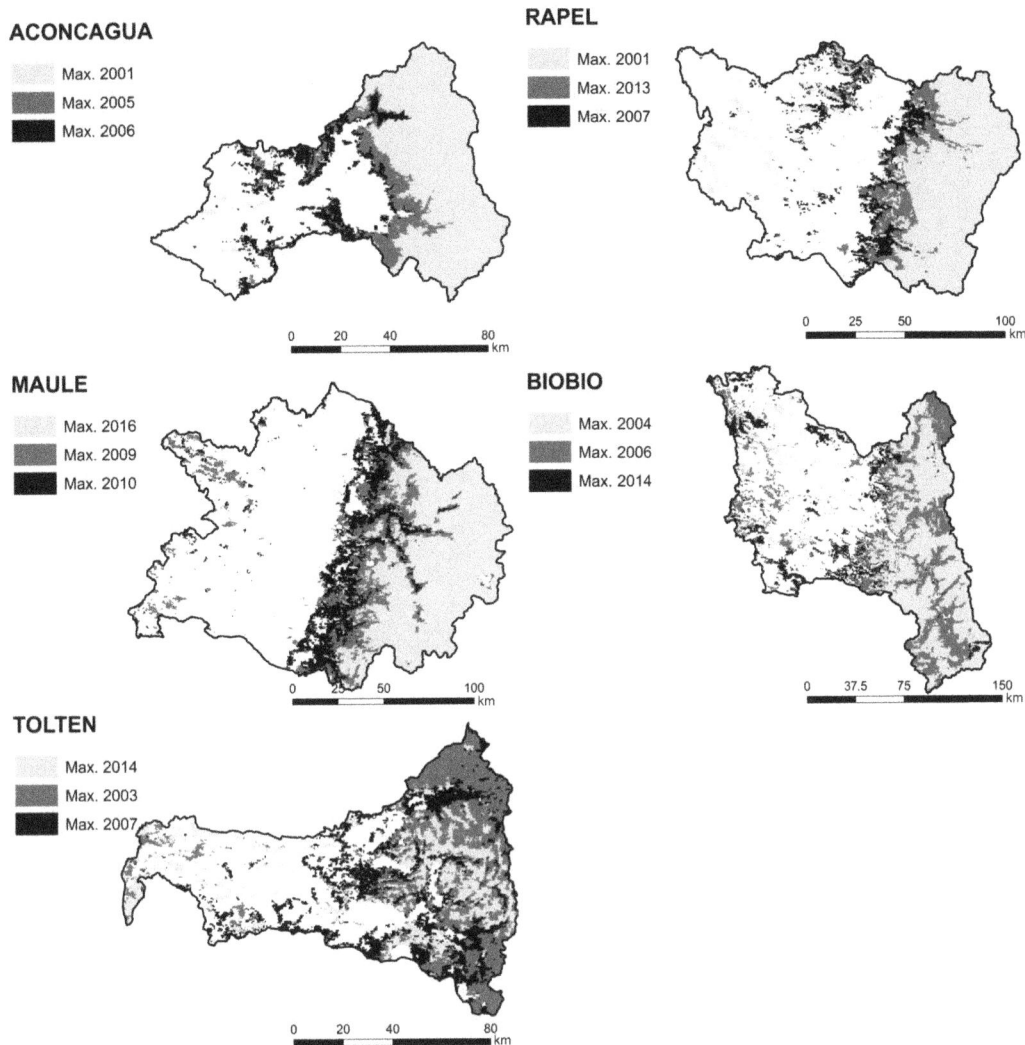

Figure 7. Spatial representation of SCA for the period 2000–2016 in Aconcagua, Rapel, Maule, Biobío and Toltén watersheds. Black indicates the year with maximum SCA, dark grey the year with average SCA and grey the year with minimum SCA.

came to their attention that, in contrast to the decreasing trend in snowpack observed across western North America, in the region they studied the average regional maximum value of snow water equivalent shows a positive to non-significant trend. In our case, we have only analysed SCA, obtaining a decreasing trend in the watersheds located in the northern part (32.0–36° S) and, to the south, a non-significant trend (35–39.5° S). These results are in agreement with the ones obtained by Falvey and Garreaud (2009), who noticed that, in central and northern Chile (17–37° S), in situ temperature observations confirm warming in the central valley and western Andes (+0.25 °C decade^{-1}). In southern Chile (38–48° S) temperature trends over land are weak (insignificant at 90 % confidence).

In the watersheds located south of 35° S we observed no relation between SCA and annual precipitation, which is in agreement with the results of Cortés et al. (2011), who found

that all watersheds they studied that were located north of 35° S had a high and significant correlation with ONI, and that precipitation during El Niño and La Niña episodes seems to be the most important factor controlling the water year hydrograph centre of timing, but with no influence in the southern ones.

5 Conclusions

The first validation of MODIS snow product MOD10A2 for estimation of snow covered areas (SCAs) via remote sensing in watersheds located in the Southern Hemisphere was presented. Ground observations of SCA were conducted during the years 2010 and 2011 at six study sites including six meteorological stations, 124 1-day single-observation points, and 11 snow courses. The SCA was determined for 636 days from MODIS snow products and compared with the SCA

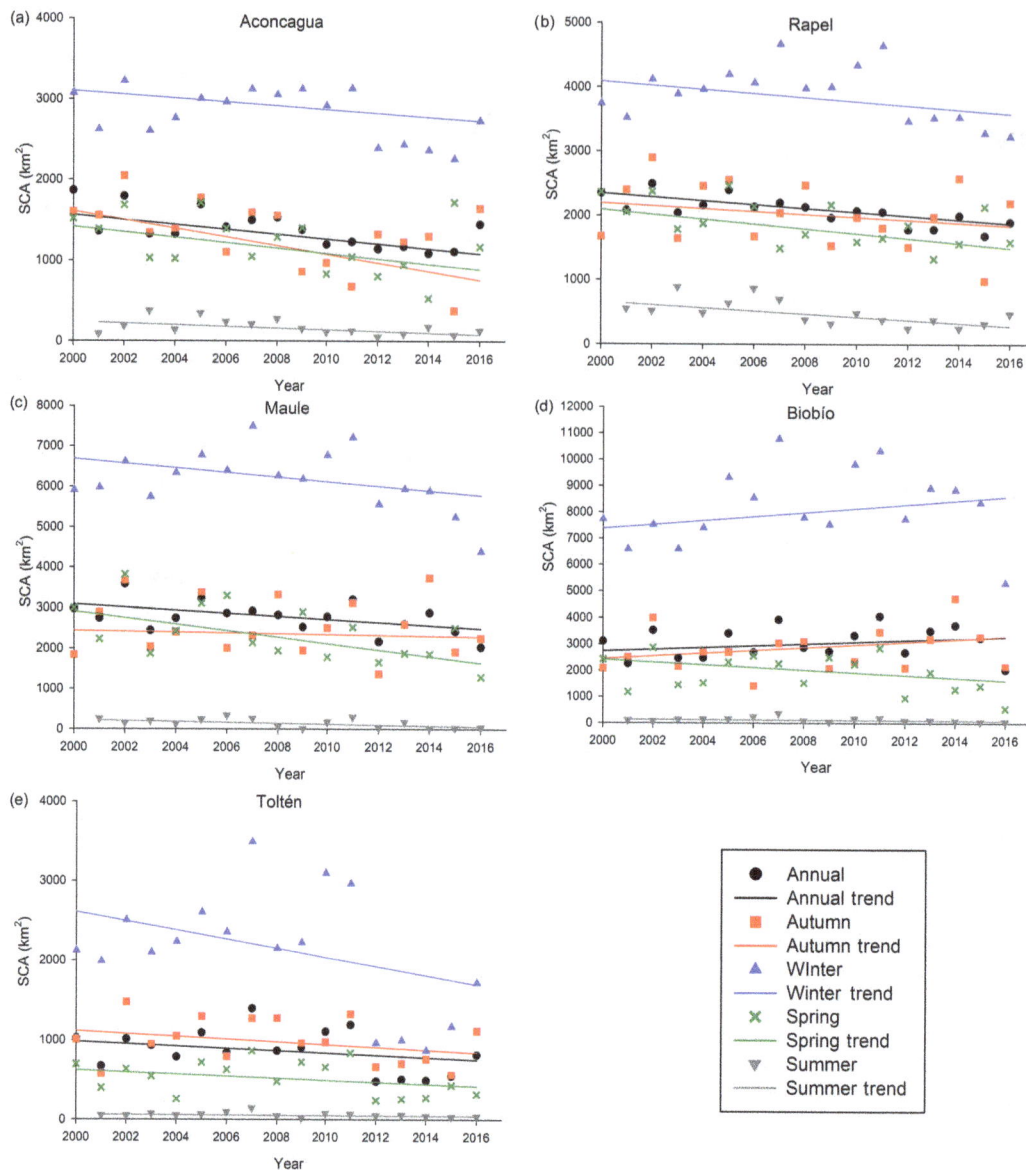

Figure 8. Annual and seasonal trends of SCA at Aconcagua, Rapel, Maule, Biobío and Toltén watersheds.

measured in situ. The SCA estimated from MOD10A2 presented an overall agreement from 81 to 98 % with SCA determined from ground observations, showing that the MODIS snow product can be taken as a feasible remote sensing tool for SCA estimation in southern–central Chile.

On the other side, we have analysed SCA trends for the period 2000–2016 in five of the most important watersheds of southern–central Chile in the context of water use and inhabitants, covering a longitudinal gradient from 32°14′ to 39°38′ S, which implies different climates regarding precipitation, drier in the northern part than in the southern part. Results indicate that all watersheds have the same SCD with more snow during wintertime, which decreases during spring and summer. Furthermore, a significant negative trend in an-

nual mean SCA is observed for the Aconcagua and Rapel watersheds, which can have implications for water availability for summertime. In general, we can see that there is no significant reduction in SCA for 2000–2016, with the exception of the Aconcagua watershed. From the data we can appreciate an important decline in SCA for the period of 2012 to 2016, which is coextensive with the rainfall deficit that occurred during the same years.

Results were compared with the ENSO episode during 2000–2016. From the latter comparison we can conclude that in the previously mentioned period, Niña years are coincident with maximum SCA at wintertime in all watersheds, with similar amounts of SCA between different Niña episodes.

In summary, the results presented in this work are highly relevant and can be used as one feasible approximation to obtain SCA, particularly in the Chilean Andes because of the lack of hydrological data such as discharge data, snow courses, and snow depths. These can become an outstanding tool to improve the analysis of important hydrological processes and the validation of water quantity prediction models.

Competing interests. The authors declare that they have no conflict of interest.

Acknowledgements. The present research was conducted in the framework of the FONDECYT 11100119 project.

Edited by: Stefan Uhlenbrook

References

Aulta, T. W., Czajkowski, K. P., Benko, T., Coss, J., Struble, J., Spongberg, A., Templin, M., and Gross, C.: Validation of the MODIS snow product and cloud mask using student and NWS cooperative station observations in the Lower Great Lakes Region, Remote Sens. Environ., 105, 341–353, 2006.

Ayala, A., McPhee, J., and Vargas, X.: Altitudinal gradients, midwinter melt, and wind effects on snow accumulation in semiarid midlatitude Andes under La Niña conditions, Water Resour. Res., 50, 3589–3594, 2014.

Azmat, M., Umar Liaqat, U. W., Muhammad Uzair Qamar, M. U., and Awan, U. K.: Impacts of changing climate and snow cover on the flow regime of Jhelum River, Western Himalayas, Reg. Environ. Change, 17, 813–825, 2017.

Barnett, T. P., Adam, J. C., and Lettenmaier, D. P.: Potential impacts of a warming climate on water availability in snow-dominated regions, Nature, 438, 303–309, 2005.

Boegh, E., Thorsen, M., Butts, M. B., Hansen, S., Christiansen, J. S., Abrahamsen, P., Hasager, C. B., Jensen, N. O., van der Keur, P., Refsgaard, J. C., Schelde, K., Soegaard, H., and Thomsen, A.: Incorporating remote sensing data in physically based distributed agro-hydrological modeling, J. Hydrol., 287, 279–299, 2004.

Campbell, J. B.: Introduction to remote sensing, 2nd Edn., London, Taylor and Francis, 1996.

Canters, F.: Evaluating the uncertainty of area estimates derived from fuzzy land-cover classification, Photogramm. Eng. Rem. S., 63, 403–414, 1997.

Congalton, R.: A review of assessing the accuracy of classifications of remotely sensed data, Remote Sens. Environ., 37, 35–46, 1991.

Congalton, R. and Mead, R.: A quantitative method to test for consistency and correctness in photointerpretation, Photogramm. Eng. Rem. S., 49, 69–74, 1983.

Cornwell, E., Molotch, N. P., and McPhee, J.: Spatio-temporal variability of snow water equivalent in the extra-tropical Andes Cordillera from distributed energy balance modeling and remotely sensed snow cover, Hydrol. Earth Syst. Sci., 20, 411–430, 2016.

Cortés, G., Vargas, X., and McPhee, J.: Climatic sensitivity of streamflow timing in the extratropical western Andes Cordillera, J. Hydro, 405, 93–109, 2011.

Demaria, E. M. C., Maurer, E. P., Thrasher, B., Vicuña, S., and Meza, F. J.: Climate change impacts on an alpine watershed in Chile: Do new model projections change the story?, J. Hydrol., 502, 128–138, 2013.

DGA: Manual de Cálculo de Crecidas y Caudales Mínimos en Cuencas Sin Información Fluviométrica, 1995.

DGA: Información pluviométrica, fluviométrica, estado de embalses y aguas subterráneas, ssd no. 6423872, Boletín no. 416 diciembre, 2012.

DGA: Información pluviométrica, fluviométrica, estado de embalses y aguas subterráneas, ssd no. 7429039, Boletín no. 428 diciembre, 2013.

DGA: Información pluviométrica, fluviométrica, estado de embalses y aguas subterráneas, ssd no. 8441530, Boletín no. 440 diciembre, 2014.

DGA: Información pluviométrica, fluviométrica, estado de embalses y aguas subterráneas, ssd no. 9490165, Boletín no. 452 diciembre, 2015.

DGA: Información pluviométrica, fluviométrica, estado de embalses y aguas subterráneas, ssd no. 10557841, Boletín no. 464 diciembre, 2016.

Dietz, A., Wohner, C., and Kuenzer, C.: European Snow Cover Characteristics between 2000 and 2011 Derived from Improved MODIS Daily Snow Cover Products, Remote Sens., 4, 2432–2454, https://doi.org/10.3390/rs4082432, 2012.

DMC: Dirección General de Aeronáutica Civil Dirección Meteorológica e Chile. Subdepartamento Climatología y Met. Aplicada, Anuario Climatológico 2012 Santiago–Chile 2013, 2013.

DMC: Dirección General de Aeronáutica Civil Dirección Meteorológica e Chile. Subdepartamento Climatología y Met. Aplicada, Anuario Climatológico 2013 Santiago–Chile 2014, 2014.

Er-Raki, S., Chehbouni, A., and Duchemin, B.: Combining Satellite Remote Sensing Data with the FAO-56 Dual Approach for Water Use Mapping In Irrigated Wheat Fields of a Semi-Arid Region, Remote Sens., 2, 375–387, 2010.

Escobar, F. and Aceituno, P.: Influencia del fenómeno ENSO sobre la precipitación nival en el sector andino de Chile central durante el invierno, Bull. Inst. Fr. Etudes Andines, 27, 753–759, http://documentos.dga.cl/MET1074.pdf, 1998.

Falvey, M. and Garreaud, R.: Regional cooling in a warming world: Recent temperature trends in the southeast Pacific and along the west coast of subtropical South America (1979–2006), J. Geophys. Res., 114, D04102, https://doi.org/10.1029/2008JD010519, 2009.

Fassnacht, S. R. and Hultstrand, M.: Snowpack variability and trends at long-term stations in northern Colorado, USA, Proceedings of the International Association of Hydrological Sciences, 371, 131–136, 2015.

Foody, G. M.: Status of Land Cover Classification Accuracy Assessment, Remote Sens. Environ., 80, 185–201, https://doi.org/10.1016/S0034-4257(01)00295-4, 2002.

Garreaud, R.: Estimación de la línea de nieve en cuencas de Chile Central, Rev. Soc. Chilena Ing. Hidráulica, Revista de la Sociedad Chilena de Ingeniería Hidráulica, 7, 21–32, 1992.

Garreaud, R.: Warm winter storms in Central Chile, J. Hydrometeorol., 14, 1515–1534, 2013.

Gilbert, R. O.: Statistical methods for environmental pollution monitoring, Van Nostrand Reinhold, New York, 1987.

Gurung, D. R., Maharjan, S. B., Shrestha, A. B., Shrestha, M. S., Bajracharya, S. R., and Murthy, M. S. R.: Climate and topographic controls on snow cover dynamics in the Hindu Kush Himalaya, Int. J. Climatol., 37, 3873–3992, https://doi.org/10.1002/joc.4961, 2017.

Hall, D. K. and Riggs, G. A.: Accuracy assessment of the MODIS snowcover products, Hydrol. Process., 21, 1534–1547, 2007.

Hall, D. K., Riggs. G. A., and Salomonson, V. V.: Development of methods for mapping global snow cover using bloderate Resolution Imaging Spectroradiometer (MODIS) data, Remote Sew. Environ., 54, 127–140, 1995.

Hall, D. K., Riggs, G. A, Salomonson, V. V., Di Girolamo, N. E., and Bayr, K. J.: MODIS snow-cover products, Remote Sens. Environ., 83, 181–194, 2002.

Hall, D. K., Riggs, G. A., and Salomonson, V.: Updated weekly.MODIS/Terra Snow Cover 8-day L3 Global 500 m Grid 15 V005, 24 February 2000–31 December 2016, Boulder, Colorado USA, National Snow and Ice Data Center, Digital media, 2016.

Harpold, A. A., Brooks, P. D., Rajogopal, S., Heidebuchel, I., Jardine, A., and Stielstra, C.: Changes in snowpack accumulation and ablation in the intermountain west, Water Resour. Res., 48, W11501, https://doi.org/10.1029/2012WR011949, 2012.

Huang, X. D., Liang, T. G., Zhang, X. T., and Guo, Z. G.: Validation of MODIS snow cover products using Landsat and ground measurements during the 2001–2005 snow seasons over northern Xinjiang, China, Int. J. Remote Sens., 32, 133–155, 2011.

Klein, A. G. and Barnett, A. C.: Validation of daily MODIS snow cover maps of the Upper Rio Grande River Basin for the 2000–2001 snow year, Remote Sens. Environ., 86, 162–176, 2003.

Kuchment, L. S., Romanov, P., Gelfan, A. N., and Demidov, V. N.: Use of satellite-derived data for characterization of snow cover and simulation of snowmelt runoff through a distributed physically based model of runoff generation, Hydrol. Earth Syst. Sci., 14, 339–350, https://doi.org/10.5194/hess-14-339-2010, 2010.

Kunkel, K., Robinson, D. A., Champion, S., Yin, X., Estilow, T., and Frankson, R.: Trends and Extremes in Northern Hemisphere Snow Characteristics, Curr. Clim. Change Rep., 2, 65–73, https://doi.org/10.1007/s40641-016-0036-8, 2016.

Krajcí, P., Holko, L., and Parajka, J.: Variability of snow line elevation, snow cover area and depletion in the main Slovak basins in winters 2001–2014, J. Hydrol. Hydromech., 64, 12–22, 2016.

Lee, S., Klein, A. G., and Over, T. M.: A Comparison of MODIS and NOHRSC Snow-Cover Products for Simulating Streamflow using the Snowmelt Runoff Model, Hydrol. Process., 19, 2951–2972, 2006.

Li, H.-Y. and Wang, J.: Simulation of snow distribution and melt under cloudy conditions in an Alpine watershed, Hydrol. Earth Syst. Sci., 15, 2195–2203, https://doi.org/10.5194/hess-15-2195-2011, 2011.

Liang, T., Huang, X., Wu, C., Liu, X., Li, W., Guo, Z., and Ren, J.: Application of MODIS data on snow cover monitoring in pastoral area: a case study in the Northern Xinjiang, China, Remote Sens. Environ., 112, 1514–1526, 2008a.

Marchane, A., Jarlan, L., Hanich, L., Boudhar, A., Gascoinb, S., Tavernier, A., Filali, N., Le Page, M., Hagolle, O., and Berjamy,

B.: Assessment of daily MODIS snow cover products to monitor snow cover dynamics over the Moroccan Atlas mountain range, Remote Sens. Environ., 160, 72–86, 2015.

Masiokas, M., Villalba, R., Luckman, B., Le Quesne, C., and Aravena, J. C.: Snowpack Variations in the Central Andes of Argentina and Chile, 1951–2005: Large-Scale Atmospheric Influences and Implications for Water Resources in the Region, J. Clim., 19, 6334–6352, 2006.

Maskey, S., Uhlenbrook, S., and Ojha, S.: An analysis of snow cover changes in the Himalayan region using MODIS snow products and in-situ temperature data, Climatic Change, 108, 391–40, 2011.

Melesse, A. M., Weng, Q. H., Thenkabail, P. S., and Senay, G. B.: Remote sensing sensors and applications in environmental resources mapping and modeling, Sensors, 7, 3209–3241, 2007.

Meza, F. J., Wilks, D. S., Gurovich, L., and Bambach, N.: Impacts of climate change on irrigated agriculture in the Maipo Basin, Chile: reliability of water rights and changes in the demand for irrigation, J. Water Res. Pl.-ASCE, 138, 421–430, 2012.

Milzowa, C., Kgotlhan, L., Kinzelbach, W., Meier, P., and Bauer-Gottwein, P.: The role of remote sensing in hydrological modelling of the Okavango Delta, Botswana, J. Environ. Manage., 90, 2252–2260, 2009.

Montecinos, A. and Aceituno, P.: Seasonality of the ENSO-related rainfall variability in central Chile and associated circulation anomalies, J. Clim., 16, 281–296, 2003.

Montzka, C., Canty, M., Kunkel, R., Menz, G., Vereecken, H., and Wendland, F.: Modelling the water balance of a mesoscale catchment basin using remotely sensed land cover data, J. Hydrol., 353, 322–334, 2008.

Pederson, G. T., Betancourt, J. L., and McCabe, G. J.: Regional patterns and proximal causes of the recent snowpack decline in the Rocky Mountains, US, Geophys. Res. Lett., 40, 1811–1816, 2013.

Peña, H. and Vidal, F.: Estimación Estadística de la Línea de Nieves durante los Eventos de Precipitación entre las latitudes 28 y 38 grados Sur, XI Congreso Chileno de Ingeniería Hidráulica, Concepción, Chile, 1993.

Ragettli, S., Cortés, G., McPhee, J., and Pellicciotti, F.: An evaluation of approaches for modelling hydrological processes in high-elevation, glacierized Andean watersheds, Hydrol. Process., 28, 5674–5695, 2013.

Riaño, D., Chuvieco, E., Salas, J., and Aguado, I.: Assessment of Different Topographic Corrections in Landsat-TM Data for Mapping Vegetation Types, IEEE T. Geosci. Remote, 41, 1056–1061, 2003.

Riggs, G. A., Hall, D. K., and Salomonson, V. V.: MODIS Snow Products User Guide to Collection 5, Online article, retrieved on 2 January 2007 at: http://modis-snow-ice.gsfc.nasa.gov/userguides.html, 2006.

Simic, A., Fernandes, R., Brown, R., Romanov, P., and Park, W.: Validation of Vegetation, MODIS, and GOES Plus SSM/I Snow-Cover Products Over Canada Based on Surface Snow Depth Observations, Hydrol. Process., 18, 1089–1104, 2004.

Snehmani, J. K. D., Kochhar, I., Hari Ram, R. P., and Ganju, A.: Analysis of snow cover and climatic variability in Bhaga basin located in western Himalaya, Geocarto Inter., 31, 1094–1107, 2016.

Stehr, A., Debels, P., Arumi, J. L., Romero, F., and Alcayaga, H.: Combining the Soil and Water Assessment Tool (SWAT) and MODIS imagery to estimate monthly flows in a data-scarce Chilean Andean basin, Hydrolog. Sci. J., 54, 1053–1067, 2009.

Stehr, A., Aguayo, M., Link, O., Parra, O., Romero, F., and Alcayaga, H.: Modelling the hydrologic response of a mesoscale Andean watershed to changes in land use patterns for environmental planning, Hydrol. Earth Syst. Sci., 14, 1963–1977, https://doi.org/10.5194/hess-14-1963-2010, 2010.

Story, M. and Congalton, R. G.: Accuracy assessment: a user's perspective, Photogramm. Eng. Rem. S., 52, 397–399, 1986.

Tang, Z., Wang, J., Li, H., and Yan, L.: Spatiotemporal changes of snow cover over the Tibetan plateau based on cloud-removed moderate resolution imaging spectroradiometer fractional snow cover product from 2001 to 2011, J. App. Remote Sens., 7, 1–14, 2013.

Tekeli, A. E., Akyürek, Z., Sorman, A. A., Sensoy, A., and Sorman, A. Ü.: Using MODIS snow cover maps in modeling snowmelt runoff process in the eastern part of Turkey, Remote Sens. Environ., 97, 216–230, 2005.

Thomlinson, J. R., Bolstad, P. V., and Cohen, W. B.: Coordinating methodologies for scaling landcover classification from site-specific to global: steps toward validating global maps products, Remote Sens. Environ., 70, 16–28, 1999.

Valdés-Pineda, R., Pizarro, R., García-Chevesich, P., Valdés, J. B., Olivares, C., Vera, M., and Abarza, A.: Water governance in Chile: Availability, management and climate change, J. Hydrol., 519, 2538–2567, 2014.

Verbunt, M., Gurtz, J., Jasper, K., Lang, H., Warmerdam, P., and Zappa, M.: The hydrological role of snow and glaciers in alpine river basins and their distributed modeling, J. Hydrol., 282, 36–55, 2003.

Vicuña, S., Garreaud, R., and McPhee, J.: Climate change impacts on the hydrology of a snowmelt driven basin in semiarid Chile, Climatic Change, 105, 469–488, 2011.

Vicuña, S., McPhee, J., and Garreaud, R.: Agriculture Vulnerability to Climate Change in a Snowmelt Driven Basin in Semiarid Chile, J. Water Res. Pl.-ASCE, 138, 431–441, 2012.

Vicuña, S., Gironás, J., Meza, F. J., Cruzat, M. L., Jelinek, M., Bustos, E., Poblete, D., and Bambach, N.: Exploring possible connections between hydrological extreme events and climate change in central south Chile, Hydrolog. Sci. J., 58, 1598–1619, 2013.

Wang, W., Huang, X., Deng, J., Xie, H., and Liang, T.: Spatio-Temporal Change of Snow Cover and Its Response to Climate over the Tibetan Plateau Based on an Improved Daily Cloud-Free Snow Cover Product, Remote Sens., 7, 169–194, 2015.

Wang, X., Xie, H., and Liang, T.: Evaluation of MODIS Snow Cover and Cloud Mask and its Application in Northern Xinjiang, China, Remote Sens. Environ., 112, 1497–1513, 2008.

Waylen, P. R. and Caviedes, C. N.: Annual and seasonal fluctuations of precipitation and streamflow in the Aconcagua River basin, J. Hydrol., 120, 79–102, 1990.

Zeinivand, H. and De Smedt, F.: Simulation of snow covers area by a physical based model, World Academy of Science, Engineering and Technology, 55, 469–474, 2009.

Zhang, G., Xie, H., Yao, T., Liang, T., and Kang S.: Snow cover dynamics of four lake basins over Tibetan Plateau using time series MODIS data (2001–2010), Water Resour. Res., 48, W10529, https://doi.org/10.1029/2012WR011971, 2012.

Zheng, W., Du, J., Zhou, X., Song, M., Bian, G., Xie, S., and Feng, X.: Vertical distribution of snow cover and its relation to temperature over the Manasi River Basin of Tianshan Mountains, Northwest China, J. Geogr. Sci., 27, 403–419, https://doi.org/10.1007/s11442-017-1384-6, 2017.

Zhou, X., Xie, H., and Hendrickx, M. H. J.: Statistical evaluation of remotely sensed snow-cover products with constraints from streamflow and SNOTEL measurements, Remote Sens. Environ., 94, 214–231, 2005.

10

The effect of GCM biases on global runoff simulations of a land surface model

Lamprini V. Papadimitriou[1], **Aristeidis G. Koutroulis**[1], **Manolis G. Grillakis**[1], **and Ioannis K. Tsanis**[1,2]

[1]Technical University of Crete, School of Environmental Engineering, Chania, Greece
[2]McMaster University, Department of Civil Engineering, Hamilton, ON, Canada

Correspondence to: Ioannis K. Tsanis (tsanis@hydromech.gr)

Abstract. Global climate model (GCM) outputs feature systematic biases that render them unsuitable for direct use by impact models, especially for hydrological studies. To deal with this issue, many bias correction techniques have been developed to adjust the modelled variables against observations, focusing mainly on precipitation and temperature. However, most state-of-the-art hydrological models require more forcing variables, in addition to precipitation and temperature, such as radiation, humidity, air pressure, and wind speed. The biases in these additional variables can hinder hydrological simulations, but the effect of the bias of each variable is unexplored. Here we examine the effect of GCM biases on historical runoff simulations for each forcing variable individually, using the JULES land surface model set up at the global scale. Based on the quantified effect, we assess which variables should be included in bias correction procedures. To this end, a partial correction bias assessment experiment is conducted, to test the effect of the biases of six climate variables from a set of three GCMs. The effect of the bias of each climate variable individually is quantified by comparing the changes in simulated runoff that correspond to the bias of each tested variable. A methodology for the classification of the effect of biases in four effect categories (ECs), based on the magnitude and sensitivity of runoff changes, is developed and applied. Our results show that, while globally the largest changes in modelled runoff are caused by precipitation and temperature biases, there are regions where runoff is substantially affected by and/or more sensitive to radiation and humidity. Global maps of bias ECs reveal the regions mostly affected by the bias of each variable. Based on our findings, for global-scale applications, bias correction of radiation and humidity, in addition to that of precipitation
and temperature, is advised. Finer spatial-scale information is also provided, to suggest bias correction of variables beyond precipitation and temperature for regional studies.

1 Introduction

In recent years, there has been a strong consensus on the changes in climate caused by increased concentrations of anthropogenic greenhouse gas emissions (King et al., 2015; O'Neill et al., 2017; Stocker et al., 2013). Under the pressing circumstances of a warming world, scientific research has focused on estimating the range of changes in the future climate and the effectiveness of different adaptation strategies. The main tool for the investigation of future climate is the utilization of global climate models (GCMs). GCMs are based on physical principles that describe the components of the climate system, such as cloud formation and water and energy flux exchanges.

Although each generation of GCMs shows improvements compared to its predecessor (Koutroulis et al., 2016), climate model outputs still contain substantial biases that are expressed as deviations of the modelled climate variables from respective historical observations. These inherent biases can emanate from misrepresentations of physical atmospheric processes (Maraun, 2012), from uncertainties regarding the boundary and initial model conditions (Bromwich et al., 2013), and from the relatively coarse resolution employed by the GCMs (Katzav and Parker, 2015). As a result, outcomes of hydrological climate change impact studies have been reported to become unrealistic without a prior adjustment of climate forcing biases (Ehret et al., 2012; Hansen et al., 2006; Harding et al., 2014; Sharma et al., 2007). To overcome this limitation, various bias correction techniques have been developed to post-process climate model data to statistically match observations. Bias correction methods are calibrated based on a historical time period for which observations are available. The adjustment is then applied to both the modelled historical period and to the period beyond the time frame of the observations.

Bias correction procedures have mainly focused on adjusting the biases of precipitation and/or temperature (Christensen et al., 2008; Li et al., 2010; Miao et al., 2016; Photiadou et al., 2016; Piani et al., 2010). These variables have traditionally been prioritized for bias correction as they are considered the most important driving variables of hydrological processes in modelling applications – even though from a physical perspective radiation is the driving force of the hydrological cycle. However, many state-of-the-art regional and global hydrological models (GHMs) and land surface models (LSMs) require – apart from precipitation and temperature – additional meteorological forcing, such as solar radiation, air humidity, surface air pressure, and wind speed (a summary of the input variables needed by various hydrological models can be found in the Supplement of Hattermann et al., 2017). For this reason, biases in variables like radiation, humidity, and wind speed can hinder the representation of hydrological fluxes such as runoff, evapotranspiration (ET), snow accumulation, and snowmelt by the impact models (Hagemann et al., 2011; Haddeland et al., 2012), indicating that bias correction should be extended to include more input variables.

Bias correction itself also has limitations, as it is a demanding process in terms of both computational cost and the involved methodological development. Moreover, the use of bias correction is challenged by conceptual pitfalls such as the disruption of the physical consistency of climate variables, the mass–energy balance and the omission of correction feedback mechanisms to other climate variables (Ehret et al., 2012). For these reasons, it is worth examining whether the effect of biases of input variables on hydrological outputs justifies the use of bias correction. Even though this information would be key for making informed decisions on the variables that should be bias corrected for a specific model application, few relevant studies can be found in the literature. Some insight is given by Haddeland et al. (2012), who investigate the combined effect of bias correcting radiation, humidity, and wind speed in addition to precipitation and temperature on hydrological simulations. However, the extent to which individual forcing variable biases affect hydrological simulations and the way that this effect varies spatially are important research questions that remain open.

Here we investigate the effect of the biases in GCM climate variables on the historical runoff output of a large-scale LSM. To this end, we firstly quantify the improvements in the representation of historical modelled runoff when bias corrected variables are used as forcing. Secondly, we examine the individual effect that the bias of each climate variable can have on runoff simulations. This way we can provide an assessment of the variables beyond precipitation and temperature that may be considered "priority" variables for bias correction, due to their possible pronounced effect on hydrological simulations.

2 Methods

2.1 The JULES land surface model

Hydrological simulations were performed with the Joint UK Land Environment Simulator (JULES) model (Best et al., 2011). JULES is a physically based model that calculates water, energy, and carbon exchanges between the land surface and the atmosphere. The science modules that comprise the model are surface energy fluxes, snow cover and surface hydrology, soil moisture and temperature, soil carbon, vegetation dynamics, and plant physiology. The model requires seven climate variables as forcing, namely, precipitation, temperature, longwave and shortwave radiation, specific humidity, surface pressure, and wind speed. Runoff production in JULES has two components. The first one is surface runoff, produced by the infiltration excess mechanism. The second one is subsurface runoff (or drainage from the bottom of the soil column), which is calculated as a Darcian flux under the assumption of zero gradient of matric potential. Calculation of potential evaporation follows the Penman–Monteith approach (Monteith, 1965). Water held at the plant canopy evaporates at the potential rate, while restrictions of canopy resistance and soil moisture are applied for the simulation of evaporation from soil and plant transpiration from potential evaporation (Best et al., 2011). For a detailed description of JULES, the reader can refer to the model description papers of Best et al. (2011) and Clark et al. (2011). Examples of recent model applications to climate change impact assessments can be found in the studies of Papadimitriou et al. (2016), where JULES is used to investigate future water availability in Europe, and Grillakis et al. (2016), who estimated the climate-induced changes in soil temperature regimes.

2.2 Model set-up and outputs

JULES was run at the global scale, with a spatial resolution of 0.5°. A daily time step was employed for all the model runs. To warm up the model, 10 spin-up cycles from 1973 to 1978 were performed before each main run. The main runs span from 1978 to 2010, but only the time period of 1981 to 2010 is used for the analysis. The model outputs are produced with a daily time resolution.

2.3   Hydrological evaluation

This study focuses on the runoff production output of JULES, hereafter denoted RF. For the assessment of model performance, RF is aggregated at the basin level to allow for comparison with discharge observations. To this end, RF is converted to discharge at the basin outlet (denoted Q) through a delay algorithm proposed by Zulkafli et al. (2013) and the use of the TRIP river routing scheme (Oki and Sud, 1998) to determine the grid boxes upstream of the basin's outlet.

For the evaluation of JULES' hydrological performance, three metrics are used: Nash–Sutcliffe efficiency (NSE), percent bias (PBIAS), and the coefficient of determination (R^2). The formulas for the calculation of NSE and PBIAS are given in Eqs. (1) and (2):

$$\text{NSE} = 1 - \left[\frac{\sum (Q_{\text{sim}} - Q_{\text{obs}})^2}{\sum (Q_{\text{obs}} - Q_{\text{mean}})^2} \right], \tag{1}$$

$$\text{PBIAS} = \left[\frac{\sum (Q_{\text{sim}} - Q_{\text{obs}}) \cdot 100}{\sum Q_{\text{obs}}} \right] \%, \tag{2}$$

where Q_{sim} is simulated discharge, Q_{obs} is observed discharge, and Q_{mean} is the mean of observed discharge data. Discharge observations were obtained from the Global Runoff Data Centre (GRDC) database for nine large-scale basins shown in Fig. 1. Information on the basin stations for model evaluation is presented in Table S1 in the Supplement of this paper.

The evaluation metrics are calculated from monthly discharge data. These are the monthly averages of daily discharge for simulations, while observations were obtained in monthly time steps. Model evaluation was based on the historical period from 1981 to 2010. The months missing from the observed discharge time series were neglected from the calculation of the evaluation metrics.

2.4   Climate data

The climate dataset used for bias correction of the GCM data and as a baseline for comparison of the results is the WATCH Forcing Data methodology applied to ERA-Interim data (WFDEI; Weedon et al., 2014). WFDEI data span from 1979 to 2012, but here only the time period from 1981 to 2010 was used. The WFDEI dataset is based on its predecessor WFD (WATCH Forcing Data; Weedon et al., 2010), which was derived from the ERA-40 reanalysis product (Uppala et al., 2005). For detailed information on the derivation of the WFDEI dataset, the reader is referred to Weedon et al. (2014).

Data from three GCMs participating in the fifth phase of the Coupled Model Intercomparison Project (CMIP5; Taylor et al., 2012) were used as forcing. Information on the ensemble members can be found in Table 1. Climate model outputs were interpolated to the 0.5° spatial resolution of the WFDEI dataset, using the nearest-neighbour method.

2.5   Bias correction method

The bias correction methodology presented by Grillakis et al. (2013), namely multi-segment statistical bias correction (MSBC), is used to adjust the biases in precipitation. MSBC follows the principles of quantile mapping correction techniques and was originally designed and tested for GCM precipitation adjustment. According to the method, the cumulative distribution function (CDF) space is split into discrete segments and then the individual quantile mapping correction is applied to each segment, achieving a better fit of the parametric equations on the data and thus better correction, especially on the CDF edges. The optimal number of segments is estimated by the Schwarz Bayesian information criterion to balance between complexity and performance. A modification of the methodology is used for bias adjustment of the rest of the variables that were used. The modified methodology uses linear functions instead of the gamma functions that were used in the original methodology. This change allows for the facilitation of negative variable values that the gamma functions cannot simulate. Hence, the methodology becomes more universal, to be used in different variable types and distributions. An additional methodological change is performed to the highest and lowest segments' corrections, which are explicitly corrected using only the difference between the historical period model data and the observations. This provides rigidity to the correction, avoiding unrealistic temperature values at the edges of the corrected data CDF. A detailed description and technical details of the modification can be found in Grillakis et al. (2017). As MSBC methodology belongs to the parametric quantile mapping techniques, it shares their advantages and drawbacks. A comprehensive analysis of advantages and disadvantages of the methods that follow the quantile mapping compared to others can be found in Maraun et al. (2010) and Themeßl et al. (2012). The methodology has already been used in in the framework of the ECLISE FP7 (265240) and HELIX FP7 (603864) projects and in a number of climate change impact studies (Grillakis et al., 2016; Papadimitriou et al., 2016). In addition, MSBC has participated in the Bias Correction Intercomparison Project (BCIP) (Nikulin et al., 2015), where it was found to compare well to the other methodologies and was ranked high in performance.

As the bias adjustment involves only the reference period of the GCM data using the same period's observations, its effect is simply limited to the equalization of the cumulative density functions of the raw GCM data towards the WFDEI data. A number of parameter checks were performed on the corrected data, such as prevention of unrealistic values (e.g. negative values to positively constrained variables) and the avoidance of extreme values beyond or below the historical record of WFDEI. The correction was performed separately for each calendar month, keeping physical coherence of the bias adjusted variables, as they are adjusted for their season-

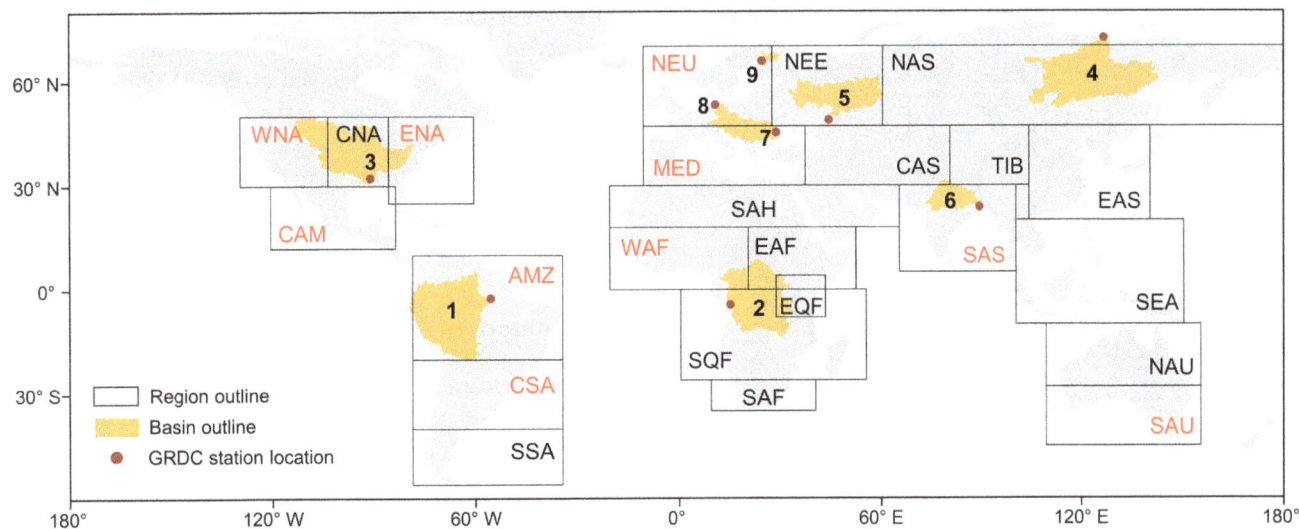

Figure 1. Outlines of study focus regions and hydrological basins and locations of the GRDC gauging stations. With red colour are denoted the regions selected for more detailed analysis. The hydrological basins have been numbered in decreasing order according to their area: (1) Amazon, (2) Congo, (3) Mississippi, (4) Lena, (5) Volga, (6) Ganges, (7) Danube, (8) Elbe, and (9) Kemijoki.

Table 1. Information on the GCMs used for this study.

Modelling group	Institute ID	Model name	°Lon × °Lat	Key reference
Institut Pierre-Simon Laplace	IPSL	IPSL-CM5A-LR	3.75 × 1.88	Dufresne et al. (2013)
Japan Agency for Marine-Earth Science and Technology, Atmosphere and Ocean Research Institute (The University of Tokyo), and the National Institute for Environmental Studies	MIROC	MIROC-ESM-CHEM	2.81 × 2.81	Watanabe et al. (2011)
US Dept. of Commerce/NOAA/Geophysical Fluid Dynamics Laboratory	GFDL-NOAA	GFDL-ESM2M	2.50 × 2.00	Dunne et al. (2012)

ality in a coherent way according to the observational dataset that is used.

2.6 Experimental design

In order to examine the effect of each forcing variable's bias on runoff we designed and implemented an experiment comprised of two parts (bias assessment and partial correction bias assessment) and nine sets of JULES' runs in total. A graphical description of the performed experiment is shown in Fig. 2. Climate data from three GCMs and the WFDEI dataset are used as JULES' forcing. The sets of runs forced with GCM data include three model runs – one per GCM. Then the analysis progresses using the ensemble mean. The time span of this analysis is the historical period 1981–2010. This is also the time span of the period used for bias correction of the GCM output.

2.7 Bias assessment

The first part of the experiment is to assess initial and remaining biases in the forcing data and in simulated runoff. Initial bias refers to the difference between raw GCM variables and the respective WFDEI variables. Remaining bias is the bias in the forcing variables after the bias correction, i.e. the difference between bias corrected GCM variables and the respective WFDEI variables. Referring to runoff, "initial" and "remaining" biases are defined as the difference between runoff simulations forced with raw and bias corrected forcing respectively from simulations forced with the WFDEI dataset. This definition is employed to shorten and simplify the expressions used in this paper (i.e. "initial bias in runoff" instead of "the difference between runoff forced with raw GCM data and WFDEI data"). In this part of the experiment, three sets of JULES' runs were conducted:

    i. forced with WFDEI (WFDEI);

    ii. forced with uncorrected climate data (raw); and

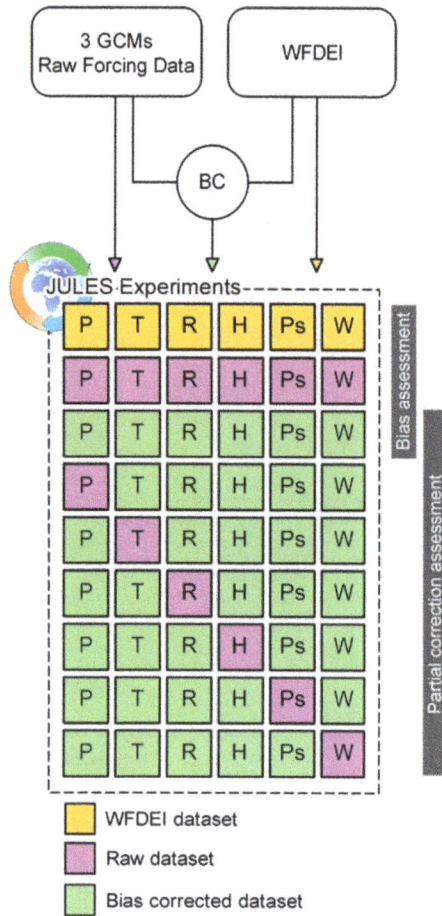

Figure 2. Graphical description of the performed experiment.

iii. forced with bias corrected climate data (BC).

2.8 Partial correction bias assessment

For the second part of the experiment – the partial correction bias assessment – six more sets of JULES' runs were performed. In each of these runs, one of the six forcing variables (precipitation, temperature, radiation, humidity, surface pressure, and wind speed) is used in its raw form, while the rest of the input forcing is bias corrected. The partial correction assessment runs are symbolized as NobcV (NOt Bias Corrected variable V), where V is one of the six forcing variables: precipitation (P), temperature (T), radiation (R), specific humidity (H), surface pressure (Ps), and wind (W). It has to be noted here that downward longwave radiation (Rl) and downward shortwave (Rs) were examined together; hence, in the respective NobcR run, both downward shortwave and downward longwave radiation were forced in uncorrected form. Partial correction assessment is composed as a tool to quantify the individual effect of each forcing variable on runoff, but is not designed to suggest and assess run formats.

The simulated runoff of each partially corrected input is compared to the respective simulation in which all input variables are bias corrected (denoted as BC). This comparison allows us to assess the "loss" of the performance of simulations when a variable is neglected from the bias correction procedure. It must be noted however that the "loss of performance" concept bears the assumption that the BC simulation is closer to the WFDEI simulation compared to a partially corrected set.

2.9 Categorization of individual variable bias effects

A new framework for the classification of the effects of forcing variables' biases on modelled runoff is developed and implemented. The classification employs the comparison of the bias in each forcing variable (ΔV) and the corresponding relative effect in simulated runoff (ΔRF), discretizing four different categories (Fig. 3). To facilitate the comparison among the different forcing variables, ΔV and ΔRF are expressed as percentages. More specifically, ΔV and ΔRF are defined as follows.

ΔV is the difference between the raw and bias corrected variable value, divided by the bias corrected variable value. ΔV is estimated by Eq. (3).

$$\Delta V = \frac{\text{raw variable} - \text{BC variable}}{\text{BC variable}} \cdot 100\% \qquad (3)$$

As an exception, for temperature ΔV refers to the absolute difference between raw and bias corrected temperature (in K).

ΔRF expresses the effect of a variable's bias on runoff and is calculated from the difference between runoff forced with all bias corrected variables except for the examined variable V (NobcV) and runoff forced with all bias corrected variables (BC), divided by the runoff of all bias corrected variables (BC). ΔRF is estimated by Eq. (4).

$$\Delta \text{RF} = \frac{\text{RF from NobcV} - \text{RF from BC}}{\text{RF from BC}} \cdot 100\% \qquad (4)$$

Sensitivity of runoff to changes in forcing variables (S) is the fraction of runoff change over the forcing variable change and serves as a measure to assess the relative magnitude of ΔRF compared to ΔV. When ΔRF is sensitive to ΔV, relatively smaller changes in the variable should cause relatively larger changes in runoff and vice versa. Sensitivity is in general dimensionless, but for temperature has units of K^{-1}. S is estimated by

$$S = \Delta \text{RF} / \Delta V. \qquad (5)$$

In total, there are six sets of ΔVs and six sets of ΔRFs, one for each examined variable and experiment respectively, and six sets of sensitivities (S). The absolute values of ΔV, ΔRF, and S denoted $|\Delta V|$, $|\Delta \text{RF}|$, and $|S|$ are used to avoid dealing with the sign of the changes and rather focus on their magnitude.

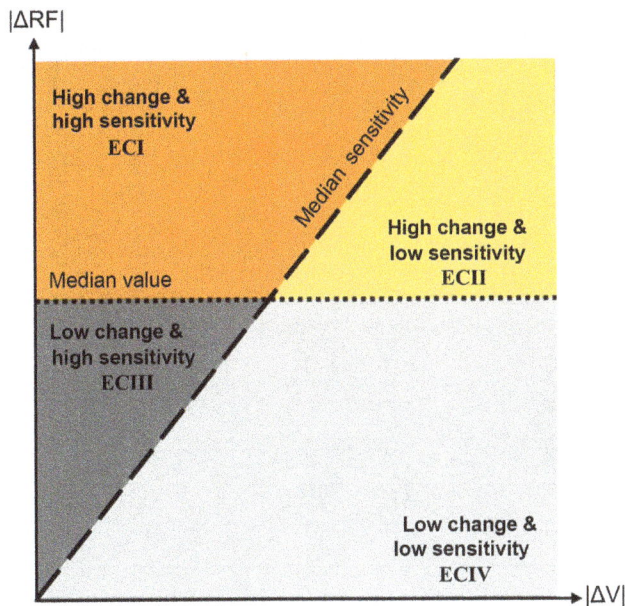

Figure 3. Categorization of the effect of changes in forcing variables (V) on runoff (RF). The four areas correspond to the four defined effect categories. The x axis corresponds to relative changes in forcing variables and the y axis to relative changes in runoff. For all changes, the absolute value is considered.

As shown in Fig. 3, the effect of each variable's bias ($|\Delta V|$) on runoff ($|\Delta RF|$) is separated into four different categories according to two rules. The first rule is the characterization of $|\Delta RF|$ among all the experiments as "low" or "high" relative to its median value, shaping the ordinate $y = \text{median}(|\Delta RF|)$. Median($|\Delta RF|$) is derived considering the $|\Delta RF|$ values of all land grid boxes and for all the experiments. The second rule is the characterization of sensitivity $|S|$ as high or low relative to its median value. The latter forms a bisectrix $s = \text{median}(|S|)$. Median($|S|$) is, accordingly to median($|\Delta RF|$), derived from the $|S|$ values of all grid boxes and for all the experiments apart from temperature. In the case of temperature, median($|S|$) is explicitly recalculated from the values of all the land grid boxes of this specific experiment. These two rules form the four categories of Fig. 3. Combinations of the two rules result in four different effect categories (ECs) presented in decreasing order of the effect of a variable's bias on runoff:

i. High change and high sensitivity (ECI);

ii. high change and low sensitivity (ECII);

iii. low change and high sensitivity (ECIII); and

iv. low change and low sensitivity (ECIV).

2.10 Regional-scale bias assessment

Regional focus is given in 24 regions and 9 hydrological basins. The regions were selected from the 26 regions pre-

Table 2. 24 regions of the globe, selected from Giorgi and Bi (2005).

Region name	Abbreviation
North Europe	NEU
Mediterranean Basin	MED
Northeast Europe	NEE
North Asia	NAS
Central Asia	CAS
Tibet	TIB
Eastern Asia	EAS
Southeast Asia	SEA
Northern Australia	NAU
Southern Australia	SAU
Sahara	SAH
Western Africa	WAF
Eastern Africa	EAF
East Equatorial Africa	EQF
South Equatorial Africa	SQF
Southern Africa	SAF
Western North America	WNA
Central North America	CNA
Eastern North America	ENA
Central America	CAM
Amazon	AMZ
Central South America	CSA
Southern South America	SSA
South Asia	SAS

sented in Giorgi and Bi (2005) (in our study Alaska and Greenland are excluded from the analysis). The hydrological basins were selected to cover different hydro-climatic regimes, in conjunction with GRDC data availability. The selected regions and basins are shown in Fig. 1. The abbreviations of the regions' names can be found in Table 2.

3 Results and discussion

3.1 Long-term annual biases in forcing variables at the global scale

Global maps of the initial and remaining annual biases of the forcing variables are shown in Fig. 4. Respective information on the seasonal biases is presented in Figs. S1 and S2 of the Supplement of this paper. In general terms the remaining annual biases are smaller than the initial ones by 1 to 2 orders of magnitude. For precipitation (Fig. 4a), the largest initial wet biases are observed for regions with high mountain ranges (the Andes in South America, the Alaska Range and the Rocky Mountains in North America, and the Himalayas in Asia) and for the tropical African and Indonesian regions. Only a very small percentage (0.75 %) of the land surface has small biases (-0.01 to $0.01\,\text{mm}\,\text{day}^{-1}$), while the largest biases (> 5 or $< -5\,\text{mm}\,\text{day}^{-1}$) occupy 31.18 % of the land surface. The remaining biases in precipitation are small (up

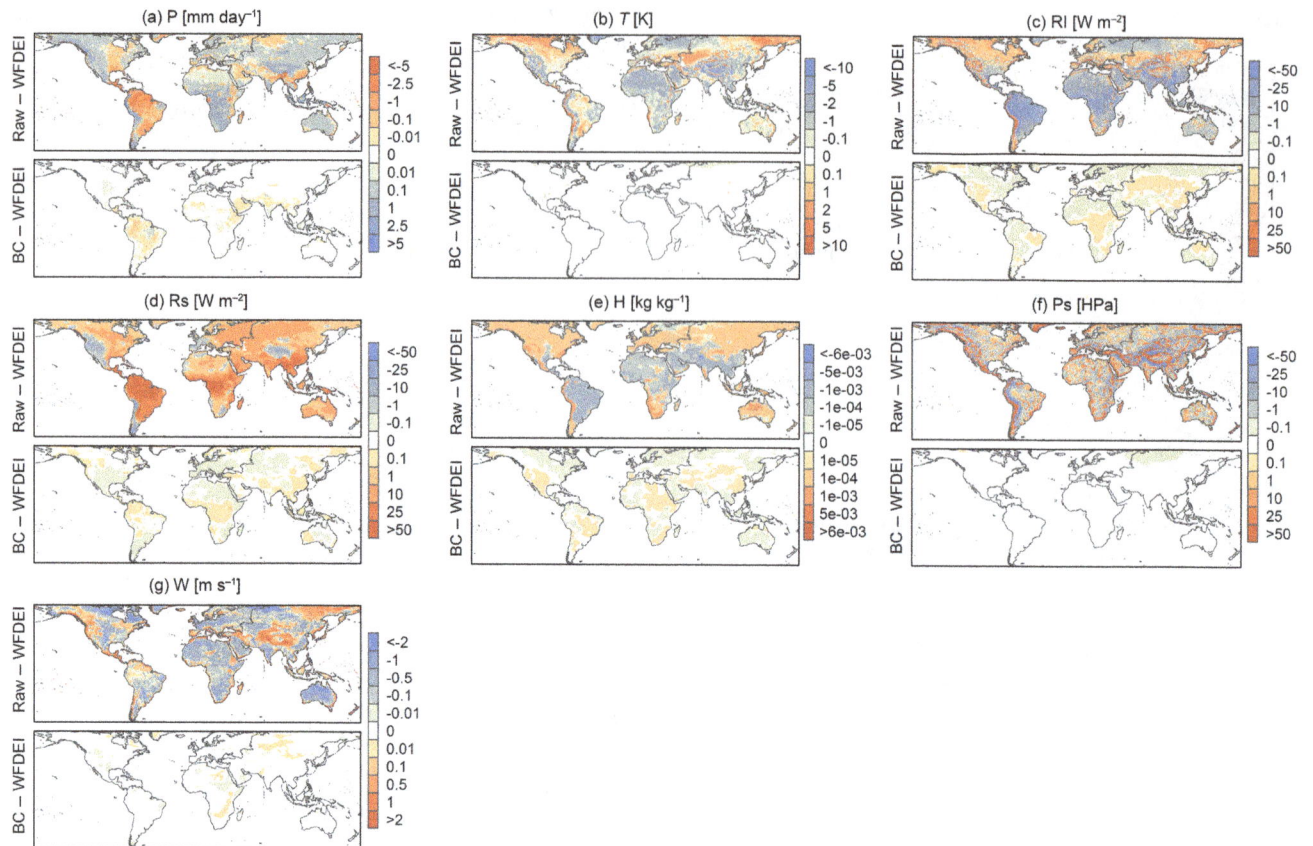

Figure 4. Difference maps, showing initial (raw-WFDEI) and remaining (BC-WFDEI) biases of the GCM ensemble forcing variables: **(a)** precipitation, **(b)** temperature, **(c)** longwave downward radiation, **(d)** shortwave downward radiation, **(e)** specific humidity, **(h)** surface pressure, and **(g)** wind. Differences are calculated between the long-term annual averages (ANN) of the 1981–2010 period.

to 0.01 mm day^{-1} in absolute terms, for 80.32 % of the land surface) and located in the tropics. The initial biases in temperature are cold biases for 57.82 % of the land surface, while warm biases (mainly found in the Alaskan, Greenland, and northern and central Asia regions, as well as in the Mediterranean and the Andes) occupy 42.12 % of the land surface (Fig. 4b). Initial biases greater than 2 K in absolute terms cover approximately one-third of the land surface (34.74 %). After bias adjustment, the remaining temperature bias is less than 0.1 K for the vast majority of the land surface (97.27 %).

The initial biases of longwave and shortwave radiation (Fig. 4c and d respectively) exhibit similar spatial variations but have different signs. Shortwave radiation shows a greater extent of large biases (> 50 W m^{-2} in absolute terms) compared to longwave radiation (8.16 % as opposed to 2.95 % of the land surface). Initial biases in specific humidity are greater than 10^{-3} kg kg^{-1} (1 g kg^{-1}), in absolute terms, for one-quarter of the land surface (23.65 %) (Fig. 4e). The largest biases in surface pressure (> 50 or < −50 HPa) occupy 10.01 % of the land surface and are found in the areas where high mountain ranges are located (Rocky Mountains, Andes, Himalayas) (Fig. 4f). The remaining bias in surface pressure is less than 0.1 HPa (in absolute terms) for most of the land

surface (96.50 %). For more than half of the land surface (55.79 %), the wind's initial biases are larger than 0.5 m s^{-1} or smaller than −0.5 m s^{-1} (Fig. 4g). The remaining biases of the wind variable range between −0.01 and 0.01 m s^{-1} for the majority of the land surface (87.71 %).

Generally, the initial GCM biases in precipitation and temperature are more pronounced over high mountainous regions and the tropics. Recent studies argue for a dependency between biases and altitude. According to the study of Haslinger et al. (2013), temperature and precipitation biases of a GCM tested over the Alpine region both show increasing trends with height. Regarding the tropics, various studies show increased GCM biases in these regions compared to model performance in other climate zones (Koutroulis et al., 2016; Randall et al., 2007; Solman et al., 2013). The initial surface pressure biases are also linked to altitude, as surface pressure heavily depends on elevation. Initial biases in surface pressure have a similar elevation pattern and could be a result of the different spatial resolutions of the elevation model in the GCMs and WFDEI. The WFDEI dataset resolution is 0.5°, while the original GCM spatial resolution is considerably lower (around 2.5°). GCM surface pressure is simulated taking into account a relatively low-resolution elevation model. Although GCM surface pressure is interpolated to the WFDEI resolution, this does not correct the elevation-induced error in the GCM simulations.

The remaining biases in precipitation in the tropical regions were also identified and discussed extensively by Grillakis et al. (2013) and are related to the error in the CDF approximation during bias correction. For the rest of the variables, the remaining bias, although not actually zero, is very close to zero (well below the smallest positive and above the smallest negative rank in the legend, e.g. below −0.1 K and below 0.1 K for temperature). The colour scale in Fig. 4 was selected with the intention of showing the remaining biases, but this does not mean that their values are accountable. They are rather trace errors occurring due to truncation numerical errors during the bias correction process. Hence the remaining biases (except for precipitation) could not be attributed to a specific mechanism.

3.2 Regional and seasonal biases in forcing variables

Figure 5 illustrates the initial biases of the GCM ensemble, spatially aggregated over 24 regions of the globe. To account for possible seasonality variations, the biases are calculated for the annual mean (ANN) and for the December–January–February (DJF) and June–July–August (JJA) means. The remaining biases are not shown because their regionally aggregated values are negligible and would be indistinguishable in the figure. Additionally, an insight into the behaviour of each ensemble member, in comparison to the ensemble mean and WFDEI, is given by Table S2. Table S2 provides the values of raw input variables for each ensemble member, the ensemble mean value, and the respective WFDEI value, averaged for the 24 study regions.

Precipitation biases are less pronounced in Europe (NEU, MED, and NEE) and in central and northern Asian regions (CAS and NAS). The wettest precipitation biases are encountered in equatorial and southern Africa (EQF, SQF, and SAF) and concern DJF precipitation (Fig. 5). The driest biases are found for the CAM, AMZ, and SAS regions, for JJA precipitation. Temperature displays cold biases in most regions. A notable exception is the warm bias in DJF temperature in the NAS region, which is the most pronounced temperature bias found. Generally the DJF temperature biases are the largest, followed by ANN, while the JJA season has the smallest temperature biases.

The two radiation components, longwave (Rl) and shortwave (Rs) radiation, show an inverse behaviour in their biases (Fig. 5). That is to say, in regions where Rl has negative biases, Rs exhibits positive biases and vice versa. According to Demory et al. (2014), overestimation of shortwave radiation is a common issue amongst the GCMs. Negative biases are dominant for Rl, in contrast to the Rs variable, which mostly shows positive biases. Specific humidity has negative biases over the northern part of the African continent (SAH, WAF, EAF, and EQF), Central and South America (CAM,

AMZ, and CSA), and South Asia (SAS). Positive humidity biases are identified in the southern part of Africa (SQF and SAF) and North America (WNA, CNA, and ENA).

Surface pressure shows almost exclusively positive biases (Fig. 5). The regions that distinguish for the largest biases are MED, SEA, SAH, SAF, CAM, CSA, and SSA. The most dominant negative wind speed bias is found in NAU. Most of the African continent (SAH, WAF, EAF, EQF, and SQF) and of South America (AMZ and CSA) also have negative biases in wind. The largest positive biases are encountered in the southern part of South America (SSA) for the JJA season and for the DJF season in regions of North America (WNA and CAM), Europe (MED), and Asia (CAS, TIB, and SEA).

3.3 Model evaluation

In order to assess JULES' performance, we compare discharge modelled with WFDEI and with the raw GCM dataset to discharge observations for nine study basins. Figure 6 shows the seasonality of observed and modelled discharge and the evaluation metrics of the two sets of simulations (WFDEI and raw GCM) are presented in Table 3.

For seven out of the nine basins (Amazon, Congo, Volga, Ganges, Danube, Elbe, and Kemijoki) seasonality is captured well by the WFDEI simulation (Fig. 6). In contrast, the raw GCM simulation exhibits significant positive and negative biases for these seven basins. For the two remaining basins, however (Mississippi and Lena), seasonality is better captured by the raw GCM simulation. The WFDEI run results in positive NSE values (0.24 to 0.94) for all the basins. By contrast, the raw GCM run results in negative NSE values for six out of the nine basins. PBIAS indicates that the raw GCM simulation exhibits greater deviations from observations than the WFDEI run for most basins (exceptions are the Mississippi, Lena, Ganges, and Danube). Finally, the R^2 metric shows that the linear correlation between simulations and observations is stronger for the WFDEI run for seven out of the nine basins (exceptions are the Mississippi and Elbe). For both simulations the lowest R^2 value is reported for the Congo basin (0.45 and 0.2 for the WFDEI and raw GCM runs respectively). The best correlations per simulation are found for the Ganges for the WFDEI run (0.99) and for the Amazon for the raw GCM run (0.94).

The shown persistent departure from the mean climatology of discharge could include three types of errors. The first is the error stemming from the insufficient description of the runoff processes by the land surface model and from the routing algorithm (Blyth et al., 2011). The second type of error is a result of errors in the forcing datasets (either observational or GCM output) with regards to depicting the real climatic drivers (Elsner et al., 2014; Mizukami et al., 2014). A third possible error comes from the comparison of naturalized discharge of the simulations with measured discharge due to influences like abstractions and dams regulating the natural river flow (Müller Schmied et al., 2014). An extra

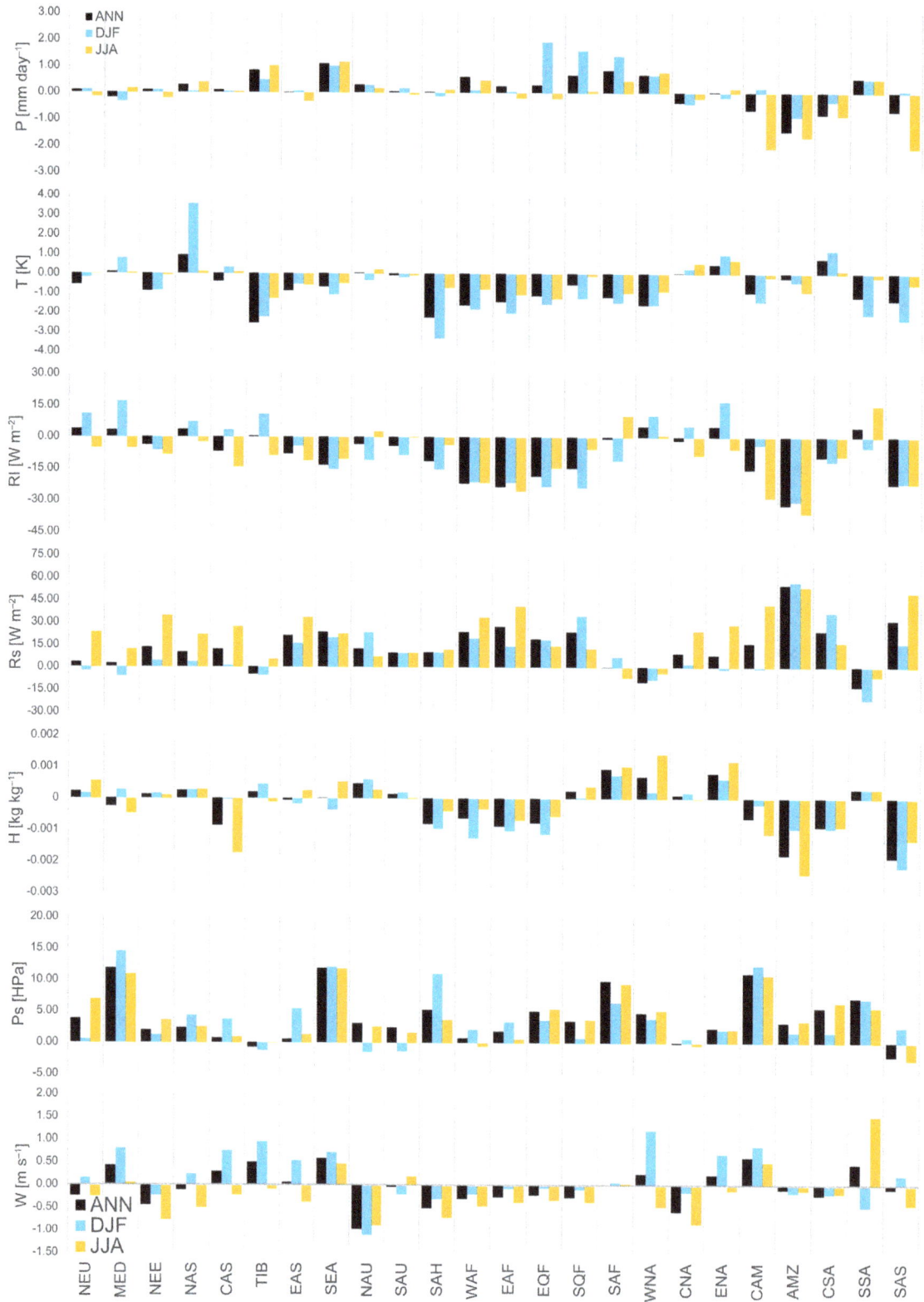

Figure 5. Initial biases (raw-WFDEI) of the GCM ensemble forcing variables, spatially averaged for 24 Giorgi regions. Biases are calculated between long-term annual averages (ANN) and December–January–February (DJF) and June–July–August (JJA) averages of the period 1981–2010.

Table 3. Evaluation metrics derived from monthly discharge data. Metrics are calculated for JULES' simulations from WFDEI data (WFDEI) and the ensemble mean of raw GCM data (raw EM).

Indices	NSE		PBIAS		R^2	
Basins	WFDEI	Raw EM	WFDEI	Raw EM	WFDEI	Raw EM
Amazon	0.48	−2.66	−18.68	−51.84	0.96	0.94
Congo	0.39	−36.40	4.06	116.77	0.45	0.20
Mississippi	0.24	0.90	21.56	−4.46	0.73	0.92
Lena	0.56	0.82	−39.32	32.14	0.98	0.89
Volga	0.82	−1.42	−17.09	35.12	0.95	0.66
Ganges	0.94	0.80	19.48	−9.51	0.99	0.91
Danube	0.28	−1.51	15.20	1.14	0.88	0.19
Elbe	0.67	−26.04	8.28	179.83	0.81	0.86
Kemijoki	0.91	−0.98	8.55	66.50	0.94	0.89

error component, which is not considered here, could result from the uncertainty in discharge measurements (Coxon et al., 2015).

The model evaluation has revealed two basins (Mississippi and Lena) for which raw GCM forced discharge simulations outperform the WFDEI simulations. For the Mississippi, the WFDEI run gives higher discharge than the observations throughout the year, revealing a deficiency of the model in capturing the water balance of this basin. Most of the Mississippi extent is in the CNA region, where negative precipitation biases have been documented (Fig. 5). Thus, the raw GCM run is forced with less precipitation compared to WFDEI and less discharge is produced, masking the model deficiency in this basin and improving the metrics of model performance. It is also important to note that the range of the raw GCM simulations is quite broad, especially for a three-member ensemble. The upper range of the GCM ensemble exceeds the WFDEI-simulated runoff during almost half the seasonal cycle. This indicates that the individual ensemble members would not necessarily outperform the WFDEI run and that, for this specific basin, the ensemble averaging has possibly produced a "false positive" in model performance. In this particular basin, model performance may also be hindered due to the comparison of naturalized and actual discharge, as the Mississippi is a heavily regulated river. For the Lena, the WFDEI run underestimates measured discharge by about 40 %. The Lena basin falls into the extent of the NAS region, for which positive precipitation biases have been documented (Fig. 5). The extra water in the raw GCM run counteracts the tendency of the model to underestimate discharge in the Lena basin, resulting in an improved model performance. In the context of the present study we are not able to identify the exact reasons why model performance is hindered in some basins. It is unrealistic for a global LSM to achieve top performance around the world (Hattermann et al., 2017), as, due to its global nature, some fixes in some regions could result in deteriorations in performance in other parts of the land surface. Thus, the interpretation of the following

analysis of the present study should consider the model deficiencies revealed in this section.

3.4 Long-term biases in runoff at the global scale

Figure 7 shows the initial and remaining biases in runoff, derived from ANN, DJF, and JJA long-term means. As with the biases in the input forcing variables, the remaining bias in runoff is 1 to 2 orders of magnitude smaller than the initial bias. Hence, the use of bias corrected data led to an improved representation of runoff by the model, compared to the baseline of the WFDEI run. Accordingly, the studies of Teutschbein and Seibert (2012) and Rojas et al. (2011) found that hydrological simulations are substantially improved with the use of bias corrected forcing.

Regarding the raw GCM run, the largest runoff underestimation biases ($< -5\,\text{mm day}^{-1}$) are encountered in Central and North America, the central–eastern part of South America, and East Asia. The most pronounced runoff overestimation biases are found in the western part of North and South America, in equatorial and southern Africa, northern Europe, the Tibetan region, and Indonesia. Initial runoff biases are larger than $1\,\text{mm day}^{-1}$ in absolute terms for 16.26, 14.85, and 20.18 % of the land surface respectively for ANN, DJF, and JJA. The differences between the seasonal means (DJF, JJA) and the annual mean (ANN) are in general subtle. However, the increases in runoff overestimation biases in DJF in southern equatorial Africa and in JJA in the Tibetan plateau are worth noting. Large initial biases ($> 5\,\text{mm day}^{-1}$ in absolute terms) in seasonal means occupy a greater percentage of the land surface compared to the annual mean (0.70 % for ANN, compared to 1.25 and 1.97 % for DJF and JJA respectively).

The remaining biases in runoff range from −0.1 to $0.1\,\text{mm day}^{-1}$ for the majority of the land surface (95.19, 87.40, and 80.30 % for ANN, DJF, and JJA respectively). Negligible biases (smaller than $0.01\,\text{mm day}^{-1}$ in absolute terms) are found for more than one-third of the land surface (specifically for 38.06 % of the land area for ANN, 37.60 %

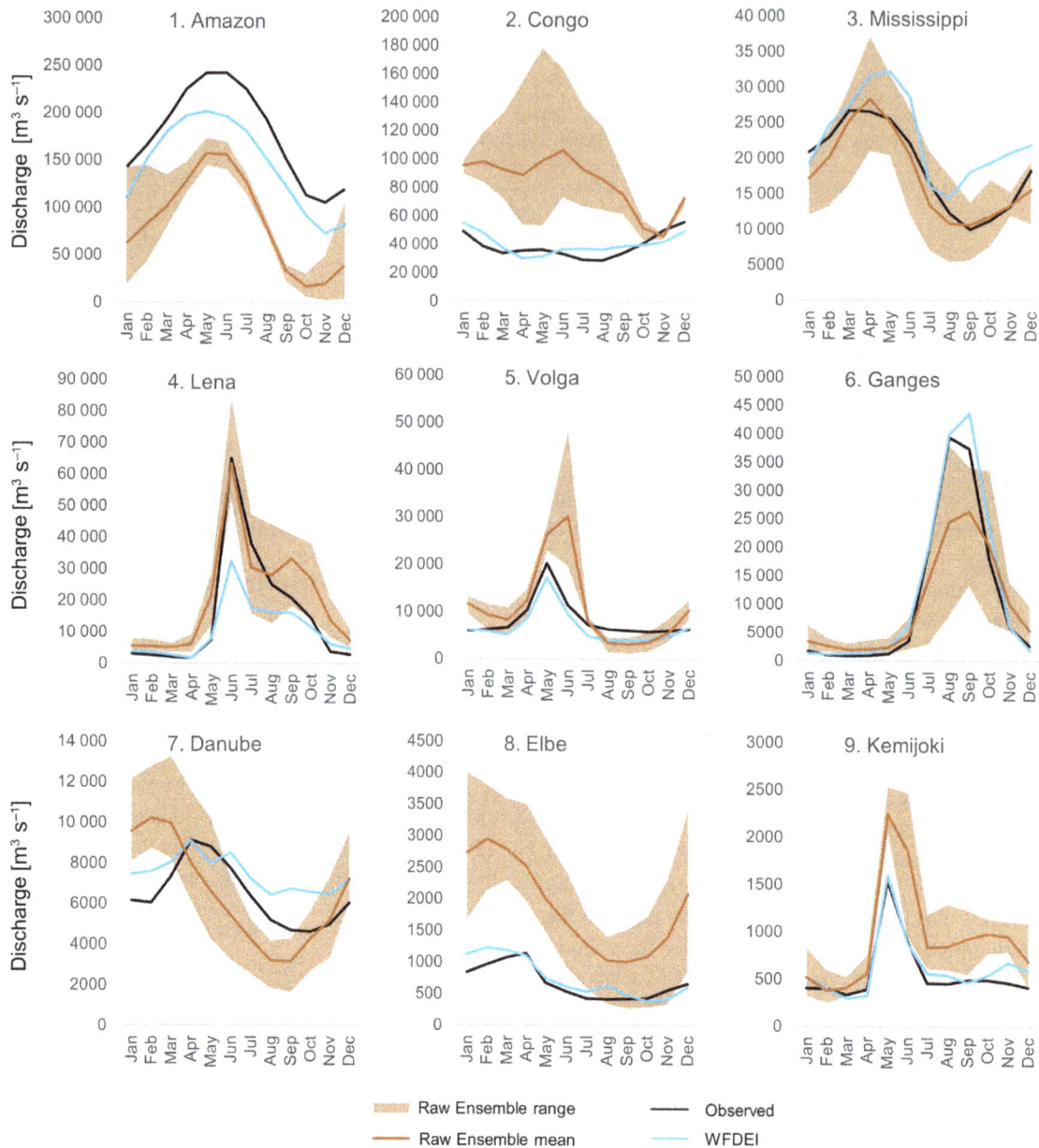

Figure 6. Discharge seasonality (m^3 s^{-1}) derived from the period 1981–2010 for nine study basins. Each panel shows observed discharge (GRDC measurements) compared to JULES' simulated discharge from WFDEI data and raw GCM data (the mean and the range of the ensemble are shown).

for DJF, and 34.42 % for JJA). The (negative) remaining bias in ANN runoff is more pronounced in the western Amazonian region. This probably corresponds to the remaining bias in precipitation identified for the Amazonian region (Fig. 4). In addition to the significant reduction of the biases in runoff forced with bias corrected data, it can be observed that the remaining biases have switched signs compared to the initial biases. This means that in regions where the initial bias in runoff is positive (negative), the raw GCM forced runoff is larger (smaller) than runoff forced with WFDEI, and the use of bias corrected forcing results in runoff slightly lower

(higher) than WFDEI runoff. A respective behaviour was not observed in the initial and remaining biases of the most impacting forcing variables (P and T), but it was, to an extent, present for other variables (Rl, Rs, and H). Thus, the "overcorrection" manifested for bias corrected runoff compared to WFDEI runoff cannot be attributed to remaining biases in precipitation and temperature. Instead, it could plausibly be associated with the compound effect of the remaining biases in some (or in all other) forcing variables.

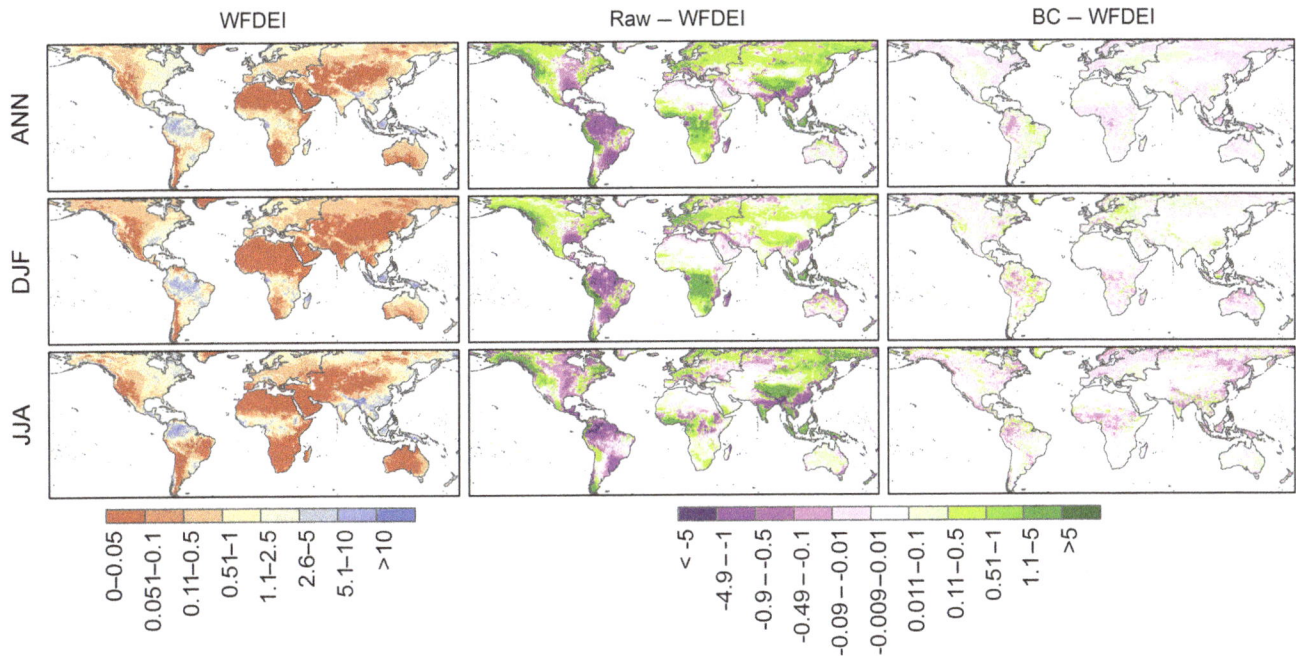

Figure 7. Runoff (mm day^{-1}) from WFDEI data (left column). Initial (raw-WFDEI) and remaining (BC-WFDEI) biases in runoff are shown in the middle and right columns respectively. Results are shown for long-term annual averages (ANN) and for December–January–February (DJF) and June–July–August (JJA) averages of the 1981–2010 period.

3.5 Effect of each forcing variable's bias on runoff

The effect that the bias of each forcing variable can have on runoff is investigated here, by comparing runoff from the bias corrected run to the partial correction assessment runs. The results are shown in Fig. 8, for ANN, DJF, and JJA averages.

First, we discuss the runoff differences calculated from the ANN period. Precipitation and temperature are the only two variables that cause runoff differences larger than 5 mm day^{-1} (in absolute terms) when neglected from bias correction. However, these differences regard a very small percentage of the land surface: 0.61 % for precipitation and only 0.02 % for temperature. Moreover, precipitation bias causes changes in runoff greater than 1 mm day^{-1} (in absolute terms) for 14.28 % of the land area. Such changes for the other variables occupy a significantly smaller fraction of the land area (ranging from 1.21 % for temperature to 0.05 % for wind). Based on the above it can be stated that precipitation is the variable that most affects runoff response. Precipitation bias causes both wet and dry biases in different regions of the land surface, with a pattern that closely resembles the effect of the initial GCMs' biases on runoff (Fig. 7). A similar pattern between precipitation and runoff biases was also observed by Teng et al. (2015), who noted that precipitation errors are magnified in modelled runoff. Temperature biases result in runoff overestimation for around 60 % of the land surface (e.g. over western and eastern North America, the Amazon region, equatorial Africa, northern Europe, and parts of Asia) and runoff underestimation for around 40 % (example

regions: parts of Central and South America and of central Asia). Temperature biases correspond to small changes in runoff (up to 0.01 mm day^{-1} in absolute terms) over about one-third of the land area. Excluding the radiation components from the bias correction procedure produces negative runoff changes for the majority of the land surface (67.60 %), while for around 80 % of the land surface the differences in runoff range between −0.1 and 0.1 mm day^{-1}. The bias in the specific humidity variable corresponds to runoff overestimations for 64 % of the land area. The areas of runoff overestimation are mainly located at the higher latitudes (northern part of North America, Europe, and northern Asia). For 36.43 % of the land surface, changes in runoff due to specific humidity biases span between 0.1 and 0.5 in absolute terms. Surface pressure and wind are the variables that show the smaller effect on the hydrological output, as their exclusion from bias correction corresponds to small changes in runoff (less than 0.1 mm day^{-1} in absolute terms) for the vast majority of the land surface (around 94 and 92 % of the land surface respectively for surface pressure, and wind speed). The most pronounced differences in runoff due to surface pressure biases are negative and are encountered over the high mountain range regions of South America and Asia (Andes and Himalayas respectively).

The patterns of runoff changes due to the biases of the forcing variables derived from annual (ANN) and seasonal (DJF, JJA) averages show only subtle variations. In general the above analysis of the ANN runoff differences applies also

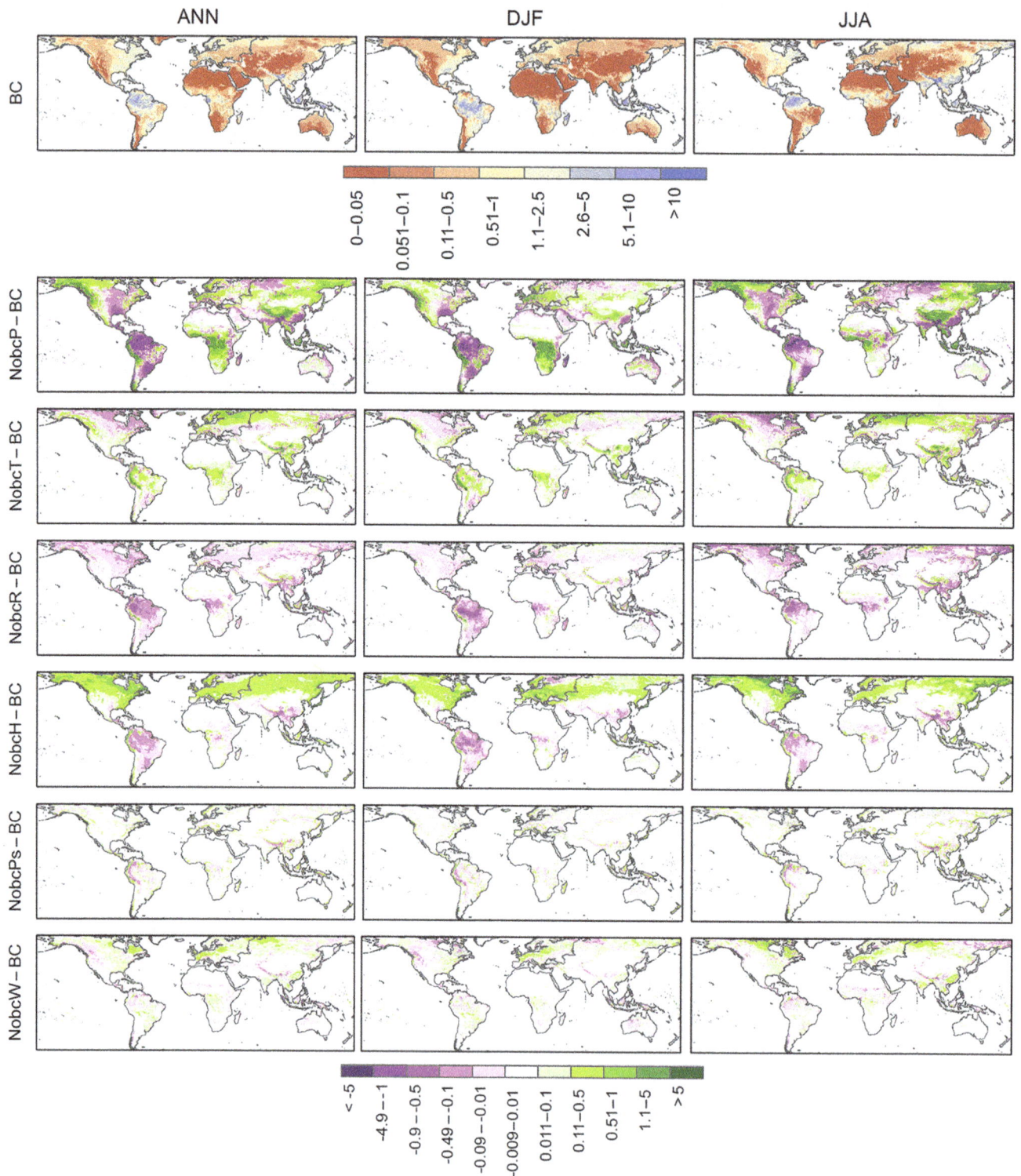

Figure 8. (Top row) Runoff (mm day^{-1}) from bias corrected GCM ensemble forcing (BC) and (second to last row) runoff differences between the bias corrected run (BC) and the partially corrected runs (NobcV, where V is one of the forcing variables P, T, R, H, Ps, or W). Results are shown for long-term annual averages (ANN) and for December–January–February (DJF) and June–July–August (JJA) averages of the 1981–2010 period.

to the seasonal values, with small variations on the land fractions that show a specific response to forcing biases.

From this analysis it can be deduced that apart from the main hydrological cycle drivers (precipitation and temperature), radiation and specific humidity can also have a substantial effect on runoff, especially for specific regions. These findings will be further investigated and discussed in the following sections. Other studies also advocate the considerable effect that biases in radiation (Mizukami et al., 2014) and humidity (Masaki et al., 2015) can have on hydrological fluxes.

3.6 Runoff sensitivities to forcing variables

Sensitivity of runoff changes to the biases of the forcing variables is examined by exploring the relationship between the input forcing biases (ΔV) and the corresponding changes in runoff (ΔRF). The regional variation of this relationship is also investigated. Figure 9 shows scatterplots of ΔRF versus ΔV for each examined variable, for 10 selected regions. The dots in each scatterplot correspond to the land grid boxes of each region. The presented regions are selected as representative of different parts of the land surface, as the number of the regions shown in the manuscript had to be reduced for clarity of the results. Scatterplots of the 24 examined regions can be found in the Supplement of this paper (Fig. S3). The median values of ΔV, ΔRF, and S of the land grid boxes of each region, for the 24 examined regions, are shown in Table 4.

The correlation between the six ΔVs and respective ΔRFs differs substantially between the examined regions. Generally, the correlations show a non-uniform behaviour, identified by the highly scattered data clouds. This implies a high spatial variability of runoff sensitivity to the examined variables.

For precipitation, the ΔRF over ΔP relationship exhibits a non-linear behaviour, indicating that the relative change in runoff is not proportional to precipitation bias, but also depends on the magnitude of precipitation bias. Renner et al. (2012) also identified non-linearities in the relationship between relative changes in streamflow and changes in precipitation, and argued that non-linear behaviour is a result of the combined effects of water and energy balances. Temperature biases have an inversely proportional and highly non-linear relationship with changes in runoff. The ΔRF over ΔT relationship is also variant for different regions. For example, the scatterplots for NEU and WNA indicate that small temperature biases may correspond to large changes in runoff. In contrast, the scatterplot for CAM indicates that larger temperature biases correspond to smaller changes in runoff compared to the other regions. Radiation biases are small but can correspond to high changes in runoff for some regions (WNA, SAS, WAF, and AMZ). For specific humidity it can be observed that small positive biases correspond to high changes in runoff for some regions (NEU, MED, WNA, and ENA). A different behaviour is observed for CAM, SAS,

AMZ, and CSA, where the data cloud is more scattered on the x axis (meaning larger biases in specific humidity) and less scattered on the y axis (i.e. changes in runoff are smaller). Surface pressure has smaller biases compared to the other forcing variables and its effect on runoff also appears reduced. Wind has a wide range of both positive and negative biases which, however, do not seem to affect runoff accordingly.

The variation of the ΔRF over ΔV relationships across the different regions can be attributed to a number of factors. First, it depends on the magnitude and signal of the biases in the forcing variables. As previously shown, these can have significant spatial variations (Fig. 4). For example, according to the median values of relative changes in Table 3, some regions are dominated by negative precipitation biases (MED, SAS, AMZ, and CSA) and others by positive biases (NEU, WNA, ENA, CAM, WAF, and SAU). Second, it reflects the climatology of each region. The same biases would affect differently regions with different runoff (and evapotranspiration) fractions of each region. The precipitation partitioning to runoff and evapotranspiration is a climate characteristic and is controlled by either water or energy limitations, depending on the region. Additionally, we should consider that although we assess the effect of long-term annual biases on long-term annual runoff, the results are still dependent on the seasonal cycles of the variables and/or runoff, especially if the seasonality of precipitation in the region is strong. For example, the same annual bias in temperature would translate differently to runoff changes in a region with precipitation evenly dispersed throughout the year and in another region where most of the annual precipitation happens during the summer months. Finally, as this is a model-based experiment, we should consider whether high sensitivities of some variables for specific regions are a result of over-sensitivity of the model. Vano et al. (2012) documented considerable differences in the spatial distribution of sensitivities to precipitation modelled by five LSMs.

3.7 Spatial distribution of bias effect categories

Figure 10 shows global maps of bias ECs for each forcing variable, derived according to the methodology described in Sect. 2.8. The land area fraction corresponding to each EC is tabulated in Table 5.

Precipitation is the variable whose biases have the largest effect on runoff, with the vast majority of the land surface (92 %) corresponding to the high change categories ECI (67.80 %) and ECII (24.20 %). Radiation has the second largest land fraction in ECI, but temperature has the second largest land fraction in the high change categories (ECI and ECII). Radiation also has the largest land fraction in the high sensitivity categories (ECI and ECIII). As discussed in Sect. 3.6, this is possibly a result of combining shortwave and longwave radiation for the calculation of the radiation biases. For specific humidity, the most affected areas (ECI)

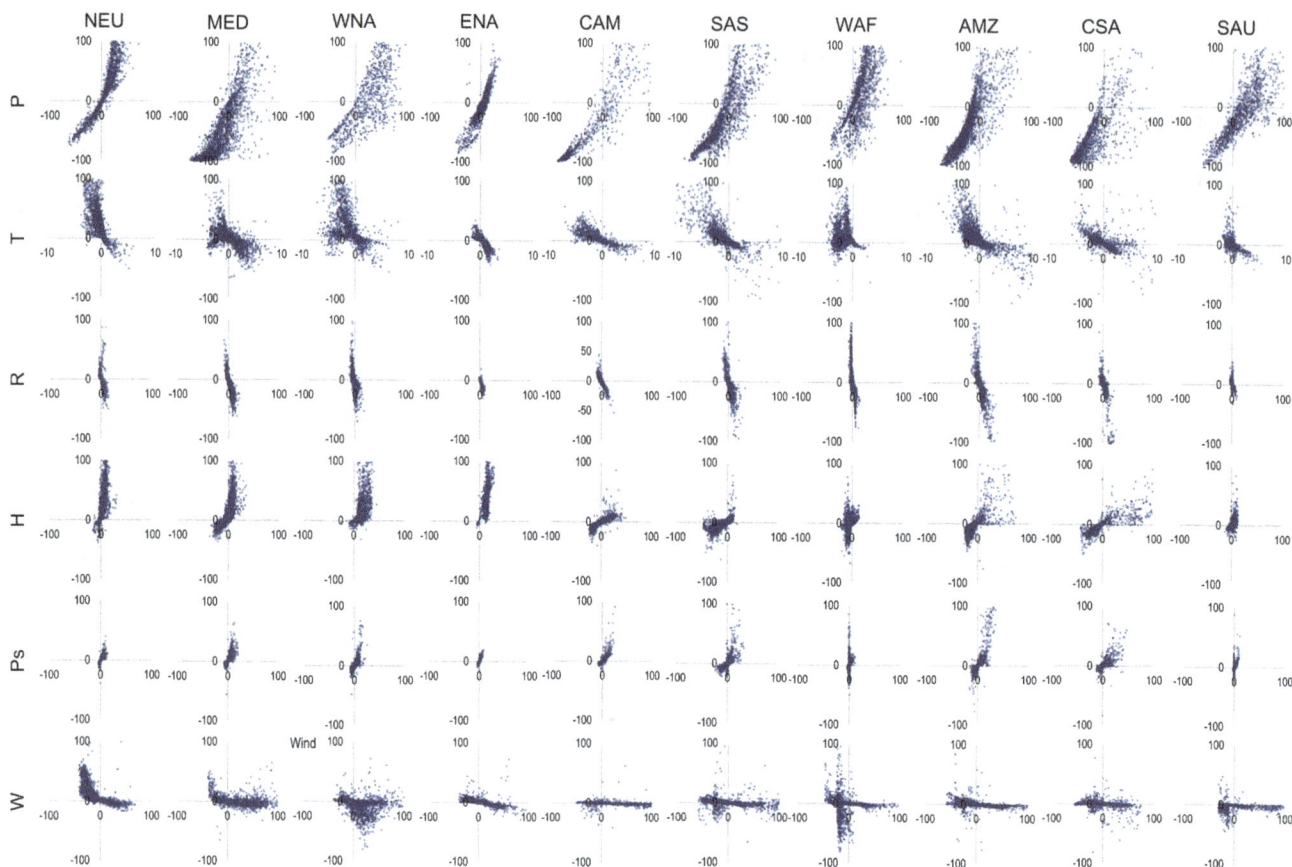

Figure 9. Scatterplots of relative changes in the forcing variable (ΔV, x axis) and corresponding relative changes in runoff (ΔRF, y axis), for all the forcing variables and for selected regions. In each panel, each dot represents the ΔRF / ΔV relationship of each land grid box in the examined region.

show a significant spatial coherence and are clustered at the higher latitudes of the globe. Surface pressure biases belong to ECI for around one-tenth of the land surface. The highly affected areas mainly correspond to regions with high mountain ranges. For wind the majority of the land surface corresponds to ECIV. Still, around one-quarter of the land surface belongs to the high change categories (ECI and ECII).

3.8   Discussion of runoff sensitivities

Here we compare our findings to the respective literature to assess the realism of JULES' sensitivity. We use the median sensitivity value of the grid boxes of each region (Table 4) as the representative sensitivity S for each region. Moreover, we discuss issues of possible model over-sensitivity in particular regions and the caveats of this study.

3.8.1   Sensitivity of runoff to precipitation

Most studies have examined the sensitivity (also reported as elasticity) of runoff (or discharge) to precipitation. A number of studies have examined sensitivity to precipitation for regions or basins in the United States. Values of runoff sensi-

tivity (S) to precipitation between 1.5 and 2.5 were reported by Sankarasubramanian and Vogel (2003) for the US (WNA, CNA, and ENA). Fu et al. (2007) reported values of 1.5 to 1.67 for the Spokane River basin (located in WNA). Vano et al. (2012) found that S to precipitation ranged from 2.2 to 3.3 for different LSMs for the Colorado River basin (also located in WNA). For the Mississippi River basin (mainly located in CNA), Renner et al. (2012) found that S of streamflow to precipitation is 2.38 and 2.55 using two different methods for sensitivity estimation. For another basin located in CNA, Brikowski (2015) reported runoff S to precipitation to be 2.64. For the US region, the S values found in this study compare very well with the literature values. Runoff S to precipitation is 2.12 for WNA, 2.54 for CNA, and 1.69 for ENA. Many studies report S to precipitation for regions or basins of China. Reported values of runoff S to precipitation in the Yellow River basin (located in EAS) are 1.4 to 1.69 (Fu et al., 2007), 1.6 to 3.9 for 89 catchments of the EAS region (Yang and Yang, 2011), and 1.71 and 1.74 (estimates of two different methods) for the headwaters of the Yellow River (Renner et al., 2012). Again, the value found in our study is in good

Table 4. Relative change (%) in forcing variable (ΔV), corresponding relative change (%) in runoff (ΔRF), and sensitivities ($S = \Delta RF/\Delta V$) per region, for each variable. For each region, the median of the ΔV, ΔRF, and S values of all land grid boxes is shown.

	Variables	P	T^*	R	H	Ps	W
GLOBAL	ΔV	14.46	−0.57	1.73	0.91	−0.02	−5.86
	ΔRF	2.49	3.38	−3.71	2.04	−0.04	0.21
	S	1.76	−0.05	−2.12	0.81	1.18	−0.06
NEU	ΔV	14.6	−0.46	1.86	4.1	−0.05	−9.79
	ΔRF	27.97	22.68	−5.25	25.49	−0.02	3.62
	S	2.10	−0.31	−3.31	5.24	2.90	−0.36
MED	ΔV	−14.39	−0.15	0.55	−1.34	0.41	14.94
	ΔRF	−58.56	1.55	−1.51	4.07	0.44	−0.47
	S	2.02	−0.04	−2.52	0.77	1.08	−0.08
NEE	ΔV	4.89	−1.44	2.44	3.32	0.1	−11.77
	ΔRF	5.75	47.11	−5.39	32.73	0.26	5.98
	S	2.28	−0.32	−2.64	9.58	3.31	−0.50
NAS	ΔV	26.05	0.67	3.53	8.05	−0.06	−1.08
	ΔRF	59.36	11.8	−10.08	63.98	0.02	4.06
	S	2.35	−0.07	−2.95	7.58	2.43	−0.29
CAS	ΔV	6.44	−0.03	1.37	−13.00	−0.41	8.09
	ΔRF	−9.94	1.31	−0.44	−0.19	−0.36	−1.29
	S	2.49	−0.05	−3.50	0.31	0.88	−0.09
TIB	ΔV	128.47	−2.94	−1.14	7.69	−0.12	12.59
	ΔRF	1017.17	5.38	0.97	0.81	0.02	0.06
	S	7.27	−0.02	−2.07	0.18	0.40	0.00
EAS	ΔV	19.25	−0.94	2.51	2.92	−0.2	−3.55
	ΔRF	4.36	5.54	−2.96	3.66	−0.05	0.76
	S	1.70	−0.06	−1.53	0.82	1.07	−0.09
SEA	ΔV	19.76	−0.87	1.11	0.89	0.23	34.57
	ΔRF	43.92	5.97	−3.2	1.66	0.32	−1.04
	S	2.07	−0.08	−2.68	1.16	1.54	−0.05
NAU	ΔV	41.15	−0.04	1.43	7.71	0.1	−28.46
	ΔRF	−5.13	1.02	−1.16	1.38	0.09	−0.44
	S	0.37	−0.03	−0.75	0.31	0.56	0.00
SAU	ΔV	18.92	−0.28	0.85	2	−0.13	−11.2
	ΔRF	−9.29	1.07	−0.11	1.4	0.06	−0.49
	S	0.82	−0.05	−0.88	0.67	1.00	−0.03
SAH	ΔV	54.11	−2.73	−0.47	−8.96	0.22	−13.59
	ΔRF	−2.59	−0.68	0.64	−0.32	0	0.08
	S	0.94	0.00	−0.25	0.04	0.04	−0.01
WAF	ΔV	26.74	−1.51	−0.88	−5.79	−0.1	−15.13
	ΔRF	58.24	5.61	−1.57	−0.71	−0.13	0.09
	S	2.78	−0.04	−2.61	0.22	1.28	−0.04
EAF	ΔV	23.22	−1.68	−0.06	−5.76	−0.25	−12.11
	ΔRF	42.13	7.24	−1.51	−3.74	−0.28	0.09
	S	2.12	−0.05	−1.95	0.48	0.95	0.00
EQF	ΔV	5.64	−1.55	−0.25	−2.15	−0.2	−10.09
	ΔRF	−0.14	6.21	0.92	−1.29	0	0.07
	S	2.26	−0.05	−1.73	0.49	0.92	−0.01

Table 4. Continued.

	Variables	P	T^*	R	H	Ps	W
SQF	ΔV	36.45	−0.9	0.9	0.89	−0.03	−15.6
	ΔRF	−73.18	−82.26	−85.07	−84.68	−84.2	−84.18
	S	2.94	−0.07	−1.91	0.59	1.10	−0.04
SAF	ΔV	89.8	−1.41	−0.38	14.28	0.68	−4.74
	ΔRF	85.47	5.5	0.54	5.33	0.42	−0.02
	S	1.35	−0.04	−1.66	0.45	0.72	−0.05
WNA	ΔV	65.92	−1.75	−1.23	13.55	0.14	10.23
	ΔRF	112.66	17.94	−0.48	9.85	0.16	−2.5
	S	2.12	−0.13	−2.01	0.77	0.98	−0.17
CNA	ΔV	−12.84	0.11	1.68	2.29	−0.08	−14.79
	ΔRF	−50.86	1.53	−2.06	6.57	−0.05	1.96
	S	2.54	−0.07	−1.47	1.08	1.09	−0.13
ENA	ΔV	4.08	0.49	2.71	13.4	0.1	5.47
	ΔRF	−0.38	−0.38	−5.18	39.72	0.13	0.86
	S	1.69	−0.07	−1.92	3.17	1.54	−0.11
CAM	ΔV	11.43	−0.98	−0.4	−6.16	0.15	25.27
	ΔRF	−7.73	3.65	−0.1	−2.55	0.14	−0.52
	S	1.32	−0.04	−1.58	0.49	0.77	−0.02
AMZ	ΔV	−26.58	−0.35	4.06	−13.19	−0.19	−4
	ΔRF	−40.52	4.88	−9.34	−6.01	−0.23	0.03
	S	1.42	−0.05	−2.37	0.53	1.44	−0.04
CSA	ΔV	−32.8	0.7	3.05	−11.53	−0.23	−7.5
	ΔRF	−63.21	−1.49	−3.22	−5.75	−0.13	0.38
	S	1.59	−0.04	−1.16	0.53	0.83	−0.04
SSA	ΔV	72.07	−1.22	−1.77	5.07	0.08	9.91
	ΔRF	84.32	10.06	−0.47	12.05	0.34	−2.44
	S	1.53	−0.09	−0.50	1.48	1.29	−0.04
SAS	ΔV	−9.19	−1.08	1.39	−13.11	−0.05	−6.81
	ΔRF	−26.35	5.2	−4.07	−2.53	−0.09	0.51
	S	1.62	−0.05	−2.46	0.29	0.90	−0.05

$^*\Delta V$ for temperature is the absolute change in temperature.

Table 5. Percent of land area (%) under each of the four effect categories (ECs).

Variables/ECs	I	II	III	IV
P	67.80	24.20	1.82	6.18
T	45.15	22.03	2.46	30.35
R	48.74	1.30	26.16	23.80
H	40.80	13.76	5.58	39.86
Ps	12.17	1.83	38.48	47.52
W	6.09	19.19	2.35	72.37

3.8.2 Sensitivity of runoff to temperature and other variables

A number of studies have examined runoff sensitivity to temperature changes. Vano et al. (2012) reported S to temperature values ranging from −2 to −9 C^{-1} between five LSMs for the Colorado River basin (WNA) and Brikowski (2015) reported a value of −0.41 C^{-1} for S to temperature in a basin in CNA. Our values for these regions are substantially lower (−0.13 K^{-1} for WNA and −0.07 K^{-1} for CNA). This divergence could be attributed to two factors. First, to an extent it could be connected to possible non-sensitivities of our model to temperature changes for these regions. Second, the differences could arise from the inclusion (or not) of the physical link between temperature and other variables in the analysis. Vano et al. (2012) use different LSMs to calculate sensitiv-

agreement with the literature (S to precipitation for EAS is 1.70).

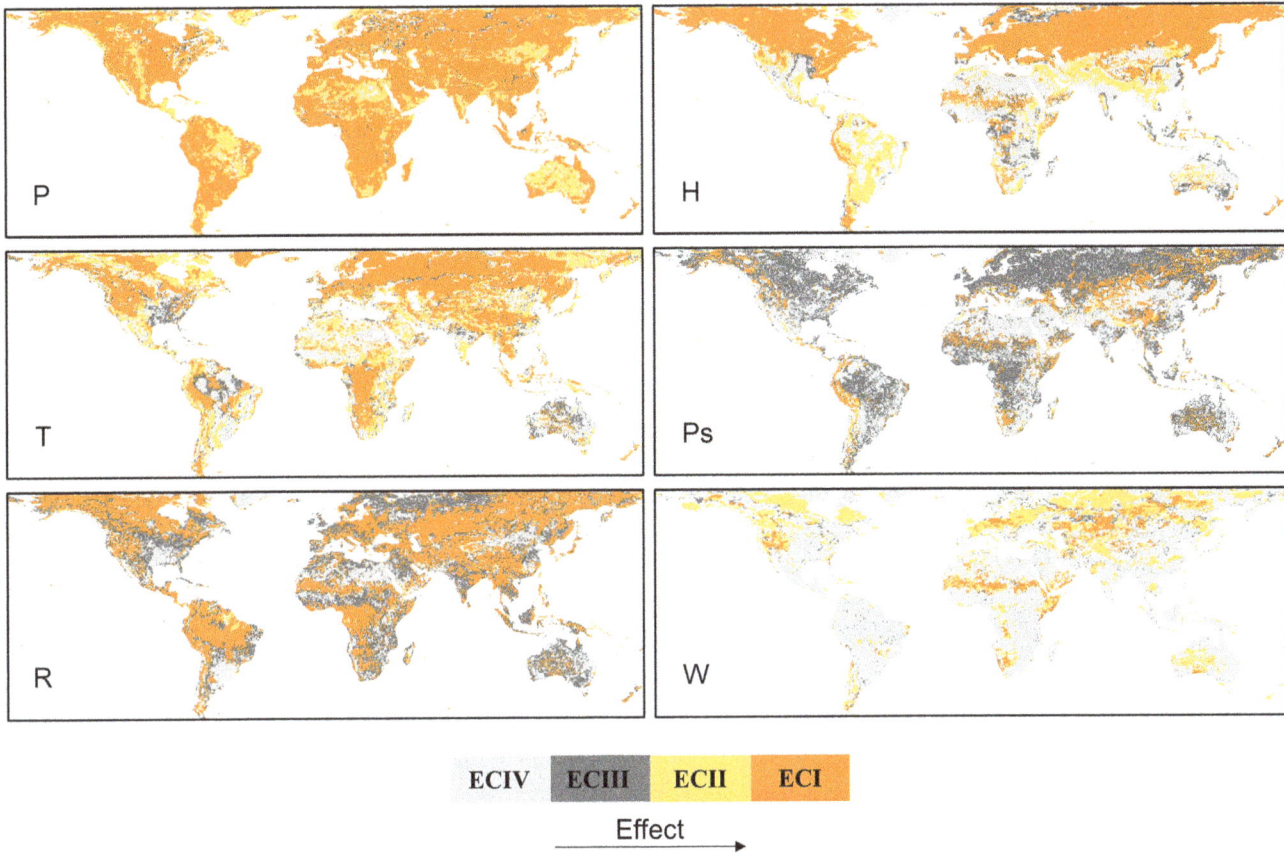

ECIV | ECIII | ECII | ECI

Effect →

Figure 10. Global maps of bias effect categories (ECs) for each forcing variable.

ities by perturbing daily temperature maxima and minima. These changes also affect the downward longwave radiation and humidity, which are then used by the evapotranspiration routines of the LSMs. In our case, the change in temperature does not interact with radiation and humidity, as those are read as input variables by the model. When temperature is allowed to interact with humidity, increased temperature will increase the water vapour capacity of the air, and more water will be evaporated. The lack of this physical link in our simulations could, to an extent, explain the decreased sensitivity of runoff to temperature changes compared to Vano et al. (2012). In the analysis of Brikowski (2015), sensitivities of runoff to precipitation and temperature are derived from the respective historical data. Thus, sensitivity to temperature will also include the changes caused by the interaction of temperature with other meteorological variables. In a study with a different approach, Yang and Yang (2011) separated the effect of precipitation, temperature, net radiation, relative humidity, and wind speed on runoff and calculated sensitivities for each variable. They reported values of S to temperature ranging from -0.11 to $-0.02\,\text{C}^{-1}$ between 89 catchments of the EAS region. For the same region, we have computed S to temperature as $-0.06\,\text{K}^{-1}$, which is included in the stated range in the literature. Moreover, our S val-

ues for radiation, humidity, and wind speed are also in good agreement with Yang and Yang (2011). According to Yang and Yang (2011), S to radiation ranges from -1.9 to -0.3, S to humidity from 0.2 to 1.9, and S to wind speed from -0.8 to -0.1. The range refers to values computed for 89 catchments in the EAS region. Our respective values for this region are -1.53 for radiation, 0.82 for humidity, and -0.09 for wind speed. This supports the argument that the large deviations of the sensitivity to temperature between our study and the studies of Vano et al. (2012) and Brikowski (2015) result from interactions in the forcing variables included in the referenced studies.

3.8.3 Sensitivity of runoff to radiation

The reported S to radiation values are higher in absolute terms than S to precipitation values for many of the examined regions and also globally (Table 4). However, according to the findings presented in Sect. 3.5, precipitation and temperature correspond to higher changes in runoff compared to radiation. That is because high S to radiation results from relatively low ΔV values, rather than from relatively high ΔRF values (compared e.g. to precipitation). Small ΔV for radiation is possibly the consequence of combining shortwave and longwave radiation to calculate the total bias in radiation, as

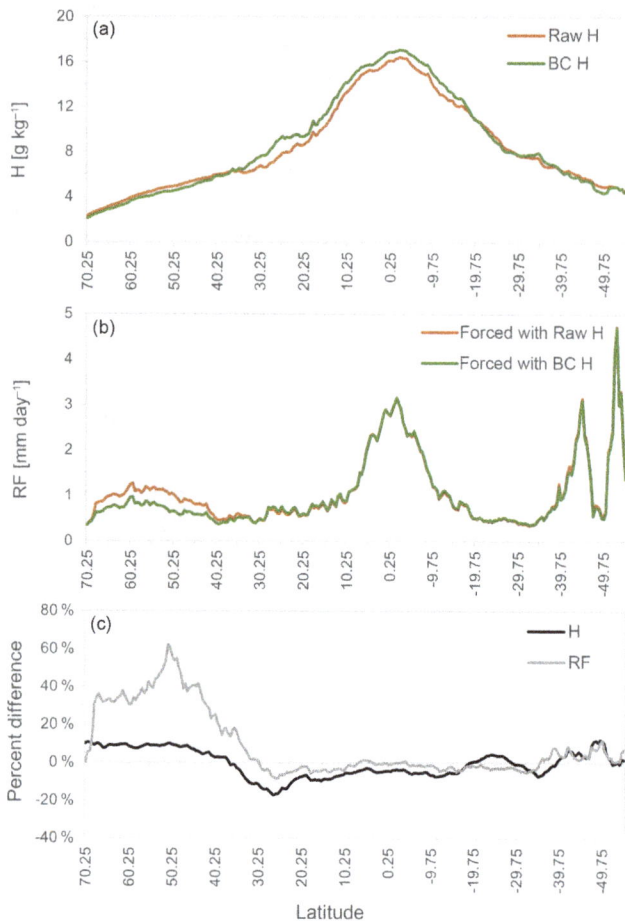

Figure 11. (a) Latitudinal means of raw and bias corrected specific humidity (g kg^{-1}), **(b)** latitudinal means of JULES' runoff forced with raw and bias corrected specific humidity (mm day^{-1}), and **(c)** percent differences of the latitudinal means in **(a)** H and **(b)** RF. The latitudinal means are calculated from the 1981–2010 period.

the two radiation components have inverse signs for most regions (Fig. 5).

3.8.4 Sensitivity of runoff to specific humidity in high-latitude regions

Although S to humidity for EAS compares well with the literature, unexpectedly high values of S to humidity are found for other regions (5.24 for NEU, 9.58 for NEE, and 7.58 for NAS). We performed an extra analysis to investigate this issue and the basic findings are included in Fig. 11 and the Supplement of this paper. Figure 11 examines the differences between the latitudinal mean of raw and bias corrected specific humidity and the resulting runoff. Very high sensitivity of runoff to H is observed for a specific area, the zone between 70 and 40° N latitudes. In that zone, a difference of about 10 % in H corresponds to an increase of 40 to 60 % in runoff. Investigation of the different fluxes related to runoff production in the model revealed two mechanisms

that explain this behaviour. First, due to higher humidity, the water vapour deficit of the air is reduced and evapotranspiration is decreased, thus allowing more of the precipitated water available as runoff. This mechanism explains around one-third of the magnitude of reported changes in runoff (Fig. S4). The second mechanism happens due to supersaturation of the air, especially during the colder months of the year when the dew point is lower, and includes the condensation and deposition of water vapour (direct transition from vapour to ice). Deposited water accumulates as snow mass. Snow mass is higher for the raw H run (H has positive biases), which results in increased snowmelt and thus increased runoff (Fig. S5).

A comparison of supersaturated air conditions for the different sets of data (WFDEI, raw, BC, and NobcH) can help us identify the origin of the aforementioned behaviour. From the input specific humidity H, we estimated the respective relative humidity (this transformation also requires temperature T and surface pressure Ps as input to the Clausius–Clapeyron equation). Then we calculated the fraction of time (based on a daily time step) in which supersaturated conditions occur, for the historical period 1981–2010. The estimation was performed for (a) the WFDEI H, T, and Ps, (b) the raw H, T, and Ps, (c) the bias corrected H, T, and Ps, and (d) for a combination of data corresponding to the NobcH run (raw H combined with bias corrected T and Ps). The results are presented in Fig. 6 of the Supplement of this paper. The analysis reveals that the higher-latitude regions – that display high sensitivity of runoff to H – are under supersaturated conditions for more than 10 % of the time (Fig. S6). The length of supersaturated conditions estimated for the WFDEI, raw, or BC data do not exhibit a respective spatial pattern, although supersaturation is found in all three datasets (Fig. S6). Thus, the high runoff sensitivity over the high-latitude regions is not a result of supersaturated conditions in the raw GCM H, and it rather stems from (1) raw GCM H being higher than BC H and (2) the calculation of relative humidity within JULES, done by combining raw GCM H with bias corrected T and Ps. This inconsistency strengthens the argument for the need for bias correction of more forcing variables – in addition to P and T. Specific humidity is a variable that is often left uncorrected, a practice that could possibly result in runoff overestimations at the northern latitudes based on our findings, in cases where hydrological models which account for deposition and condensation are used.

Since this experiment was performed with a single LSM, it cannot be concluded whether this behaviour is common between the LSMs or is an over-sensitivity of the JULES model. However, it highlights the importance of bias correction for specific humidity for specific regions, where runoff would have been highly overestimated using raw specific humidity as forcing.

3.9 Study caveats

An issue that must be considered for the interpretation of the results of this study is that they have been based on a single impact model. As the uncertainty stemming from the selection of the impact model is large (Gudmundsson et al., 2012; Hagemann et al., 2013), it is preferable to use multiple models in order to capture a wide range of possible results. The effect of the meteorological forcing on a hydrological output is heavily model dependent, as different models employ different concepts and/or equations for the representation of key hydrological processes. This concern has also been discussed by other single model studies on meteorological variables' effects on hydrological outputs (Mizukami et al., 2014; Masaki et al., 2015). Nonetheless, the results of single model studies are useful in giving indicative answers on the issues they examine and set a basis for the methodology needed for the respective multi-model applications.

4 Summary and conclusions

The present study examined the effect of the biases in GCM output variables on historical runoff simulations, using the JULES LSM. The effects of biases were studied for each forcing variable separately, for a total of six meteorological variables (precipitation, temperature, radiation, specific humidity, surface pressure, and wind speed). Biases of each variable and the respective effect of runoff were quantified at the global and regional scales. A framework for the categorization of the effects of biases of the different variables was developed and implemented, leading to global maps of bias ECs.

We found that bias correction of GCM outputs results in substantially improved representation of historical runoff. For this reason, our study adds to the numerous studies that advocate the use of some kind of bias correction of GCM data prior to their use as impact model forcing. Precipitation and temperature biases were identified as causing the largest changes in runoff. Radiation and specific humidity can also have a substantial effect on runoff, especially for specific regions. The sensitivity of runoff to the different forcing variables exhibits a high spatial variability. Depending on the region, runoff can be more sensitive to radiation or humidity compared to precipitation or temperature. The produced EC maps show that all variables can potentially affect runoff to a high extent, depending on the region. The fraction of the land surface occupied by the high effect category ECI (high changes in runoff and high sensitivity of runoff to the variable's changes) ranges between the variables from 67.80 % for precipitation to 6.09 % for wind.

The produced maps of ECs aid the identification of the regions most affected by the bias of each variable. Thus, they could serve as a decision tool in cases when an informed decision needs to be made on the variables that would need to be bias corrected or could be neglected from bias correction, according to the planned model application. Moreover, when raw forcing is used in model applications, EC maps could provide guidance towards the areas where the results would need more careful interpretation.

Based on the findings of this study, we suggest that the widely used concept of bias correcting precipitation and temperature should be extended to include more input variables. Radiation and specific humidity should be added to the priority variables for bias correction in hydrological applications, along with precipitation and temperature.

Due to the heavily model-dependent nature of runoff sensitivity to forcing variables, generalized conclusions for the behaviour of other impact models to GCM biases cannot be drawn from the present single model assessment. Nevertheless, this study aims to initiate a discussion of the effect of GCM biases on hydrological output, as the consideration of these sensitivities is crucial to understanding the uncertainty spectrum of hydrologically relevant climate change assessments.

Competing interests. The authors declare that they have no conflict of interest.

Acknowledgements. We acknowledge the World Climate Research Programme's Working Group on Coupled Modelling, which is responsible for CMIP, and we thank the climate modelling groups (listed in Table 1 of this paper) for producing and making available their model output. For CMIP the US Department of Energy's Program for Climate Model Diagnosis and Intercomparison provides coordinating support and led development of software infrastructure in partnership with the Global Organization for Earth System Science Portals.

The research leading to these results has received funding from the HELIX project of the European Union's Seventh Framework Programme for research, technological development and demonstration under grant agreement no. 603864.

Edited by: Stacey Archfield

References

Best, M. J., Pryor, M., Clark, D. B., Rooney, G. G., Essery, R. L. H., Ménard, C. B., Edwards, J. M., Hendry, M. A., Porson, A., Gedney, N., Mercado, L. M., Sitch, S., Blyth, E., Boucher, O., Cox, P. M., Grimmond, C. S. B., and Harding, R. J.: The Joint UK Land Environment Simulator (JULES), model description – Part 1: Energy and water fluxes, Geosci. Model Dev., 4, 677–699, https://doi.org/10.5194/gmd-4-677-2011, 2011.

Blyth, E., Clark, D. B., Ellis, R., Huntingford, C., Los, S., Pryor, M., Best, M., and Sitch, S.: A comprehensive set of benchmark tests for a land surface model of simultaneous fluxes of water and carbon at both the global and seasonal scale, Geosci. Model Dev., 4, 255–269, https://doi.org/10.5194/gmd-4-255-2011, 2011.

Brikowski, T. H.: Applying multi-parameter runoff elasticity to assess water availability in a changing climate: An example from Texas, USA, Hydrol. Process., 29, 1746–1756, https://doi.org/10.1002/hyp.10297, 2015.

Bromwich, D. H., Otieno, F. O., Hines, K. M., Manning, K. W., and Shilo, E.: Comprehensive evaluation of polar weather research and forecasting model performance in the Antarctic, J. Geophys. Res.-Atmos., 118, 274–292, https://doi.org/10.1029/2012JD018139, 2013.

Christensen, J. H., Boberg, F., Christensen, O. B., and Lucas-Picher, P.: On the need for bias correction of regional climate change projections of temperature and precipitation, Geophys. Res. Lett., 35, L20709, https://doi.org/10.1029/2008GL035694, 2008.

Clark, D. B., Mercado, L. M., Sitch, S., Jones, C. D., Gedney, N., Best, M. J., Pryor, M., Rooney, G. G., Essery, R. L. H., Blyth, E., Boucher, O., Harding, R. J., Huntingford, C., and Cox, P. M.: The Joint UK Land Environment Simulator (JULES), model description – Part 2: Carbon fluxes and vegetation dynamics, Geosci. Model Dev., 4, 701–722, https://doi.org/10.5194/gmd-4-701-2011, 2011.

Coxon, G., Freer, J., Westerberg, I. K., Wagener, T., Woods, R., and Smith, P. J.: A novel framework for discharge uncertainty quantification applied to 500 UK gauging stations, Water Resour. Res., 51, 5531–5546, https://doi.org/10.1002/2014WR016532, 2015.

Demory, M. E., Vidale, P. L., Roberts, M. J., Berrisford, P., Strachan, J., Schiemann, R., and Mizielinski, M. S.: The role of horizontal resolution in simulating drivers of the global hydrological cycle, Clim. Dynam., 42, 2201–2225, https://doi.org/10.1007/s00382-013-1924-4, 2014.

Dufresne, J.-L., Foujols, M.-A., Denvil, S., Caubel, A., Marti, O., Aumont, O., Balkanski, Y., Bekki, S., Bellenger, H., Benshila, R., Bony, S., Bopp, L., Braconnot, P., Brockmann, P., Cadule, P., Cheruy, F., Codron, F., Cozic, A., Cugnet, D., de Noblet, N., Duvel, J.-P., Ethé, C., Fairhead, L., Fichefet, T., Flavoni, S., Friedlingstein, P., Grandpeix, J.-Y., Guez, L., Guilyardi, E., Hauglustaine, D., Hourdin, F., Idelkadi, A., Ghattas, J., Joussaume, S., Kageyama, M., Krinner, G., Labetoulle, S., Lahellec, A., Lefebvre, M.-P., Lefevre, F., Levy, C., Li, Z. X., Lloyd, J., Lott, F., Madec, G., Mancip, M., Marchand, M., Masson, S., Meurdesoif, Y., Mignot, J., Musat, I., Parouty, S., Polcher, J., Rio, C., Schulz, M., Swingedouw, D., Szopa, S., Talandier, C., Terray, P., Viovy, N., and Vuichard, N.: Climate change projections using the IPSL-CM5 Earth System Model: from CMIP3 to CMIP5, Clim. Dynam., 40, 2123–2165, https://doi.org/10.1007/s00382-012-1636-1, 2013.

Dunne, J. P., John, J. G., Adcroft, A. J., Griffies, S. M., Hallberg, R. W., Shevliakova, E., Stouffer, R. J., Cooke, W., Dunne, K. A., Harrison, M. J., Krasting, J. P., Malyshev, S. L., Milly, P. C. D., Phillipps, P. J., Sentman, L. T., Samuels, B. L., Spelman, M. J., Winton, M., Wittenberg, A. T., and Zadeh, N.: GFDL's ESM2 Global Coupled Climate–Carbon Earth System Models. Part I: Physical Formulation and Baseline Simulation Characteristics, J. Climate, 25, 6646–6665, https://doi.org/10.1175/JCLI-D-11-00560.1, 2012.

Ehret, U., Zehe, E., Wulfmeyer, V., Warrach-Sagi, K., and Liebert, J.: HESS Opinions "Should we apply bias correction to global and regional climate model data?", Hydrol. Earth Syst. Sci., 16, 3391–3404, https://doi.org/10.5194/hess-16-3391-2012, 2012.

Elsner, M. M., Gangopadhyay, S., Pruitt, T., Brekke, L. D., Mizukami, N., Clark, M. P., Elsner, M. M., Gangopadhyay, S., Pruitt, T., Brekke, L. D., Mizukami, N., and Clark, M. P.: How Does the Choice of Distributed Meteorological Data Affect Hydrologic Model Calibration and Streamflow Simulations?, J. Hydrometeorol., 15, 1384–1403, https://doi.org/10.1175/JHM-D-13-083.1, 2014.

Fu, G., Charles, S. P., and Chiew, F. H. S.: A two-parameter climate elasticity of streamflow index to assess climate change effects on annual streamflow, Water Resour. Res., 43, 1–12, https://doi.org/10.1029/2007WR005890, 2007.

Giorgi, F. and Bi, X.: Updated regional precipitation and temperature changes for the 21st century from ensembles of recent AOGCM simulations, Geophys. Res. Lett., 32, L21715, https://doi.org/10.1029/2005GL024288, 2005.

Grillakis, M. G., Koutroulis, A. G., and Tsanis, I. K.: Multisegment statistical bias correction of daily GCM precipitation output, J. Geophys. Res.-Atmos., 118, 3150–3162, https://doi.org/10.1002/jgrd.50323, 2013.

Grillakis, M. G., Koutroulis, A. G., Papadimitriou, L. V., Daliakopoulos, I. N., and Tsanis, I. K.: Climate-Induced Shifts in Global Soil Temperature Regimes, Soil Sci., 181, 264–272, 2016.

Grillakis, M. G., Koutroulis, A. G., Daliakopoulos, I. N., and Tsanis, I. K.: A method to preserve trends in quantile mapping bias correction of climate modeled temperature, Earth Syst. Dynam. Discuss., https://doi.org/10.5194/esd-2017-53, in review, 2017.

Gudmundsson, L., Tallaksen, L. M., Stahl, K., Clark, D. B., Dumont, E., Hagemann, S., Bertrand, N., Gerten, D., Heinke, J., Hanasaki, N., and Voss, F.: Comparing large-scale hydrological model simulations to observed runoff percentiles in Europe, J. Hydrometeorol., 13, 604–620, 2012.

Haddeland, I., Heinke, J., Voß, F., Eisner, S., Chen, C., Hagemann, S., and Ludwig, F.: Effects of climate model radiation, humidity and wind estimates on hydrological simulations, Hydrol. Earth Syst. Sci., 16, 305–318, https://doi.org/10.5194/hess-16-305-2012, 2012.

Hagemann, S., Chen, C., Haerter, J. O., Heinke, J., Gerten, D., and Piani, C.: Impact of a Statistical Bias Correction on the Projected Hydrological Changes Obtained from Three GCMs and Two Hydrology Models, J. Hydrometeorol., 12, 556–578, https://doi.org/10.1175/2011JHM1336.1, 2011.

Hagemann, S., Chen, C., Clark, D. B., Folwell, S., Gosling, S. N., Haddeland, I., Hanasaki, N., Heinke, J., Ludwig, F., Voss, F., and Wiltshire, A. J.: Climate change impact on available water resources obtained using multiple global climate and hydrology models, Earth Syst. Dynam., 4, 129–144, https://doi.org/10.5194/esd-4-129-2013, 2013.

Hansen, J. W., Challinor, A., Ines, A. V. M., Wheeler, T., and Moron, V.: Translating climate forecasts into agricultural terms: advances and challenges, Clim. Res., 33, 27–41, 2006.

Harding, R. J., Weedon, G. P., van Lanen, H. A. J., and Clark, D. B.: The future for Global Water Assessment, J. Hydrol., 518, 186–193, https://doi.org/10.1016/j.jhydrol.2014.05.014, 2014.

Haslinger, K., Anders, I., and Hofstätter, M.: Regional climate modelling over complex terrain: an evaluation study of COSMO-CLM hindcast model runs for the Greater Alpine Region, Clim. Dynam., 40, 511–529, https://doi.org/10.1007/s00382-012-1452-7, 2013.

Hattermann, F., Krysanova, V., Gosling, S. N., Dankers, R., Daggupati, P., Donnelly, C., Flörke, M., Huang, S., Motovilov, Y., Buda, S., Yang, T., Muller, C., Leng, G., Tang, Q., Portmann, F. T., Hagemann, S., Gerten, D., Wada, Y., Masaki, Y., Alemayehu, T., Satoh, Y., and Samaniego, L.: Cross-scale intercomparison of climate change impacts simulated by regional and global hydrological models in eleven large river basins, Climatic Change, 141, 561–576, https://doi.org/10.1007/s10584-016-1829-4, 2017.

Katzav, J. and Parker, W. S.: The future of climate modeling, Climatic Change, 132, 475–487, https://doi.org/10.1007/s10584-015-1435-x, 2015.

King, A. D., Donat, M. G., Fischer, E. M., Hawkins, E., Alexander, L. V., Karoly, D. J., Dittus, A. J., Lewis, S. C., and Perkins, S. E.: The timing of anthropogenic emergence in simulated climate extremes, Environ. Res. Lett., 10, 94015, https://doi.org/10.1088/1748-9326/10/9/094015, 2015.

Koutroulis, A. G., Grillakis, M. G., Tsanis, I. K., and Papadimitriou, L.: Evaluation of precipitation and temperature simulation performance of the CMIP3 and CMIP5 historical experiments, Clim. Dynam., 47, 1881–1898, https://doi.org/10.1007/s00382-015-2938-x, 2016.

Li, H., Sheffield, J., and Wood, E. F.: Bias correction of monthly precipitation and temperature fields from Intergovernmental Panel on Climate Change AR4 models using equidistant quantile matching, J. Geophys. Res., 115, D10101, https://doi.org/10.1029/2009JD012882, 2010.

Maraun, D.: Nonstationarities of regional climate model biases in European seasonal mean temperature and precipitation sums, Geophys. Res. Lett., 39, L06706, https://doi.org/10.1029/2012GL051210, 2012.

Maraun, D., Wetterhall, F., Ireson, A. M., Chandler, R. E., Kendon, E. J., Widmann, M., Brienen, S., Rust, H. W., Sauter, T., Themeßl, M., Venema, V. K. C., Chun, K. P., Goodess, C. M., Jones, R. G., Onof, C., Vrac, M., and Thiele-Eich, I.: Precipitation downscaling under climate change: Recent developments to bridge the gap between dynamical models and the end user, Rev. Geophys., 48, RG3003, https://doi.org/10.1029/2009RG000314, 2010.

Masaki, Y., Hanasaki, N., Takahashi, K., and Hijioka, Y.: Propagation of biases in humidity in the estimation of global irrigation water, Earth Syst. Dynam., 6, 461–484, https://doi.org/10.5194/esd-6-461-2015, 2015.

Miao, C., Su, L., Sun, Q., and Duan, Q.: A nonstationary bias-correction technique to remove bias in GCM simulations, J. Geophys. Res.-Atmos., 121, 5718–5735, https://doi.org/10.1002/2015JD024159, 2016.

Mizukami, N., Clark, M. P., Slater, A. G., Brekke, L. D., Elsner, M. M., Arnold, J. R., and Gangopadhyay, S.: Hydrologic Implications of Different Large-Scale Meteorological Model Forcing Datasets in Mountainous Regions, J. Hydrometeorol., 15, 474–488, https://doi.org/10.1175/JHM-D-13-036.1, 2014.

Monteith, J. L.: Evaporation and environment. The state and movement of water in living organisms, Symposium of the society of experimental biology, Vol. 19, 205–234, 1965.

Müller Schmied, H., Eisner, S., Franz, D., Wattenbach, M., Portmann, F. T., Flörke, M., and Döll, P.: Sensitivity of simulated global-scale freshwater fluxes and storages to input data, hydrological model structure, human water use and calibration, Hydrol. Earth Syst. Sci., 18, 3511–3538, https://doi.org/10.5194/hess-18-3511-2014, 2014.

Nikulin, G., Bosshard, T., Yang, W., Bärring, L., Wilcke, R., Vrac, M., Vautard, R., Noel, T., Gutiérrez, J. M., Herrera, S., Fernández, J., Haugen, J. E., Benestad, R., Landgren, O. A., Grillakis, M., Tsanis, I., Koutroulis, A., Dosio, A., Ferrone, A., and Switanek, M.: Bias Correction Intercomparison Project (BCIP): an introduction and the first results, in EGU General Assembly Conference Abstracts, p. 2250, 2015.

Oki, T. and Sud, Y. C.: Design of Total Runoff Integrating Pathways (TRIP) – A Global River Channel Network, 2, 7–22, 1998.

O'Neill, B. C., Oppenheimer, M., Warren, R., Hallegatte, S., Kopp, R. E., Pörtner, H. O., Scholes, R., Birkmann, J., Foden, W., Licker, R., Mach, K. J., Marbaix, P., Mastrandrea, M. D., Price, J., Takahashi, K., van Ypersele, J.-P., and Yohe, G.: IPCC reasons for concern regarding climate change risks, Nat. Publ. Gr., 7, 28–37, https://doi.org/10.1038/NCLIMATE3179, 2017.

Papadimitriou, L. V., Koutroulis, A. G., Grillakis, M. G., and Tsanis, I. K.: High-end climate change impact on European runoff and low flows – exploring the effects of forcing biases, Hydrol. Earth Syst. Sci., 20, 1785–1808, https://doi.org/10.5194/hess-20-1785-2016, 2016.

Photiadou, C., van den Hurk, B., van Delden, A., and Weerts, A.: Incorporating circulation statistics in bias correction of GCM ensembles: hydrological application for the Rhine basin, Clim. Dynam., 46, 187–203, https://doi.org/10.1007/s00382-015-2578-1, 2016.

Piani, C., Weedon, G. P., Best, M., Gomes, S. M., Viterbo, P., Hagemann, S., and Haerter, J. O.: Statistical bias correction of global simulated daily precipitation and temperature for the application of hydrological models, J. Hydrol., 395, 199–215, https://doi.org/10.1016/j.jhydrol.2010.10.024, 2010.

Randall, D. A., Wood, R. A., Bony, S., Colman, R., Fichefet, T., Fyfe, J., Kattsov, V., Pitman, A., Shukla, J., Srinivasan, J., Stouffer, R. J., Sumi, A., and Tayler, K. E.: Climate Models and Their Evaluation, Clim. Chang. 2007 Phys. Sci. Basis, edited by: Solomon al., S., Qin, D., Manning, M., Chen, Z., Marquis, M., Averyt, K. B., Tignor, M., and Miller, H. L., Cambridge Univ. Press, 589–662, 2007.

Renner, M., Seppelt, R., and Bernhofer, C.: Evaluation of water-energy balance frameworks to predict the sensitivity of streamflow to climate change, Hydrol. Earth Syst. Sci., 16, 1419–1433, https://doi.org/10.5194/hess-16-1419-2012, 2012.

Rojas, R., Feyen, L., Dosio, A., and Bavera, D.: Improving pan-European hydrological simulation of extreme events through statistical bias correction of RCM-driven climate simulations, Hydrol. Earth Syst. Sci., 15, 2599–2620, https://doi.org/10.5194/hess-15-2599-2011, 2011.

Sankarasubramanian, A. and Vogel, R. M.: Hydroclimatology of the continental United States, Geophys. Res. Lett., 30, 1–4, https://doi.org/10.1029/2002GL015937, 2003.

Sharma, D., Das Gupta, A., and Babel, M. S.: Spatial disaggregation of bias-corrected GCM precipitation for improved hydrologic simulation: Ping River Basin, Thailand, Hydrol. Earth Syst.

Sci., 11, 1373–1390, https://doi.org/10.5194/hess-11-1373-2007, 2007.

Solman, S. A., Sanchez, E., Samuelsson, P., da Rocha, R. P., Li, L., Marengo, J., Pessacg, N. L., Remedio, A. R. C., Chou, S. C., Berbery, H., Le Treut, H., de Castro, M., and Jacob, D.: Evaluation of an ensemble of regional climate model simulations over South America driven by the ERA-Interim reanalysis: model performance and uncertainties, Clim. Dynam., 41, 1139–1157, https://doi.org/10.1007/s00382-013-1667-2, 2013.

Stocker, T., Qin, D., Plattner, G.-K., Tignor, M., Allen, S., Boschung, J., Nauels, A., Xia, Y., Bex, V., and Migley, P.: IPCC, 2013: Summary for Policymakers, in: Climate Change 2013: The Physical Science Basis, Contribution of Working Group I to the Fifth Assessment Report of the Intergovernmental Panel on Climate Change, Cambridge University Press, Cambridge, United Kingdom and New York, NY, USA, 2013.

Taylor, K. E., Stouffer, R. J., and Meehl, G. A.: An Overview of CMIP5 and the Experiment Design, B. Am. Meteorol. Soc., 93, 485–498, https://doi.org/10.1175/BAMS-D-11-00094.1, 2012.

Teng, J., Potter, N. J., Chiew, F. H. S., Zhang, L., Wang, B., Vaze, J., and Evans, J. P.: How does bias correction of regional climate model precipitation affect modelled runoff?, Hydrol. Earth Syst. Sci., 19, 711–728, https://doi.org/10.5194/hess-19-711-2015, 2015.

Teutschbein, C. and Seibert, J.: Bias correction of regional climate model simulations for hydrological climate-change impact studies: Review and evaluation of different methods, J. Hydrol., 456–457, 12–29, https://doi.org/10.1016/j.jhydrol.2012.05.052, 2012.

Themeßl, M. J., Gobiet, A., and Heinrich, G.: Empirical-statistical downscaling and error correction of regional climate models and its impact on the climate change signal, Climatic Change, 112, 449–468, https://doi.org/10.1007/s10584-011-0224-4, 2012.

Uppala, S. M., KÅllberg, P. W., Simmons, A. J., Andrae, U., Bechtold, V. D. C., Fiorino, M., Gibson, J. K., Haseler, J., Hernandez, A., Kelly, G. A., Li, X., Onogi, K., Saarinen, S., Sokka, N., Allan, R. P., Andersson, E., Arpe, K., Balmaseda, M. A., Beljaars, A. C. M., Berg, L. Van De, Bidlot, J., Bormann, N., Caires, S., Chevallier, F., Dethof, A., Dragosavac, M., Fisher, M., Fuentes, M., Hagemann, S., Hólm, E., Hoskins, B. J., Isaksen, L., Janssen, P. A. E. M., Jenne, R., Mcnally, A. P., Mahfouf, J.-F., Morcrette, J.-J., Rayner, N. A., Saunders, R. W., Simon, P., Sterl, A., Trenberth, K. E., Untch, A., Vasiljevic, D., Viterbo, P., and Woollen, J.: The ERA-40 re-analysis, Q. J. Roy. Meteor. Soc., 131, 2961–3012, https://doi.org/10.1256/QJ.04.176, 2005.

Vano, J. A., Das, T., and Lettenmaier, D. P.: Hydrologic Sensitivities of Colorado River Runoff to Changes in Precipitation and Temperature, J. Hydrometeorol., 13, 932–949, https://doi.org/10.1175/JHM-D-11-069.1, 2012.

Watanabe, S., Hajima, T., Sudo, K., Nagashima, T., Takemura, T., Okajima, H., Nozawa, T., Kawase, H., Abe, M., Yokohata, T., Ise, T., Sato, H., Kato, E., Takata, K., Emori, S., and Kawamiya, M.: MIROC-ESM 2010: model description and basic results of CMIP5-20c3m experiments, Geosci. Model Dev., 4, 845–872, https://doi.org/10.5194/gmd-4-845-2011, 2011.

Weedon, G. P., Gomes, S., Viterbo, P., Österle, H., Adam, J. C., Bellouin, N., Boucher, O., and Best, M.: The WATCH forcing data 1958–2001: A meteorological forcing dataset for land surface and hydrological models, Watch. Ed. Watch Tech. Rep., 22, 2010.

Weedon, G. P., Balsamo, G., Bellouin, N., Gomes, S., Best, M. J., and Viterbo, P.: The WFDEI meteorological forcing data set: WATCH Forcing Data methodology applied to ERA-Interim reanalysis data, Water Resour. Res., 50, 7505–7514, https://doi.org/10.1002/2014WR015638, 2014.

Yang, H. and Yang, D.: Derivation of climate elasticity of runoff to assess the effects of climate change on annual runoff, Water Resour. Res., 47, 1–12, https://doi.org/10.1029/2010WR009287, 2011.

Zulkafli, Z., Buytaert, W., Onof, C., Lavado, W., and Guyot, J. L.: A critical assessment of the JULES land surface model hydrology for humid tropical environments, Hydrol. Earth Syst. Sci., 17, 1113–1132, https://doi.org/10.5194/hess-17-1113-2013, 2013.

11

Constraining frequency–magnitude–area relationships for rainfall and flood discharges using radar-derived precipitation estimates: example applications in the Upper and Lower Colorado River basins, USA

Caitlin A. Orem and Jon D. Pelletier

Department of Geosciences, The University of Arizona, 1040 E. 4th Street, Tucson, AZ 85721, USA

Correspondence to: Caitlin A. Orem (oremc@email.arizona.edu)

Abstract. Flood-envelope curves (FECs) are useful for constraining the upper limit of possible flood discharges within drainage basins in a particular hydroclimatic region. Their usefulness, however, is limited by their lack of a well-defined recurrence interval. In this study we use radar-derived precipitation estimates to develop an alternative to the FEC method, i.e., the frequency–magnitude–area-curve (FMAC) method that incorporates recurrence intervals. The FMAC method is demonstrated in two well-studied US drainage basins, i.e., the Upper and Lower Colorado River basins (UCRB and LCRB, respectively), using Stage III Next-Generation-Radar (NEXRAD) gridded products and the diffusion-wave flow-routing algorithm. The FMAC method can be applied worldwide using any radar-derived precipitation estimates. In the FMAC method, idealized basins of similar contributing area are grouped together for frequency–magnitude analysis of precipitation intensity. These data are then routed through the idealized drainage basins of different contributing areas, using contributing-area-specific estimates for channel slope and channel width. Our results show that FMACs of precipitation discharge are power-law functions of contributing area with an average exponent of 0.82 ± 0.06 for recurrence intervals from 10 to 500 years. We compare our FMACs to published FECs and find that for wet antecedent-moisture conditions, the 500-year FMAC of flood discharge in the UCRB is on par with the US FEC for contributing areas of $\sim 10^2$ to $10^3 \, \text{km}^2$. FMACs of flood discharge for the LCRB exceed the published FEC for the LCRB for contributing areas in the range of $\sim 10^3$ to $10^4 \, \text{km}^2$. The FMAC method retains the power of the FEC

method for constraining flood hazards in basins that are ungauged or have short flood records, yet it has the added advantage that it includes recurrence-interval information necessary for estimating event probabilities.

1 Introduction

1.1 Flood-envelope curves

For nearly a century, the flood-envelope curves (FECs), i.e., curves drawn slightly above the largest measured flood discharges on a plot of discharge vs. contributing area for a given hydroclimatic region (Enzel et al., 1993), have been an important tool for predicting the magnitude of potential future floods, especially in regions with limited stream-gauge data. FECs assume that, within a given hydroclimatic region, maximum flood discharges for one drainage basin are similar to those of other drainage basins of the same area, despite differences in relief, soil characteristics, slope aspect, etc. (Enzel et al., 1993). This assumption enables sparse and/or short-duration flood records over a hydroclimatic region to be aggregated in order to provide more precise constraints on the magnitude of the largest possible (i.e., long-recurrence-interval) floods.

FECs reported in the literature have a broadly similar shape across regions of widely differing climate and topography. For example, FECs for the Colorado River Basin (Enzel et al., 1993), the central Appalachian Mountains (Miller, 1990; Morrison and Smith, 2002), the 17 hydrologic regions within the US defined by Crippen and Bue (1977), the US as a whole (Costa, 1987; Herschy, 2002), and China (Herschy, 2002) are all concave-down when plotted in log-log space, with maximum recorded flood discharges following a power-law function of contributing area for small contributing areas and increasing more slowly at larger contributing areas (i.e., the curve "flattens").

Traditional FECs also have the potential problem that the maximum flood associated with smaller drainage basins may be biased upward (or the floods of larger drainage basins biased downward) because there are typically many more records of floods in smaller drainage basins relative to larger drainage basins (because there are necessarily fewer large drainage basins in any hydroclimatic region). That is, the largest flood on record for small drainage basins within a hydroclimatic region likely corresponds to a flood of a larger recurrence interval compared with the largest flood on record for larger drainage basins. In this paper we present a method that includes recurrence-interval information and avoids any sample-size bias that might exist as a function of contributing area.

The use of FECs to quantify flood regimes is limited by the lack of recurrence-interval information (Wolman and Costa, 1984; Castellarin et al., 2005) and by the short length, incomplete nature, and sparseness of many flood discharge records. Without recurrence-interval information, the data provided by FECs are difficult to apply to some research and planning questions related to floods. In the US for example, the 100- and 500-year flood events are the standard event sizes that define flood risk for land planning and engineering applications (FEMA, 2001).

Previously published studies have looked at new approaches to approve upon the FEC method. Castellarin et al. (2005) took a probabilistic approach to estimating the exceedance probability of the FEC for synthetic flood data. The authors were able to relate the FECs of certain recurrence intervals to the correlation between sites, the number of flood observations, and the length of each observation. Later, Castellarin (2007) and Castellarin et al. (2009) applied these methods to real flood record data and extreme rainfall events for basins within northern–central Italy. Castellarin et al. (2009) also created depth–duration envelope curves of precipitation to relate extreme precipitation events to mean annual precipitation. This group of studies was successful in incorporating recurrence-interval information into the traditional FEC method. However, most of the models presented in these studies were completed with synthetic data or created for design storm processes and require additional analysis. Also, most of the precipitation data used in these past studies were collected using rain gauges (point sources), while only a small subset of data in Castellarin et al. (2009) was sourced from radar-derived precipitation estimates. In contrast to these studies we formulate a simplified method (i.e., the FMAC method) that is readily applicable to any region of interest and can be directly compared to already exist-

ing FECs. Also, we favor the use of spatially complete radar-derived precipitation estimates in order to apply our methods to ungauged basins.

To mitigate the uncertainty caused by short and incomplete flood discharge records, this study uses a space-for-time substitution (e.g., regionalization) to lengthen the record for a given contributing area. Previous studies have employed similar methods, including the index-flood procedure first described by Dalyrymple (1960) and expanded upon by many subsequent authors. The index-flood method uses data from multiple sites within a region to construct more accurate flood-quantile estimates than would be possible using a single site (Stedinger et al., 1993; Hosking and Wallis, 2005). This method can also be used on precipitation data, where it is referred to as the station-year method (Buishand, 1991). The index-flood method is based on two major assumptions: (1) that observations from two or more basins are independent; and (2) that observations follow the same distribution (Wallis et al., 2007).

Here we use a regionalization method similar to the index-flood method in order to calculate rainfall-intensity values associated with specific recurrence intervals. The assumption of statistical independence of rainfall (and associated flood) observations is one that we assume in this study but understand may not be true for all samples in our natural dataset. This assumption is difficult to definitively prove with natural data (Hosking and Wallis, 2005). For example, a large rainfall event may affect two basins in a similar way and therefore create correlated maximum rainfall-intensity values. This spatial correlation is difficult to avoid and may cause biased results. However, it has been shown that the index-flood method can be used in the absence of fully statistically independent observations and still give robust results (Hosking and Wallis, 1988; Hosking and Wallis, 2005). The assumption that observations are sampled from the same distribution is also somewhat difficult to prove with natural data, but by knowing the study areas well a researcher can identify regions with similar rainfall and flood mechanisms. Many examples of this type of area analysis can be found in the literature, including Soong et al. (2004), who separated rural streams in Illinois into hydrological regions based on basin morphology and soil characteristics. Soong et al. (2004) used regionalization in their study to increase the amount of flood data available for frequency analysis. Wallis et al. (2007) employed a similar regionalization method to identify hydroloclimatic regions in their study of precipitation frequency in Washington. It should be noted that FECs in general use this type of regionalization approach to analyze maximum flood data for hydroclimatic regions with similar flood mechanisms. In this study we similarly attempt to analyze regions based on their basic rainfall mechanisms, in this case by separating the Upper and Lower Colorado River basins.

In this study, a new method for estimating flood discharges associated with user-specified recurrence intervals is introduced that uses radar-derived precipitation estimates (in this case rainfall only), combined with the diffusion-wave flow-routing algorithm, to create frequency–magnitude–area curves (FMACs) of flood discharge. Our method (i.e., the FMAC method) retains the power of the FEC approach in that data from different drainage basins within a hydroclimatic region are aggregated by contributing area, thereby enabling large sample sizes to be obtained within each contributing-area class in order to more accurately constrain the frequencies of past extreme flood events and hence the probabilities of future extreme flood events within each class. The method improves upon the FEC approach in that the complete spatial coverage of radar-derived precipitation estimates provides for large sample sizes of most classes of contributing area (larger contributing areas have fewer samples). The radar-derived precipitation estimates include only rainfall and therefore snow and other types of precipitation are not included in the study. The precipitation estimates are then used to predict flood discharges associated with specific recurrence intervals by first accounting for water lost to infiltration and evapotranspiration using runoff coefficients appropriate for different contributing areas and antecedent-moisture conditions, and then routing the available water using a flow-routing algorithm. Predicted flood discharges are presented as FMACs on log-log plots, similar to traditional FECs, except that the method predicts a family of curves, one for each user-defined recurrence interval. These plots are then compared to FECs for the study region (Enzel et al., 1993) and the US (Costa, 1987).

1.2 Study area

This study focuses on the Upper and Lower Colorado River basins (UCRB and LCRB, respectively; Fig. 1) as example applications of the FMAC method. Although the methods we develop are applied to the UCRB and LCRB in the western US in this study, the methods are applicable to any region of interest where radar-derived precipitation estimates are available (i.e., the entire US and at least 22 countries around the world; Li, 2013; RadarEU, 2014). We focus on the UCRB and LCRB because they have been a focus of flood-hazard assessment studies in the western US and hence the FECs available for them are of especially high quality. In addition, the distinctly different hydroclimatic regions of the UCRB and LCRB (Sankarasubramanian and Vogel, 2003) make working in these regions an excellent opportunity to test and develop the new methods of this study on different precipitation patterns and storm types.

Precipitation and flooding in the LCRB are caused by convective-type storms, including those generated by the North American Monsoon (NAM), and frontal-type and tropical storms sourced from the Pacific Ocean and the Gulf of California (House and Hirschboeck, 1997; Etheredge et al., 2004). In the UCRB, the influence of the NAM and tropical storms is diminished and floods are generally caused by

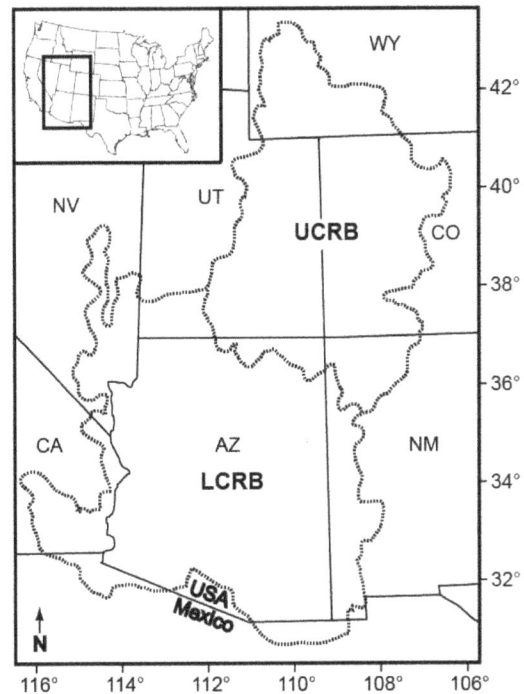

Figure 1. Map showing the locations of the Upper and Lower Colorado River basins (UCRB and LCRB, respectively) outlined by the dotted line.

Pacific frontal-type storms (Hidalgo and Dracup, 2003). In both regions, the El Niño–Southern Oscillation (ENSO) alters the frequency and intensity of the NAM, tropical storms, and the Pacific frontal systems, and can cause annual variations in precipitation and flooding (House and Hirschboeck, 1997; Hidalgo and Dracup, 2003). Winter storms in both regions are also intensified by the occurrence of atmospheric rivers (Dettinger et al., 2011), which can cause total winter precipitation to increase up to approximately 25 % (Rutz and Steenburgh, 2012). The radar-derived precipitation estimates used in this study record this natural variability in precipitation in the two regions.

The methods used in this study to calculate rainfall and flood discharges of specified recurrence intervals from radar-derived precipitation estimates require a few main assumptions. The first assumption is that of climate stationarity; i.e., the parameters that define the distribution of floods do not change through time (Milly et al., 2008). Climate is changing and these changes pose a challenge to hazard predictions based on the frequencies of past events. Nevertheless, stationarity is a necessary assumption for any probabilistic analysis that uses past data to make future predictions. The results of such analyses provide useful starting points for more comprehensive analyses that include the effects of future climate changes. The second assumption is that the sample time interval is long enough to correctly represent the current hydroclimatic state (and its associated precipitation patterns and flood magnitudes and risks) of the specified study area. Our study uses data for the 1996 to 2004 water years and therefore may be limited by inadequate sampling of some types of rare weather patterns and climate fluctuations within that time interval. To address whether or not the sample time interval used in this study includes major changes in circulation and weather patterns, and therefore is a good representation of climate in the CRB, we investigated the effect of the ENSO on rainfall intensity within the UCRB and LCRB. ENSO is a well-known important influence on the hydroclimatology of the western US (Hidalgo and Dracup, 2003; Cañon et al., 2007). In general, winter precipitation in the southwestern US increases during El Niño events and decreases during La Niña events (Hidalgo and Dracup, 2003). The opposite effects are found in the northwestern portions of the US (including the UCRB; Hidalgo and Dracup, 2003). The last assumption of the method is that all basins of similar contributing area respond similarly to input rainfall, i.e., that they have similar flood-generating and flow-routing mechanisms. Specifically, the method assumes that basins of similar contributing area have the same runoff coefficient, flow-routing parameters, basin shape, and channel length, width, and slope. This assumption is necessary in order to aggregate data into discrete contributing-area classes so that the frequency of extreme events can be estimated from relatively short-duration records. In this study, high-recurrence-interval events (i.e., low-frequency events) can be considered despite the relatively short length of radar-derived-precipitation-estimate records because the number of samples in the radar-derived record is extremely large, especially for small contributing areas and short-duration floods. For example, for a 1 h time-interval-of-measurement and a contributing area of $4096\,km^2$ event in the UCRB, there are approximately 40 (number of spatial-scale samples) times 55 000 (number of temporal-scale samples in 9 years of data) samples of rainfall-intensity values (and associated modeled discharges obtained via flow routing). As contributing area and time intervals of measurement increase there are successively fewer samples, within any particular hydroclimatic region, thus increasing the uncertainty of the resulting probability assessment for larger areas and longer time periods.

2 Next-Generation-Radar (NEXRAD) data

The specific radar-derived precipitation estimates we use in this study come from the Stage III Next-Generation-Radar (NEXRAD) gridded product, which is provided for the entire US, Guam, and Puerto Rico. NEXRAD was introduced in 1988 with the introduction of the Weather Surveillance Radar 1988 Doppler, or WSR-88D, network (Fulton et al., 1998). The WSR-88D radars use the Precipitation Processing System (PPS), a set of automated algorithms, to produce precipitation intensity estimates from reflectivity data.

Reflectivity values are transformed to precipitation intensities through the empirical $Z-R$ power-law relationship

$$Z = \alpha R^\beta, \tag{1}$$

where Z is precipitation rate ($mm\,h^{-1}$), α and β are derived empirically and can vary depending on location, season, and other conditions (Smith and Krajewski, 1993), and R is reflectivity ($mm^6\,m^{-3}$; Smith and Krajewski, 1993; Fulton et al., 1998; Johnson et al., 1999). Precipitation intensity data are filtered and processed further to create the most complete and correct product (Smith and Krajewski, 1993; Smith et al., 1996; Fulton et al., 1998; Baeck and Smith, 1998). Further information and details about PPS processing are thoroughly described by Fulton et al. (1998).

Stage III NEXRAD gridded products are Stage II precipitation products mapped onto the Hydrologic Rainfall Analysis Project (HRAP) grid (Shedd and Fulton, 1993). Stage II data are hourly precipitation intensity products that incorporate both radar reflectivity and rain-gauge data (Shedd and Fulton, 1993) in an attempt to make the most accurate precipitation estimates possible. The HRAP grid is a polar coordinate grid that covers the conterminous US, with an average grid size of 4 km by 4 km, although grid size varies from approximately 3.7 km (north to south) to 4.4 km (east to west) in the southern and northern US, respectively (Fulton et al., 1998).

3 Methods

3.1 NEXRAD data conversion and sampling

NEXRAD Stage III gridded products (hereafter NEXRAD products) for an area covering the Colorado River Basin from 1996 to 2005 were downloaded from the NOAA HDSG website (http://dipper.nws.noaa.gov/hdsb/data/nexrad/cbrfc_stageiii.php) for analysis. The data files were converted from archived XMRG files to ASCII format (each data file representing the mean rainfall intensity within each 1 h interval) using the xmrgtoasc.c program provided on the NOAA HDSG website. The ASCII data files were then input into a custom program written in IDL for analysis.

3.2 Rainfall sampling over space

In this study we quantified hourly rainfall intensities ($mm\,h^{-1}$) over square idealized drainage basins (i.e., not real drainage basins, but square drainage basins as shown schematically in Fig. 2a as brown squares) of a range of areas from 16 to $11\,664\,km^2$ (approximately the contributing area of the Bill Williams River, AZ, for readers familiar with the geography of the western US) by successively spatially averaging rainfall-intensity values at HRAP pixel-length scales of powers of 2 (e.g., 4, $16\,pixel^2$) and 3 (e.g., 9, $81\,pixel^2$; Fig. 2, Step 1). Spatial averaging is done by both powers of 2 and 3

(a)

Step 1: rainfall sampling over space and time

For "Time Interval" = 1, 2, 4, 8, 16, 32, 64 h
  Rainfall = Average (Rainfall) into blocks of time of length "Time Interval" (see below)
  For "Area" = 16, 64, 144, 256, 1024, 1296, 4096, 11664 km^2
    For "Basins" within drainage basin defined by user (see below)
      For "Time" 1 to end of 9-year record in increments of "Time Interval"
        Find max Rainfall in consecutive blocks of non-zero Rainfall a.k.a. storm event (see below)
        n = count of max Rainfall values
        Record max storm-event Rainfall ("Time Interval", "Area", n)
      End For "Time" Loop
    End For "Basins" Loop
  End For "Area" Loop
End For "Time Interval" Loop
Result: 7 x 8 x n Array of Rainfall (Time Interval, Area, n)

Example of gridded rainfall intensity aggregated over spatial scales

Example of rainfall intensity averaged over temporal scales

16 km^2 basins   64 km^2 basins

| 0 | 4 | 5 | **7** | 0 | 0 | 7 | 1 | 1h data |

| 2 | **6** | 0 | 4 | 2h data |

| **4** | 2 | 4h data |

| **3** | 8h data |

— Study area outline
☐ Basins included (within study area)
☐ Max rainfall in storm event

Step 2: rainfall recurrence interval calculations

For "Time Interval" = 1, 2, 4, 8, 16, 32, 64 h
  For "Area" = 16, 64, 144, 256, 1024, 1296, 4096, 11664 km^2
    Rank Rainfall from highest to lowest
    For Recurrence Interval, "RI", = 10, 50, 100, 500 years
      m = (n +1)/"RI" (see Equation 2)
      Record Rainfall at rank m ("Time Interval", "Area", "RI")
    End For "RI" Loop
  End For "Area" Loop
End For "Time Interval" Loop
Result: 7 x 8 x 4 Array of Rainfall (Time Interval, Area, RI)

(b)

Step 3: Q_{pm} calculation

For Antecedent Moisture Condition, "AMC" = dry, medium, wet
  For "Time Interval" = 1, 2, 4, 8, 16, 32, 64 h
    For "Area" = 16, 64, 144, 256, 1024, 1296, 4096, 11664 km^2
      Runoff Coefficient, RC, set as function of "Area" (see Fig. 3)
      For "RI" = 10, 50, 100, 500 years
        Q_p = Rainfall * "Time Interval" * "Area"
        Q_{pm} = Q_p * RC("AMC") (see Equation 3)
      End For "RI" Loop
    End For "Area" Loop
  End For "Time Interval" Loop
End For "AMC" Loop
Find peak Q_{pm} value from all values for different time intervals and use as max Q_{pm} value
Result: 3 x 8 x 4 Array of Q_{pm} (AMC, Area, RI)

Step 4: Q_{fd} calculation

For "AMC" = dry, medium, wet
  For "Area" = 16, 64, 144, 256, 1024, 1296, 4096, 11664 km^2
    For "RI" = 10, 50, 100, 500 years
      Slope, S, set as function of "Area" (see Figure 4A)
      Channel Width, W, set as function of "Area" (see Figure 4B)
      For "Iteration" = 1 to when water depth, h, change < 0.001
        If "Iteration" = 1, then h = 1 m
        Else when "Iteration" > 1, then h = last calculated h
        Channel velocity, V = solve Manning's Equation (see Equation 4)
        Drift velocity a = (1+a$_0$)V (see Equation 6)
        Diffusion coefficient b^2= solve Diffusion Equation (see Equation 7)
        For "Timestep" = 1 to end
          For "Pixel" = 1 to channel length L
            Q_{pm} added to 1D channel following triangular width function A(x)
            Solve Unit Impulse Function q_{fd} (see Equation 5)
            Record peak discharge Q_{fd} ("AMC", "Time Interval", "Area", "RI")
          End For "Pixel" Loop
        End For "Timestep" Loop
        If h change < 0.001, then record peak Q_{fd} ("AMC", "Time Interval", "Area", "RI")
      End for "Iteration" Loop
    End For "RI" Loop
  End For "Area" Loop
End For "AMC" Loop
Result: 3 x 8 x 4 Array of Q_{fd} (AMC, Area, RI)

Q_{pm}

Flow through main channel along diagonal axis (length of L) using diffusion-wave flow-routing algorithm

Idealized basin cross section w/ triangular area function

Q_{fd}

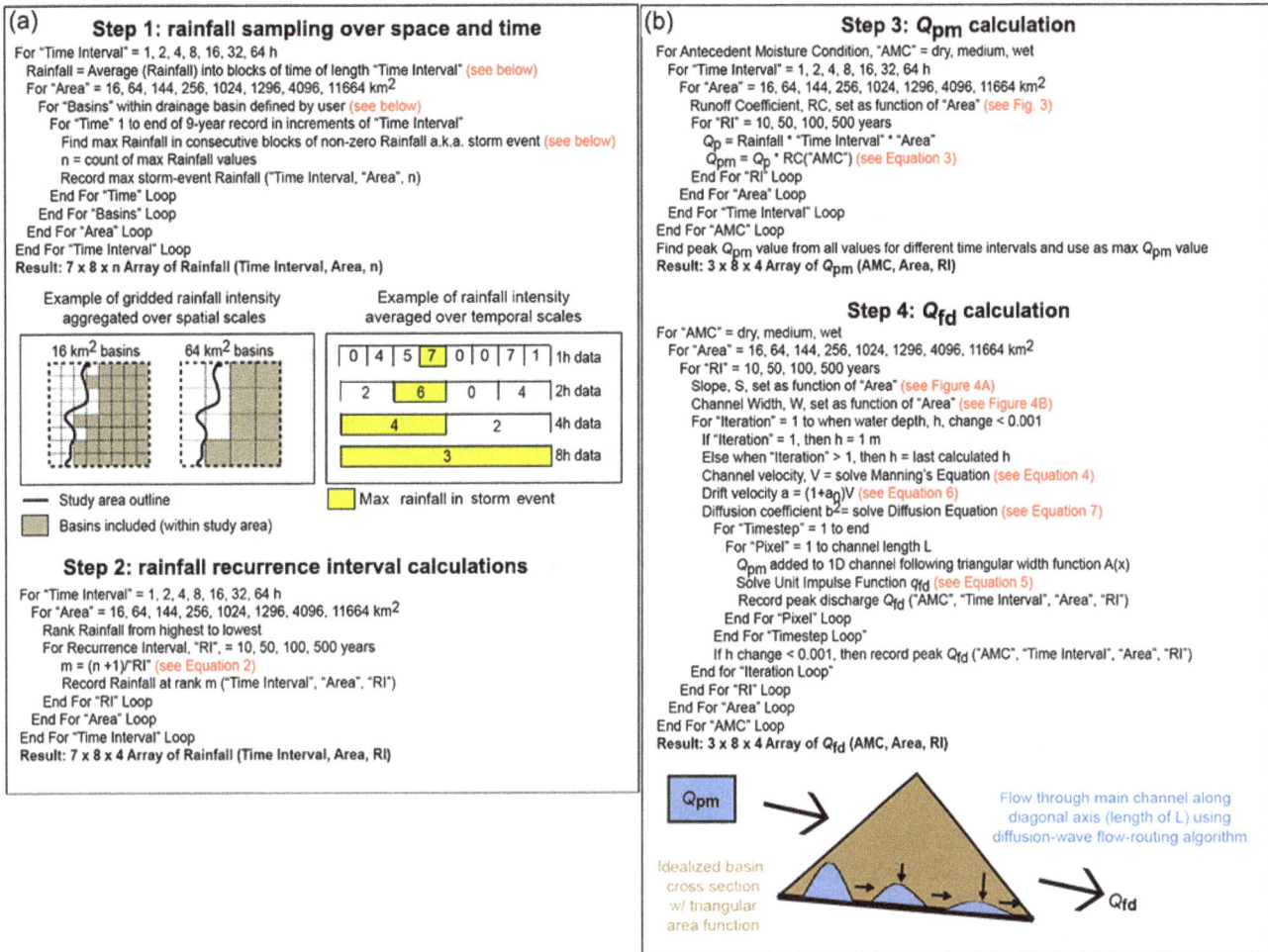

Figure 2. (a) Pseudocode describing the methods of the paper with schematic diagrams shown below pseudocode in some cases. Equations within the text and other figures are referenced in red text. **(b)** Pseudocode describing the methods of the paper with schematic diagrams shown below pseudocode in some cases. Equations within the text and other figures are referenced in red text.

simply to include more points on the FMACs than would result from using powers of 2 or 3 alone. The number of samples within each contributing area class limited the range of contributing areas used in this study; i.e., at larger contributing areas there were too few samples to successfully apply the frequency analysis.

UCRB and LCRB boundaries from GIS hydrologic unit layers created by the USGS and provided online through the National Atlas site (http://gdg.sc.egov.usda.gov) were projected to HRAP coordinates using the methods of Reed and Maidment (2006). These boundaries were used to delineate the region from which rainfall data were sampled from the NEXRAD products; i.e., when averaging rainfall data by powers of 2 and 3, a candidate square drainage basin was not included in the analysis if any portion of the square fell outside of the boundaries of the UCRB or LCRB (Fig. 2a). Throughout the analysis, the HRAP pixel size was approximated by a constant 4 km by 4 km size despite the fact that

HRAP pixel sizes vary slightly as a function of latitude (Reed and Maidment, 2006). Our study's drainage basins span latitudes between approximately 31 and 43° N, resulting in a maximum error of 15 %. However, by keeping the pixel size constant, all pixels could be treated as identical in size and shape, allowing us to sample the NEXRAD products in an efficient and automated way over many spatial scales.

For larger contributing areas, necessarily fewer samples are available within a given hydroclimatic region, thus increasing the uncertainty associated with the analysis for those larger contributing-area classes. For the UCRB and LCRB specifically, the uncertainty in the analysis becomes significant for contributing-area classes equal to and larger than $\sim 10^3$ to 10^4 km^2, depending on the recurrence interval being analyzed. Of course, if the hydroclimatic region is defined to be larger, more samples are available for each contributing-area class and hence larger basins can be analyzed with confidence.

3.3 Rainfall sampling over time

In addition to computing rainfall intensities as a function of spatial scale, we averaged rainfall intensities as a function of the time interval of measurement ranging from 1 to 64 h in powers of 2 by averaging hourly rainfall-intensity records over the entire 9-year study period (Fig. 2, Step 1). This range in time intervals was chosen in order to capture rainfall events that last on the order of ~ 1 h (convective-type storms) to days (frontal-type storms).

Rainfall data were sampled temporally by taking the maximum value of each storm event. Storm events were identified as consecutive non-zero rainfall-intensity values separated by instances of zero values in time for each temporal scale. This allows for multiple maximum rainfall values in time to be sampled within a year and throughout the entire 9-year study period. This sampling method is similar to that used in the peak over threshold (POT) method typically used on discharge data where a minimum threshold value is set and maximum peaks above the threshold value are recorded as maximum events. Here we set the minimum threshold value to zero, and hence the maximum values of all individual storm events are considered in the analysis.

3.4 Rainfall recurrence-interval calculations

To determine the rainfall-intensity values with a user-specified recurrence interval, maximum rainfall intensities of storm events sampled from the NEXRAD data for each contributing-area and time-interval-of-measurement class were first ranked from highest to lowest (Fig. 2a, Step 2). The relationship between recurrence intervals and rank in the ordered list is given by the probability-of-exceedance equation (i.e., the frequency–rank relationship):

$$\text{RI} = \frac{(n+1)}{m}, \tag{2}$$

where RI is the recurrence interval (yr), defined as the inverse of frequency (yr^{-1}) or probability of exceedance, n is the total number of samples in each contributing area and time-interval-of-measurement scaled to units in years (resulting in units of yr), and m is the rank of the magnitude ordered from largest to smallest (unitless). Here the recurrence interval is prescribed (10, 50, 100, and 500 years); then, the rank associated with this recurrence interval is computed using the frequency–rank relationship (Eq. 2). The resulting rainfall intensities associated with a user-specified recurrence interval and contributing-area and time-interval-of-measurement class were then used to calculate the Q_p value.

At the end of the calculations described above we have datasets of rainfall-intensity values for each combination of the eight contributing-area classes, the seven time-interval-of-measurement classes, and the four recurrence intervals. We then find the maximum values of rainfall intensity associated with a given contributing-area class and recurrence interval among all values of the time-interval-of-measurement

class (i.e., the values calculated for 1 to 64 h time intervals). This step is necessary in order to find the maximum values for a given contributing area class and recurrence interval independent of the time-interval-of-measurement, i.e., independent of storm durations and associated types of storms. The maximum values are used to be consistent with the methods of the traditional FECs where the points represent the largest possible storm for a given contributing area. These maximum values are used to calculate Q_p and Q_{fd} (see next section).

3.5 Rainfall and runoff calculations

The first variable calculated from the maximum rainfall intensities found for each contributing-area class and recurrence interval is the precipitation (here rainfall only) discharge, Q_p. The variable Q_p is defined as the average rainfall intensity over a basin and time interval of measurement multiplied by the contributing area, resulting in units of $\text{m}^3\,\text{s}^{-1}$. This is a simple calculation resulting in a "discharge" of rainfall to a basin. Q_p is the input value for the flow-routing algorithm that we employ to calculate the peak flood discharge (Fig. 2b, Step 3).

The flow-routing algorithm we employ does not explicitly include infiltration and other losses that can further reduce peak flood discharge relative to input to the basin, Q_p. In this study we modeled infiltration and evaporation losses by simply removing a volume of water per unit time equal to one minus the runoff coefficient, i.e., the ratio of runoff to rainfall over a specified time interval, for three antecedent-moisture scenarios (wet, med, and dry). We estimated runoff coefficients for each contributing-area class and each of three antecedent-moisture scenarios using published values for annual runoff coefficients for large basins within the UCRB and LCRB (Rosenburg et al., 2013) and published values for event-based runoff coefficients for small basins modeled with a range of antecedent-moisture conditions by Vivoni et al. (2007) (Fig. 3). On average, estimated runoff coefficients are higher for smaller and/or initially wetter basins. We found the dependence of runoff coefficients on contributing area and antecedent moisture to be similar despite the large difference in timescales between event-based and annual values. Despite the difference in geographic region between our study site and that of Vivoni et al. (2007) (they studied basins in Oklahoma), the runoff coefficients they estimated are likely to be broadly applicable to the LCRB and UCRB given that basin size and antecedent moisture are the primary controls on these values (climate and soil types play a lesser role except for extreme cases).

We applied the estimated runoff coefficients for all three antecedent-moisture scenarios by simply using them to remove a portion of the Q_p calculated for a specific time interval and basin area:

$$Q_{pm} = C \cdot Q_p, \tag{3}$$

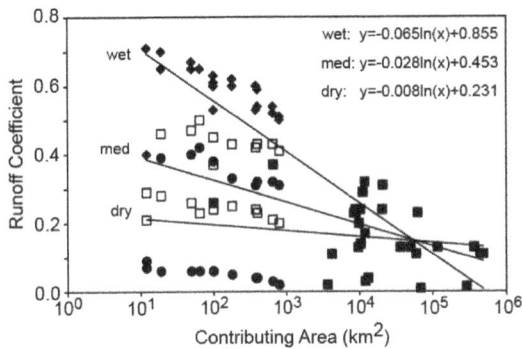

Figure 3. Logarithmic relationships between runoff coefficients and contributing area using modeled data for wet (filled diamonds), medium (open squares), and dry (filled circles) antecedent-moisture conditions (Vivoni et al., 2007) and measured data for larger contributing areas (filled squares; Rosenburg et al., 2013). The medium (open squares) and dry (filled circles) data separate into two distinct groups relating to the precipitation event used to model them, with the lower group and higher group relating to a 12 h, 1 mm h^{-1} event and 1 h, 40 mm h^{-1} event, respectively. All points were used in the least-squares weighed-regression analysis.

where C is the runoff coefficient calculated for the specific basin area and antecedent-moisture scenario under evaluation. The newly formed Q_{pm} is now the Q_p value for the wet, medium, or dry antecedent-moisture scenario under analysis for each given recurrence-interval and contributing-area class.

3.6   Flood discharge calculations

The second variable calculated in this study, and the end-result of our methods, is the peak flood discharge, Q_{fd}. The variable Q_{fd} is the peak flood discharge (m^3 s^{-1}) calculated via the diffusion-wave flow-routing algorithm for a hypothetical flood triggered by a rainfall discharge, Q_{pm}, input uniformly over the time interval of measurement to idealized square basins associated with each contributing-area class (Fig. 2b, Step 4).

The flow-routing algorithm routes flow along the main-stem channel of idealized square basins with sizes equal to the contributing area of each contributing-area class. The choice of a square basin is consistent with the square sample areas (see Sect. 3.1) and it allows for basin shape to remain the same (and therefore comparable) over the range of contributing areas used in this study. The main-stem channel, with a length of L (m), was defined as the diagonal distance from one corner to the opposite corner across the square basin (i.e., L is equal to the square root of 2 times the area of the square basin). This main-stem channel was used in conjunction with a normalized area function to represent the shape of the basin and the routing of runoff through the drainage basin network. By including the normalized area function, we can account for geomorphic dispersion (i.e., the

attenuation of the flood peak due to the fact that rainfall that falls on the landscape will take different paths to the outlet and hence reach the outlet at different times) in our analyses. The normalized area function, $A(x)$ (unitless), is defined as the portion of basin area, $A_L(x)$ (m^2), that contributes flow to the main-stem channel within a given range of distances (x) from the outlet, normalized by the total basin area, A_T (m^2; Mesa and Mifflin, 1986; Moussa, 2008). The normalized area function is assumed to be triangular in shape, with a maximum value at the midpoint of the main-stem channel from the outlet. Area functions, and related width functions, from real basins used in other studies show this triangular shape in general (Marani et al., 1994; Rinaldo et al., 1995; Veneziano et al., 2000; Rodriguez-Iturbe and Rinaldo, 2001; Puente and Sivakumar, 2003; Saco and Kumar, 2008), although not all basins show this shape. The triangular area function has been shown to approximate the average area function of basins and that the peak discharge and time to peak discharge is likely more important to the shape of the flood wave (Henderson, 1963; Rodriguez-Iturbe and Valdes, 1979).

A one-dimensional channel with simplified width and along-channel slope appropriate for channels in the CRB is used to approximate the geometry of the main-stem channel of the idealized basin in the flow-routing algorithm. In addition, values for channel slope, S (m m^{-1}), and channel width, w (m), are assigned based on the contributing area of the idealized basin and the results of a least-squares regression to channel-slope and channel-width data from the CRB. We assume here that the assigned channel slopes and widths represent the average value for the entire idealized basin. To find the best approximations for channel slope and width values, we developed formulae that predict average channel slope and channel width as a function of contributing area based on a least-squares fit of the logarithms of slope, width, and contributing area based on approximately 100 sites in the Colorado River Basin (CRB; Fig. 4). The data used in these least-squares regressions included slope, width, and contributing area information from all sites in the LCRB and southern UCRB presented in Moody et al. (2003) and additional sites from USGS stream-gauge sites from across the CRB.

The assigned channel slope and width values, together with the values of Q_{pm} modified for each antecedent-moisture scenario, were used to calculate the depth-average velocities, V (m s^{-1}), in hypothetical one-dimensional main-stem channels of idealized square drainage basins corresponding to each contributing-area and time-interval-of-measurement class. In this study, flow velocity is not modeled over space and time, but rather is set at a constant value appropriate for the peak discharge using an iterative approach that solves for the peak depth-averaged flow velocity, uses that velocity to compute the parameters of the diffusion-wave-routing algorithm, routes the flow, and then computes an updated estimate of peak depth-averaged velocity. To calculate the depth-averaged velocity, V, we used Manning's

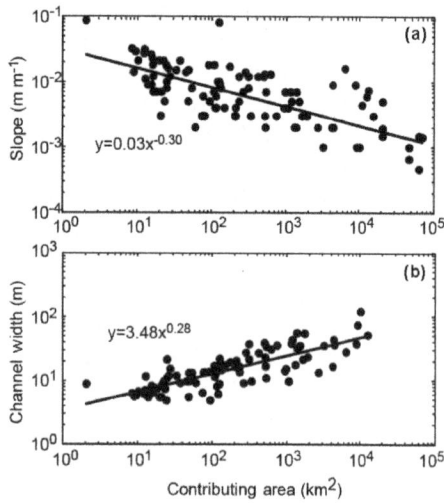

Figure 4. Power-law relationships between channel slope and contributing area **(a)** and channel width and contributing area **(b)** for the Colorado River Basin.

equation, i.e.,

$$V = \frac{1}{n_M} R^{\frac{2}{3}} S^{\frac{1}{2}}, \tag{4}$$

where n_M is Manning's n (assumed to be equal to 0.035), R is the hydraulic radius (m) calculated with the assigned channel width, and S (m m^{-1}) is the assigned channel slope. In order to calculate R, the water depth, h, of the peak discharge needed to be determined. In this study h was iteratively solved for based on the peak-flow conditions (i.e., the depth-averaged velocity, V, associated with the peak-flood discharge, Q_{fd}) with h set at 1 m for the first calculation of the flow-routing algorithm. At the end of each calculation, h is recalculated using Manning's equation. These iterations continue until the water depth converges on a value (i.e., the change from the last calculation of h to the next calculation of h is ≤ 0.1 m) corresponding to a specific recurrence interval, contributing-area class, and time-interval-of-measurement class.

The method we used to model flow through the main-stem channel is the diffusion-wave flow-routing algorithm. This approach is based on the linearized Saint-Venant equations for shallow-water flow in one dimension. To find a simpler, linear solution to Saint-Venant equations, Brutsaert (1973) removed the acceleration term from the equations, leaving the diffusion and advection terms that often provide a reasonable approximation for watershed runoff modeling (Brutsaert, 1973). Leaving the diffusion term in the flow-routing algorithm includes hydrodynamic dispersion of the flood wave in the calculation of the flood hydrograph. In the case where the initial condition is given by a unit impulse function (Dirac function), the cell response function of the channel, q_d

(units of s^{-1}), is given by

$$q_d = \frac{x}{(2\pi)^{1/2} b t_r^{3/2}} \exp\left[-\frac{(a - at_r)^2}{2b^2 t_r}\right], \tag{5}$$

where x is the distance along the channel from the location where the impulse is input to the channel, t_r is the time since the impulse was input into the channel, and the drift velocity a (m s^{-1}) and diffusion coefficient b^2 (m^2 s^{-1}) are defined as

$$a = (1 + a_0) V, \tag{6}$$

$$b^2 = \frac{V^3}{g S F r^2}\left(1 - a_0^2 F r^2\right), \tag{7}$$

where Fr is the Froude number, g is the acceleration due to gravity (m s^{-2}), and a_0 is a constant equal to $2/3$ when using Manning's equation (Troch et al., 1994). The large floods modeled in this study are assumed to have critical-flow conditions, and therefore the Froude number is set to a constant value of 1.

The unit response discharge, q_{fd} (m^2 s^{-1}), at the outlet of a drainage basin can be computed from Eqs. (3) to (5) by integrating the product of the cell response function $q_d(x, t)$ corresponding to a delta-function input of the normalized area function, $A(x)$, i.e., the spatial distribution of rainfall input. The integral is given by

$$q_{fd}(t_r) = \int_0^{t_p} \frac{Q_p}{w} dt' \int_0^L q_d\left(x, t_r - t'\right) A(x) dx, \tag{8}$$

where t_p is the time interval of measurement over which the unit impulse input (i.e., Q_p) is applied to the idealized square drainage basin, and t_r is the time after the input of the unit impulse that is long enough to capture the waxing and waning portions and the flood peak of the flood wave. The final peak discharge value, or Q_{fd} (m^3 s^{-1}), was calculated by multiplying the unit discharge q_{fd} (m^2 s^{-1}) by the channel width found through the formula derived from CRB data in Fig. 4 and then selecting the largest value from the resulting hydrograph.

3.7 Estimation of uncertainty

Confidence intervals (i.e., uncertainty estimates) were calculated to quantify the uncertainty in calculated rainfall intensities and associated Q_p and Q_{fd} values. In this study we estimated confidence intervals using a non-parametric method similar to that used to calculate quantiles for flow-duration curves (Parzen, 1979; Vogel and Fennesset, 1994). Like quantile calculations, which identify a subset of the ranked data in the vicinity of each data point to estimate expected values and associated uncertainties, we estimated confidence intervals for our predictions based on the difference in Q_p values between each point and the next largest value in the ranked list. This approach quantifies the variation in the rainfall-intensity value for a given contributing area and recurrence interval. In some cases the calculated uncertainties for rainfall intensities and associated Q_p and Q_{fd} values are infinite due to the values being past the frequency–magnitude distribution; i.e., there are not enough samples for these values to be determined and there are no finite numbers to sample. These values are not used in this study.

The resulting confidence intervals of rainfall intensity were used to calculate confidence intervals for Q_p and Q_{fd}. Confidence intervals for Q_p values were equal to the confidence intervals for rainfall intensity propagated through the calculation of Q_p (i.e., multiplying by contributing area). Confidence intervals for Q_{fd} values were calculated to be the same proportion of the Q_{fd} value as that set by the rainfall-intensity value and its confidence intervals. For example, if the upper confidence interval was 120% of a rainfall-intensity value, the upper confidence interval for the Q_{fd} value associated with the rainfall-intensity value is assumed to be 120 % of the Q_{fd} value. This approach to propagation of uncertainty treats all other variables in the calculations as constants, and additional uncertainty related to regression analyses of variables used in the flow-routing algorithm such as slope, channel width, and runoff coefficients was not included.

3.8   Testing the effects of climate variability

To quantify the robustness of our results with respect to climate variability, we separated the NEXRAD data into El Niño and La Niña months using the multivariate ENSO index (MEI). All months of data with negative MEI values (La Niña conditions) were run together to calculate the rainfall intensity and Q_p values for contributing areas of 16, 256, and 4096 km^2, time intervals of 1 to 64 h, and for 10-, 50-, 100-, and 500-year recurrence intervals. This was repeated with all months of data with positive MEI values (El Niño conditions). Figure 5 shows the distribution of negative and positive MEI values during the 1996 to 2004 water years used in this study.

4   Results

4.1   Channel characteristics and runoff coefficients

Least-squares regression of channel slopes and channel widths from the CRB vs. contributing area was used to estimate channel slope, channel width, and runoff coefficients for each idealized basin of a specific contributing-area class. Channel slope decreases as a power-law function of contributing area with an exponent of -0.30 ($R^2 = 0.39$), whereas channel width increases as a power-law function of contributing area with an exponent of 0.28 ($R^2 = 0.65$; Fig. 4). These results follow the expected relationships among channel slopes, widths, and contributing area; i.e., as

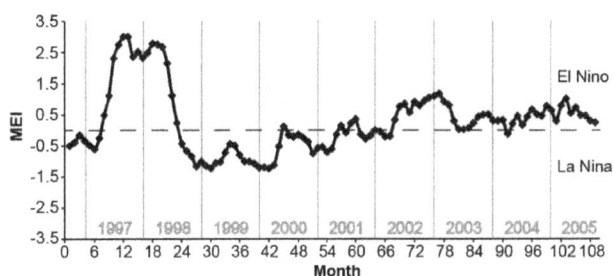

Figure 5. Multivariate ENSO index (MEI) of months included in Stage III NEXRAD gridded products. Months are numbered from September 1996 to September 2005 with years shown in gray. Dashed black line MEI equal to zero. A positive MEI indicates El Niño conditions, while a negative MEI indicates La Niña conditions.

contributing area increases, the channel slope decreases and the channel width increases.

Runoff coefficients for wet, medium, and dry antecedent-moisture conditions all decrease with increasing contributing area following a logarithmic function, with the slope of the line decreasing from wet to dry conditions. The fitness of the line to the data also decreases for the wet to dry conditions, with the R^2 values for wet, medium, and dry conditions equal to 0.78, 0.45, and 0.04, respectively. Runoff coefficients decrease with increasing contributing area due to the increased probability of water losses as basin area increases. Also, as expected, runoff coefficients are highest in basins with wet initial conditions that are primed to limit infiltration and evapotranspiration.

4.2   Trends in rainfall intensity

Maximum rainfall intensities (i.e., the maximum among all time-interval-of-measurement classes) for each contributing-area class and recurrence interval decrease systematically as power-law functions of increasing contributing area for all recurrence intervals with an average exponent of -0.18 ± 0.06 (error is the standard deviation of all calculated exponents found from a weighted least-squares regression; average coefficient of determination $R^2 = 0.78$). Note that maximum-rainfall-intensity results are not presented because they are closely related to the plots of Q_p vs. contributing area in Fig. 6; i.e., Q_p is simply the rainfall intensity multiplied by the contributing area. The decrease in maximum rainfall intensity with contributing area can be seen in Table 1, where maximum rainfall intensities over contributing areas of 11 664 km^2 are 45 to 8 % of maximum-rainfall-intensity values for basin areas of 16 km^2 in both the UCRB and LCRB (Table 1). The largest decrease in maximum-rainfall-intensity values between the smallest and largest contributing areas were found for the largest recurrence interval (e.g., 500-year) for both the UCRB and LCRB. The decrease in maximum rainfall intensity with increasing

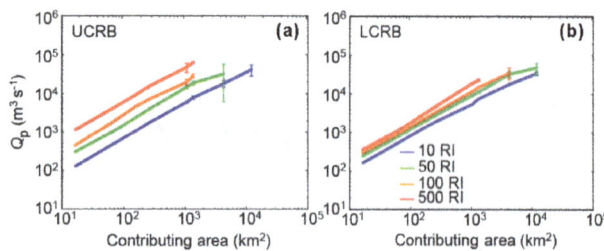

Figure 6. Frequency–magnitude–area (FMA) curves of Q_p vs. contributing area for recurrence intervals (RIs) of 10, 50, 100, and 500 years for the Upper Colorado River Basin (UCRB; **a**) and the Lower Colorado River Basin (LCRB; **b**).

contributing area suggests that there is a spatial limitation to storms of a given rainfall intensity.

Differences among maximum rainfall intensities for the four recurrence intervals as a function of contributing area are larger in the UCRB than in the LCRB (Table 1). This larger "spread" in the maximum rainfall intensities in the UCRB relative to the LCRB is also propagated throughout the maximum rainfall and flood discharge calculations. For both the UCRB and LCRB, the difference between the 50- and 100-year recurrence-interval values was the smallest (Table 1). These trends show that maximum rainfall intensities vary much more as a function of recurrence interval in the UCRB compared with the LCRB.

Maximum rainfall intensities associated with a 10-year recurrence interval are similar in the LCRB and UCRB, while intensities were higher in the UCRB than the LCRB for recurrence intervals of 50, 100, and 500 years (Table 1). The results of the comparison between the two basins suggest that common (i.e., low-recurrence-interval) rainfall events will have similar maximum rainfall intensities in the UCRB and LCRB, but that rare (i.e., high-recurrence-interval) rainfall events will have higher maximum rainfall intensities in the UCRB than in the LCRB for the same recurrence interval.

Maximum precipitation intensities associated with the four defined recurrence intervals are similar to previously published values. In general the values we calculate for the LCRB and the UCRB for the 10-, 50-, and 100-year recurrence intervals are on the order of 10s of $mm\,h^{-1}$. This is similar to the spread in values reported on precipitation intensity maps for the same duration and recurrence interval in Hershfield (1961). However, the values reported by Hershfield (1961) are slightly higher (by less than $20\,mm\,h^{-1}$) in the LCRB for the three recurrence intervals and in the UCRB for the 10-year recurrence interval than values calculated in this study. The values calculated here are also broadly consistent with presented precipitation frequency estimates for points within the LCRB and UCRB provided by the NOAA Atlas 14 Point Precipitation Frequency Estimates website (http://hdsc.nws.noaa. gov/hdsc/pfds/pfds_map_cont.html). Due to the difference in

how precipitation intensities are measured and how the frequencies are calculated, the values are expected to be slightly different but within the same order of magnitude.

4.3 Trends in Q_p

Maximum precipitation (here only rainfall) discharges (Q_p hereafter) increase with contributing area as power-law functions with an average exponent of 0.82 ± 0.06 (error is the standard deviation of all calculated exponents) based on weighed least-squares regressions on the data ($R^2 = 0.98$) for all recurrence intervals and for both the UCRB and LCRB (Fig. 6). These Q_p values for a given contributing-area class and recurrence interval are the largest values taken from the multiple values calculated for each of the seven time intervals of measurement as explained in Sect. 3.3. By taking the maximum values, the resulting Q_p FMACs approximate the upper envelope of values of a given recurrence interval. In this study the FMAC follows a power-law function that shows that Q_p increases predictably across the range in contributing areas. As with the maximum rainfall-intensity results, differences between Q_p values of different recurrence intervals for a given contributing area were larger for the UCRB than the LCRB (Fig. 6).

In general, confidence intervals for Q_p values increase with increasing contributing-area class (Table 1 and Fig. 6). The large values of the highest contributing-area classes and highest recurrence intervals show the spatial limitation of the method, meaning that at these contributing-area classes and recurrence intervals the values are sampled from the largest ranked value and have infinite confidence intervals. These values include the 50-, 100-, and 500-year recurrence intervals for the UCRB and the 100- and 500-year recurrence intervals for the LCRB at the $11\,664\,km^2$ contributing-area class. These values also include the 100- and 500-year recurrence intervals for the UCRB and the 500-year recurrence intervals for the LCRB at the $4096\,km^2$ contributing-area class. Values with infinite confidence intervals are not included in Fig. 6 due to their high uncertainties.

4.4 Trends in Q_{fd}

Maximum Q_{fd} values (hereafter Q_{fd}), i.e., the largest values taken for the multiple values calculated for each time interval of measurement for a given contributing-area class and recurrence interval, were used to plot FMACs for wet, medium, and dry conditions for both the UCRB and LCRB (Fig. 7). In general, FMACs for Q_{fd} values follow the power-law relationship shown in the Q_p FMACs until contributing areas of $\sim 1000\,km^2$, where the curves begin to very slightly flatten or decrease. As with the Q_p values, Q_{fd} values representing some of the higher recurrence intervals converge to the same value (i.e., the value corresponding to the highest rainfall intensity for the contributing-area class) at contributing areas of $\approx 10\,000\,km^2$ and the confidence intervals become infi-

Table 1. Maximum rainfall intensity and Q_p for the Upper Colorado River Basin (UCRB) and Lower Colorado River Basin (LCRB). Note that data are all sampled from time intervals of measurement ≤ 2 h.

RI	Area	Intensity ($\mathrm{mm\,h^{-1}}$)		Q_p ($\mathrm{m^3\,s^{-1}}$)	
	($\mathrm{km^2}$)	UCRB	LCRB	UCRB	LCRB
10	16	28.0 ± 0.0	36.6 ± 0.0	125 ± 0	162 ± 0
10	64	25.4 ± 0.1	32.5 ± 0.0	451 ± 1	578 ± 0
10	144	25.1 ± 1.1	29.5 ± 0.4	1004 ± 44	1182 ± 16
10	256	23.7 ± 0.2	27.3 ± 0.0	1682 ± 13	1944 ± 1
10	1024	19.8 ± 1.5	19.7 ± 0.4	5644 ± 427	5610 ± 114
10	1296	20.7 ± 2.4	21.7 ± 3.5	7439 ± 873	7820 ± 1268
10	4096	15.5 ± 3.0	15.9 ± 0.8	$17\,682 \pm 3462$	$18\,134 \pm 890$
10	11\,664	12.6 ± 1.7	11.0 ± 2.6	$40\,914 \pm 5571$	$35\,521 \pm 8586$
50	16	55.9 ± 0.7	56.2 ± 0.1	248 ± 3	250 ± 0
50	64	55.1 ± 1.2	47.7 ± 0.0	980 ± 22	847 ± 1
50	144	55.3 ± 3.5	43.3 ± 0.9	2211 ± 142	1734 ± 38
50	256	54.9 ± 1.4	40.9 ± 0.5	3901 ± 101	2908 ± 32
50	1024	50.8 ± 5.5	33.6 ± 1.4	$14\,449 \pm 1569$	9560 ± 393
50	1296	50.8 ± 25.0	32.5 ± 3.9	$18\,287 \pm 9011$	$11\,704 \pm 1410$
50	4096	27.6 ± 22.2	30.0 ± 5.2	$31\,382 \pm 25\,313$	$34\,126 \pm 5969$
50	11\,664	21.1^*	15.4 ± 8.3	$68\,434^*$	$49\,764 \pm 26\,874$
100	16	92.3 ± 0.3	68.6 ± 0.0	410 ± 1	305 ± 0
100	64	91.9 ± 2.5	54.5 ± 0.2	1635 ± 44	970 ± 3
100	144	90.1 ± 3.0	51.9 ± 1.0	3606 ± 118	2075 ± 41
100	256	88.7 ± 4.3	48.4 ± 0.4	6305 ± 307	3440 ± 27
100	1024	63.8 ± 11.0	42.5 ± 2.2	$18\,155 \pm 3139$	$12\,085 \pm 630$
100	1296	78.5 ± 50.1	43.2 ± 7.8	$28\,257 \pm 18\,022$	$15\,544 \pm 2820$
100	4096	40.8^*	32.0 ± 10.4	$46\,422^*$	$36\,425 \pm 11\,803$
100	11\,664	21.1^*	20.1^*	$68\,434^*$	$65\,011^*$
500	16	254.0 ± 0.8	81.9 ± 0.5	1129 ± 3	364 ± 2
500	64	229.0 ± 3.1	68.6 ± 1.5	4071 ± 55	1219 ± 26
500	144	219.1 ± 11.9	68.6 ± 4.7	8762 ± 476	2743 ± 187
500	256	219.4 ± 7.3	68.6 ± 3.4	$15\,600 \pm 517$	4877 ± 242
500	1024	166.0 ± 44.1	68.6 ± 3.1	$47\,229 \pm 12\,554$	$19\,507 \pm 884$
500	1296	174.6 ± 85.3	65.6 ± 31.3	$62\,862 \pm 30\,696$	$23\,624 \pm 11\,279$
500	4096	81.6^*	53.6^*	$92\,844^*$	$60\,930^*$
500	11\,664	21.1^*	20.1^*	$68\,434^*$	$65\,011^*$

* Values with infinite confidence intervals; not used in this study.

nite (Table 2). This convergence of Q_{fd} values at the largest contributing areas is due to the reduction in the range of values and the number of samples from which to calculate the associated values for each recurrence interval.

In general, the UCRB Q_{fd} FMACs (Fig. 7a, c, and e) are slightly higher in magnitude and span a larger range of magnitudes than the FMACs for the LCRB. For both basins, FMACs for the wet, medium, and dry conditions resulted in the highest, middle, and lowest magnitudes, respectively. This trend is expected due to the lowering of runoff coefficients and available water as conditions become drier.

FMACs of Q_{fd} for the LCRB plot below published FECs for the LCRB and US (Fig. 7b, d, and f) at low contributing areas, but meet and/or exceed the LCRB FEC for con-

tributing areas above ≈ 1000 and $\approx 100\,\mathrm{km^2}$ for dry and wet antecedent-moisture conditions, respectively. The FMACs for the LCRB do not exceed the US FEC. All of the FMACs of Q_{fd} for the UCRB exceed the LCRB FEC for wet conditions, with the FMACs of lower recurrence intervals exceeding the curve at higher contributing areas than the FMACs of higher recurrence intervals (Fig. 7a). The 500-year FMAC for wet conditions approximates the US FEC for contributing areas between ≈ 100 and $1000\,\mathrm{km^2}$. These results suggest that under certain antecedent-moisture conditions, and in basins of certain contributing areas, the LCRB produces floods that exceed the maximum recorded floods in the LCRB, and the UCRB produces floods of magnitudes on par with the maximum recorded floods in the US.

Table 2. Maximum Q_{fd} for the Upper Colorado River Basin (UCRB) and Lower Colorado River Basin (LCRB). Note that data are all sampled from time intervals of measurement ≤ 2 h.

RI	Area (km^2)	Wet Q_{fd} (m^3 s^{-1})		Med Q_{fd} (m^3 s^{-1})		Dry Q_{fd} (m^3 s^{-1})	
		UCRB	LCRB	UCRB	LCRB	UCRB	LCRB
10	16	65 ± 0	86 ± 0	36 ± 0	47 ± 0	20 ± 0	26 ± 0
10	64	246 ± 1	263 ± 0	137 ± 0	151 ± 0	75 ± 0	89 ± 0
10	144	465 ± 20	489 ± 7	268 ± 12	290 ± 4	156 ± 7	175 ± 2
10	256	657 ± 5	748 ± 0	388 ± 3	449 ± 0	244 ± 2	283 ± 0
10	1024	2363 ± 179	2194 ± 44	1423 ± 108	1326 ± 27	892 ± 68	820 ± 17
10	1296	2244 ± 263	2384 ± 387	1459 ± 171	1543 ± 250	1010 ± 118	1066 ± 173
10	4096	5594 ± 1095	5304 ± 260	3665 ± 718	3375 ± 166	2507 ± 491	2315 ± 114
10	11664	14603 ± 1966	11048 ± 2670	9010 ± 1213	6978 ± 1687	6105 ± 822	4942 ± 1195
50	16	131 ± 2	131 ± 0	73 ± 1	73 ± 0	41 ± 1	41 ± 0
50	64	553 ± 12	387 ± 0	307 ± 7	222 ± 0	172 ± 4	130 ± 0
50	144	1145 ± 73	720 ± 16	636 ± 41	424 ± 9	355 ± 23	259 ± 6
50	256	1772 ± 46	1119 ± 12	1043 ± 27	676 ± 7	639 ± 16	421 ± 5
50	1024	6127 ± 665	3062 ± 126	3665 ± 398	1928 ± 79	2291 ± 249	1308 ± 54
50	1296	7076 ± 3487	3562 ± 429	4265 ± 2102	2300 ± 277	2682 ± 1321	1571 ± 189
50	4096	$15\,716 \pm 12\,650$	8487 ± 1485	9451 ± 7607	5850 ± 1023	6076 ± 4890	4343 ± 760
50	11664	$44\,482^*$	$15\,700 \pm 8478$	$28\,783^*$	$10\,176 \pm 5495$	$19\,770^*$	7138 ± 3855
100	16	216 ± 1	160 ± 0	120 ± 0	89 ± 0	67 ± 0	50 ± 0
100	64	924 ± 25	442 ± 1	514 ± 14	255 ± 1	286 ± 8	150 ± 0
100	144	1807 ± 60	860 ± 17	1041 ± 35	508 ± 10	610 ± 20	309 ± 6
100	256	2888 ± 140	1324 ± 10	1706 ± 83	798 ± 6	1037 ± 50	499 ± 4
100	1024	$10\,586 \pm 1830$	3812 ± 199	6366 ± 1101	2438 ± 127	3979 ± 688	1662 ± 87
100	1296	9564 ± 6100	4713 ± 855	5752 ± 3668	3058 ± 555	3619 ± 2308	2104 ± 382
100	4096	$29\,415^*$	$10\,319 \pm 3344$	$19\,095^*$	6654 ± 2156	$13\,116^*$	4698 ± 1522
100	11664	$59\,600^*$	$18\,607^*$	$38\,667^*$	$12\,904^*$	$26\,747^*$	9609^*
500	16	594 ± 2	192 ± 1	330 ± 1	107 ± 1	184 ± 1	59 ± 0
500	64	1855 ± 25	556 ± 12	1068 ± 14	320 ± 7	628 ± 8	188 ± 4
500	144	3631 ± 197	1138 ± 77	2141 ± 116	670 ± 46	1306 ± 71	408 ± 28
500	256	6012 ± 200	1879 ± 93	3618 ± 120	1130 ± 56	2266 ± 75	709 ± 35
500	1024	$19\,049 \pm 5059$	6139 ± 278	$11\,478 \pm 3048$	3945 ± 179	7186 ± 1909	2660 ± 120
500	1296	$19\,075 \pm 9314$	7153 ± 3415	$12\,370 \pm 6041$	4656 ± 2223	8499 ± 4150	3198 ± 1527
500	4096	$43\,688^*$	$14\,892^*$	$28\,354^*$	$10\,460^*$	$19\,481^*$	7800^*
500	11664	$65\,705^*$	$23\,062^*$	$42\,738^*$	$16\,198^*$	$29\,364^*$	$12\,080^*$

* Values with infinite confidence intervals; not used in this study.

4.5 The effects of ENSO on rainfall

Definitive differences in maximum rainfall intensities and Q_p values were found between months with positive vs. months with negative MEI values (Table 3). For very small contributing areas (16 km^2) in the LCRB, maximum rainfall intensities and Q_p values are similar during negative and positive MEI conditions. Larger contributing areas (256 and 4096 km^2) show higher maximum rainfall intensities during negative MEI conditions regardless of recurrence interval. Values of Q_p show the same trend as the maximum rainfall intensity in the LCRB. In the UCRB, maximum rainfall intensities and Q_p values during negative MEI conditions are higher than those during positive MEI conditions regardless of recurrence interval.

5 Discussion

5.1 Use and accuracy of NEXRAD products

NEXRAD products are widely used as precipitation inputs in rainfall–runoff modeling studies due to the spatially complete nature of the data necessary for hydrologic and atmospheric models (Ogden and Julien, 1994; Giannoni et al., 2003; Kang and Merwade, 2011). In contrast to past studies similar in scope to this study (Castellarin et al., 2005, 2009; Castellarin, 2007), we did not use rain-gauge data and only

Table 3. Maximum rainfall intensity and Q_p values for 10-, 50-, 100-, and 500-year recurrence intervals during negative (neg) and positive (pos) multivariate ENSO index (MEI) conditions within the Lower Colorado River Basin (LCRB) and Upper Colorado River Basin (UCRB). Note that data are all sampled from time intervals of measurement ≤ 2 h.

Basin	MEI	Area	Intensity (mm h^{-1})				Q_p (m^3 s^{-1})			
		(km^2)	10 yr	50 yr	100 yr	500 yr	10 yr	50 yr	100 yr	500 yr
LCRB	neg	16	39	56	69	77	175	250	305	343
	neg	256	31	46	53	69	2206	3251	3741	4877
	neg	4096	21	32	43	54	23 856	36 425	48 363	60 930
	pos	16	40	64	74	130	179	284	330	576
	pos	256	27	38	47	52	1943	2690	3369	3721
	pos	4096	13	20*	20*	20*	15 229	22 689*	22 689*	22 689*
UCRB	neg	16	41	98	162	254	186	435	721	1129
	neg	256	33	101	155	254	2366	7172	11 012	18 055
	neg	4096	22	34	41	82	25 556	39 013	46 422	92 844
	pos	16	26	51	56	74	115	225	248	330
	pos	256	18	40	51	56	1255	2810	3601	4018
	pos	4096	10	26	27*	27*	10 822	30 034	31 044*	31 044*

* Values with infinite confidence intervals; not used in this study.

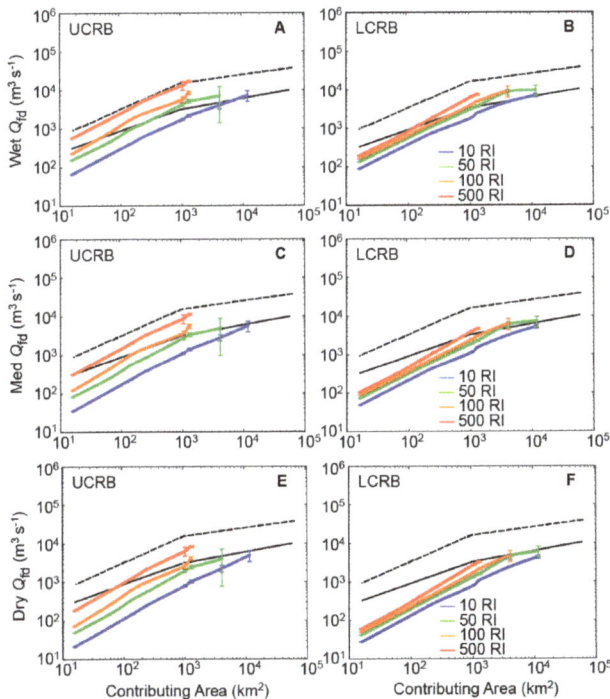

Figure 7. Q_{fd} frequency–magnitude–area curves of 10, 50, 100, and 500-year recurrence intervals (RIs) and for wet, medium, and dry conditions for the Upper Colorado River Basin (UCRB) and the Lower Colorado River Basin (LCRB). Published FECs (black lines) for the Lower Colorado River Basin (solid black line) from Enzel et al. (1993) and the US (dashed black line) from Costa (1987) are also shown.

used NEXRAD products to determine the FMACs for precipitation and flood discharges. We favor NEXRAD products due to the spatial completeness of the data.

Intuitively, NEXRAD products that are spatially complete and average precipitation over a 4 km by 4 km area would not be expected to match rain-gauge data within that area precisely (due to the multi-scale variability of rainfall), although some studies have tried to address this discrepancy (Sivapalan and Bloschl, 1998; Johnson et al., 1999). Xie et al. (2006) studied a semi-arid region in central New Mexico and found that hourly NEXRAD products overestimated the mean precipitation relative to rain-gauge data in both monsoon and non-monsoon seasons by upwards of 33 and 55 %, respectively. Overestimation of precipitation has also been noted due to the range and the tilt angle at which radar reflectivity data are collected (Smith et al., 1996). Underestimation of precipitation by NEXRAD products relative to rain-gauge data has also been observed (Smith et al., 1996; Johnson et al., 1999), however.

Under- and over-estimation of precipitation by NEXRAD products in relation to rain-gauge data is partly due to the difference in sampling between areal NEXRAD products and point data from rain gauges and partly due to sampling errors inherent to both methods. For example, NEXRAD products include problems such as the use of incorrect $Z–R$ relationships for high-intensity storms and different types of precipitation, such as snow and hail (Baeck and Smith, 1998). Also, because of its low reflectivity, snow in the NEXRAD products is measured as if it were light rain (David Kitzmiller, personal communication, 10 January 2012). This means the NEXRAD products likely underestimate snowfall and therefore snowfall is not fully accounted for in this study. Due to snowfall not being included in this study, associated snow-pack and snowmelt effects were also not accounted for. Rain gauges can also suffer from a number of measurement errors that usually result in an underestimation of rainfall (Burton Jr. and Pitt, 2001). In addition, gridded rainfall data derived from rain gauges are not spatially complete and therefore must be interpolated between point measurements to form a spatially complete model of rainfall. It is impossible to discern which product is more correct due to the differences in measurement techniques and errors, but by taking both products and combining them into one, the Stage III NEXRAD precipitation products generate the best precipitation estimate possible for this study. Moreover, it should be noted that 100-year flood magnitude predictions based on regression equations have very large relative error bars (ranging between 37 and 120 % in the western US; Parrett and Johnson, 2003) and that measurements of past extreme floods can have significant errors ranging from 25 to 130 %, depending on the method used (Baker, 1987). As such, even a ∼ 50 % bias in NEXRAD-product-derived precipitation estimates is on par or smaller than the uncertainty associated with an analysis of extreme flood events.

As stated previously, the NEXRAD precipitation estimates used here do not include snowfall and other non-rainfall precipitation types. In this study we also do not include snowpack information in our flood discharge calculations. The omission of snowpack is a reasonable assumption for our low-elevation, warm regions within most of the UCRB and LCRB. However, we acknowledge that some of our higher elevation areas at higher latitudes may underestimate the maximum flood discharge by only including rainfall-derived runoff. If the methodology in this paper were applied to a snowmelt-dominated region, snowpack would need to be added to accurately estimate the maximum flood discharge.

5.2 Comparison of FMACs to published FECs

FMACs of Q_{fd} exhibit a similar shape and similar overall range in magnitudes to previously published FECs, derived from stream-gauge and paleoflood records, for the LCRB and the US (Fig. 7). In general, the FMACs exceed or match published FECs at larger contributing areas, and are lower than or on par with published FECs at the smallest contributing areas (Fig. 7).

All FMACs except the 500-year recurrence-interval curve for the UCRB under wet conditions are positioned well below the US FEC presented by Costa (1987; Fig. 7a). The similarity between the 500-year recurrence interval Q_{fd} FMAC for the UCRB under wet conditions and the US FEC suggests that the US FEC includes floods of larger recurrence-intervals, which are similar in magnitude to the 500-year recurrence-interval floods within the UCRB. The approximation of the US FEC by the 500-year UCRB FMAC is a significant finding due to the fact that the US FEC includes

storms from other regions of the US with extreme climatic forcings (i.e., hurricanes, extreme convection storms).

The Q_{fd} FMACs for the LCRB can be directly compared to the FEC for the LCRB presented by Enzel et al. (1993). At contributing areas smaller than approximately 100 km^2, Q_{fd} FMACs for wet conditions and all recurrence intervals are positioned below the LCRB FEC, but at larger contributing areas Q_{fd} FMACs exceed or approximate the LCRB FEC. Q_{fd} FMACs calculated for medium and dry antecedent conditions show the same trend, but exceed the LCRB FEC at larger contributing areas (≥ 1000 km^2). This comparison suggests that although the FMACs overlap the overall range of flood magnitudes of the LCRB FEC, the two methods are not capturing the same trend for extreme flood discharges and the LCRB is capable of producing floods larger than those on record.

The difference in the slope of the FMACs, and specifically the exceedance of the published LCRB FEC, suggests that the two methods are not capturing the same information. This difference may be due to the difference in how the data are sourced for each method. FECs are created as regional estimates of maximum flood discharges and are based on stream-gauging station and paleoflood data. The FECs are then used to provide flood information for the region, including ungauged and unstudied drainage basins. FECs are limited to the number of stream gauges employed by public and private parties and do not include all basins within a region. In general, FECs may underestimate maximum floods in larger basins, relative to smaller basins, because there are a larger number of smaller basins to sample than larger basins. This sample-size problem introduces bias in the record where flood estimates for smaller contributing areas may be more correct than estimates for larger basins. In this study, the regional precipitation information given by the NEXRAD network is used to form the FMAC, therefore taking advantage of the entire region and using precipitation data to calculate flood discharges, rather than directly measuring flood discharges. This sampling scheme allows for much larger sample sizes for the range of contributing areas, therefore minimizing the sample bias of the traditional FEC.

This study aimed to introduce the new method of the FMAC and therefore improve upon the traditional methods of the FEC. By calculating FMACs we provide frequency and magnitude information of possible flood events for a given region, in contrast to the FECs that only provide an estimate of the largest flood on record. This information is vital for planning and infrastructure decisions and the accurate representation of precipitation and flooding in design-storm and watershed modeling. In addition, the fact that the FMACs match the FECs for large (500-year) recurrence intervals and do not exhibit the same trends suggests that the FMACs are capturing different samples than the FECs. This indicates that by using the NEXRAD products, the FMACs may provide a more inclusive flood dataset for a region (es-

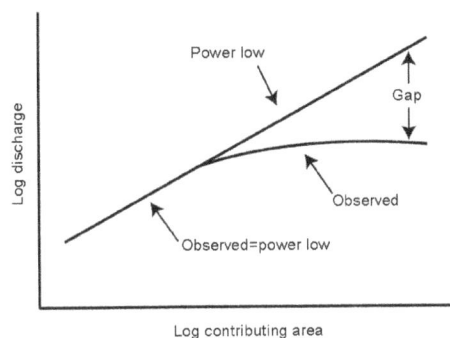

Figure 8. Conceptual diagram of the characteristic concave-down shape of the FEC (observed) shown in comparison to a power-law function between Q_p and contributing area. The "gap" between the observed curve and the predicted power law is caused by precipitation limitations and mechanisms occurring during the routing of water over the landscape.

pecially ungauged areas) than the traditional stream-gauge records.

5.3 Precipitation controls on the form of the FEC

Q_p FMACs were shown to have a strong (average $R^2 = 0.93$) power-law relationship between Q_p and contributing area for all recurrence intervals. Figure 8 shows a conceptualized FEC where the concave-down shape is created when the observed envelope curve diverges from the constant positive power-law relationship between Q_p and contributing area. This diversion creates a "gap" between the two curves and indicates that flood discharge is not a simple power-law function of contributing area. Three mechanisms have been proposed to explain the "gap" and characteristic concave-down shape of FECs: (1) integrated precipitation (i.e., total precipitation over an area) is more limited over larger contributing areas compared to smaller contributing areas (Costa, 1987), (2) a relative decrease in maximum flood discharges in larger contributing areas due to geomorphic dispersion (Rodriguez-Iturbe and Valdes, 1979; Rinaldo et al., 1991; Saco and Kumar, 2004), and (3) a relative decrease in maximum flood discharges in larger basins due to hydrodynamic dispersion (Rinaldo et al., 1991). The first explanation, proposed by Costa (1987), suggests that there is a limitation to the size of a storm and the amount of water that a storm can precipitate. The effect of precipitation limitations may be evidenced by the decreasing maximum rainfall intensities with increasing contributing area. However, the strong power-law relationship between Q_p and contributing area for all recurrence intervals indicates that Q_p is, in general, increasing predictably over the range of contributing areas used in this study. Even if precipitation limitations affect the shape of the curve, this single hypothesis does not account for all of the concave-down shape of each FEC suggesting that other mechanisms are important to creating the characteristic shape. However,

it is important to note that the importance of each mechanism may be different for different locations.

5.4 Climate variability in the NEXRAD data

The results from comparing negative and positive MEI conditions in the UCRB and LCRB are generally consistent with ideas about ENSO and how it affects precipitation in the western US. In the LCRB, during negative MEI conditions, small, frequent storms have similar or slightly higher maximum rainfall intensities and Q_p values than during positive MEI conditions. This similarity between the two conditions may be explained by the balancing of increased winter moisture during El Niño in the southwestern US (Hidalgo and Dracup, 2003) and increased summer moisture through the strengthening of the NAM system and the convective storms it produces during La Niña conditions (Castro et al., 2001; Grantz et al., 2007). In general, the strengthening of the NAM may explain the higher maximum rainfall intensities and Q_p values during negative MEI conditions in the LCRB. Strengthening of the NAM may be due in part to the large temperature difference between the cool sea surface of the eastern Pacific Ocean and the hot land surface of the southwestern US and northwestern Mexico during La Niña conditions. The large temperature gradient increases winds inland, bringing the moisture associated with the NAM (Grantz et al., 2007). In the UCRB it is during negative MEI conditions, where the highest maximum rainfall intensities and Q_p values for all recurrence intervals occur. This suggests that the UCRB is affected by ENSO much like the northwestern US, where wetter winters are affiliated with La Niña and not El Niño conditions (Cayan et al., 1999; Hidalgo and Dracup, 2003). It is important to note that this comparison is of intensity rates and not total precipitated moisture so the MEI condition resulting in wetter conditions is not known.

In addition to the ENSO analysis, by investigating previous studies we see that, along with natural yearly precipitation variability, the 1996 to 2004 water years included many atmospheric river events (Dettinger, 2004; Dettinger et al., 2011). It is important that these events were included due to their ability to greatly increase winter precipitation in the UCRB and LCRB (Rutz and Steenburgh, 2012). Atmospheric river events (sometimes known as Pineapple Express events) can also be tied to major Pacific climate modes such as the ENSO (Dettinger, 2004; Dettinger et al., 2011), the Pacific Decadal Oscillation (PDO; Dettinger, 2004), and the North Pacific Gyre Oscillation (NPGO; Reheis et al., 2012) in southern California. Unfortunately, correlations between atmospheric river events are unknown and/or less clear for the interior western US. However, all three of these Pacific climate modes shifted during the 9-year study period in $\sim$ 1998 to 1999 (Reheis et al., 2012), indicating that both positive and negative conditions of the ENSO, PDO, and NPGO exist in the NEXRAD products used in this study.

The presence of distinct trends in maximum rainfall-intensity and Q_p values calculated for negative and positive MEI conditions, as well as the information in the literature on atmospheric river events, indicates the NEXRAD products used in this study incorporate circulation-scale weather patterns. In addition, the patterns in maximum rainfall intensities and Q_p values during different MEI conditions agree with common understanding of the effects of ENSO on the western US and provide evidence that the data and methods used in this paper to analyze precipitation are reliable. This analysis shows that the NEXRAD products worked well in this location and that using radar-derived precipitation products may be useful for identifying precipitation and climatic trends in other locations where the FMAC method can be applied.

6 Conclusions

In this study we present the new FMAC method of calculating precipitation and flood discharges of a range of recurrence intervals using radar-derived precipitation estimates combined with a flow-routing algorithm. This method improves on the traditional FEC by assigning recurrence-interval information to each value and/or curve. Also, instead of relying on stream-gauge records of discharge, this method uses up-to-date and spatially complete radar-derived precipitation estimates (in this case NEXRAD products) to calculate flood discharges using flow-routing algorithms. This study presents an alternative data source and method for flood-frequency analysis by calculating extreme (high recurrence interval) event magnitudes from a large sample set of magnitudes made possible by sampling the radar-derived precipitation estimates.

The FMACs for Q_p and Q_{fd} for the UCRB were similar to those produced for the LCRB. In general, all recurrence-interval curves followed the same general trend, indicating that the mechanisms of precipitation and flood discharge are similar for the two basins. However, there were some differences between the two basins. Overall, there were larger differences between curves of different recurrence intervals for the UCRB than the LCRB, suggesting a larger range in maximum rainfall intensities, and therefore Q_p and Q_{fd}, in the UCRB relative to the LCRB. For both the UCRB and LCRB the 50- and 100-year recurrence-interval curves for all precipitation and discharge FMACs were the most similar. This similarity may mean that although historical discharge records are short, having a 50-year record may not underestimate the 100-year flood as much as one might expect. Also, for Q_p and Q_{fd}, low recurrence-interval values were slightly higher in the LCRB than in the UCRB. This relationship was opposite for high recurrence-interval values. This likely points to a general hydroclimatic difference between the two basins, with the LCRB receiving high-intensity storms annu-ally due to the NAM and the UCRB receiving more intense and rarer winter frontal storms.

Power-law relationships between maximum rainfall intensity, Q_p, and contributing area were also found in this study. Maximum rainfall intensities decreased as a power-law function of contributing area with an average exponent of -0.18 ± 0.06 for all recurrence intervals. Q_p values for all recurrence intervals increased as a power-law function of contributing area with an exponent of approximately 0.82 ± 0.06 on average. Based on the constant power-law relationship between Q_p and contributing area, the "gap" or characteristic concave-down shape of published FECs is likely not caused by precipitation limitations.

In general, the FMACs of Q_{fd} calculated in this study are lower than, and exceed, the published FECs for the LCRB at lower and higher contributing areas. All FMACs of Q_{fd} were positioned well below the US FECs except the UCRB 500-year FMAC, which approximated the US FECs during wet antecedent-moisture conditions. All FMACs of Q_{fd} for all moisture conditions in the LCRB closely approximated the same magnitudes as the published LCRB FEC, but exceeded it for larger contributing areas. The higher estimates of flood discharges at larger contributing areas may be the result of the difference of sampling methods and are likely not erroneous and may be proved true by future events.

Lastly, the approximately 9 years of NEXRAD products were found to be a good representation of climate in the CRB. This conclusion was made based on differences in precipitation between positive and negative ENSO conditions in both the UCRB and LCRB and additional data found in the literature. In general, the UCRB was found to have a hydroclimatic regime much like that of the northwestern US, where El Niño conditions result in lower maximum rainfall intensities and amounts and La Niña conditions result in higher maximum rainfall intensities. The LCRB showed a more complex trend with similar maximum rainfall intensities for both El Niño and La Niña conditions.

Here this method is applied to the UCRB and LCRB in the southwestern US, but could be applied to other regions of the US and the world with variable climate and storm types where radar-derived precipitation estimates are available. In this study we used set values for contributing area, drainage basin shape, time intervals of measurement, and recurrence intervals that can be changed based on the focus of future studies. However, it is also important to note that a number of assumptions were made in this study that simplified our analysis; most importantly, (1) space-for-time substitution, or regionalization, was used to increase the number of samples and assumed that observations were independent and sampled from the same distribution; (2) it was assumed that the time period length and the spatial and temporal sampling scales were sufficient to create a representative sample from the observations; and (3) it was assumed that similar flood-generating and flow-routing mechanisms (and related variables such as runoff coefficients) were present in each basin regardless of size or location. These assumptions allowed us to form and apply the methods described here to our study area, but may not apply to all areas. Other variables such as snowpack, elevation, land use, and climate change that were not included in this study should be explored in conjunction with this methodology to better understand controls on precipitation and flooding. The absence of these elements from the method here may limit the application of this method to other locations.

Acknowledgements. This study was supported by the Jemez River Basin and Santa Catalina Critical Zone Observatory NSF grants EAR-0724958 and EAR-1331408. We would like to thank Vic Baker, Phil Pearthree, Peter Troch, and Katie Hirschboeck for helpful discussions and suggestions.

Edited by: P. Saco

References

Baeck, M. L. and Smith, J. A.: Rainfall estimates by the WSR-88D for heavy rainfall events, Weather Forecast., 13, 416–436, 1998.

Baker, V. R.: Paleoflood hydrology and extraordinary flood events, J. Hydrol., 96, 77–99, 1987.

Brutsaert, W.: Review of Green's functions for linear open channels, J. Eng. Mech.-ASCE, 99, 1247–1257, 1973.

Buishand, T. A.: Extreme rainfall estimation by combining data from several sites, Hydrolog. Sci. J., 36, 345–365, doi:10.1080/02626669109492519, 1991.

Burton Jr., G. A. and Pitt, R. E. (Eds.): Stormwater effects handbook: a toolbox for watershed managers, scientists, and engineers, Lewis Publishers, Boca Raton, Florida, 2001.

Cañon, J., González, J., and Valdes, J.: Precipitation in the Colorado River Basin and its low frequency associations with PDO and ENSO signals, J. Hydrol., 333, 252–264, 2007.

Castellarin, A.: Probabilistic envelope curves for design flood estimation at ungauged sites, Water Resour. Res., 43, W04406, doi:10.1029/2005WR004384, 2007.

Castellarin, A., Vogel, R. M., and Matalas, N. C.: Probabilistic behavior of a regional envelope curve, Water Resour. Res., 41, W06018, doi:10.1029/2004WR003042, 2005.

Castellarin, A., Merz, R., and Bloschl, G.: Probabilistic envelope curves for extreme rainfall events, J. Hydrol., 378, 263–271, 2009.

Castro, C. L., McKee, T. B., and Pielke Sr., R. A.: The relationship of the North American Monsoon to Tropical and North Pacific surface temperatures as revealed by observational analyses, J. Climate, 14, 4449–4473, 2001.

Cayan, D. R., Redmond, K. T., and Riddle, L. G.: ENSO and hydrologic extremes in the western United States, J. Climate, 12, 2881–2893, 1999.

Costa, J. E.: A comparison of the largest rainfall-runoff floods in the United States with those of the People's Republic of China and the World, J. Hydrol., 96, 101–115, 1987.

Crippen, J. R. and Bue, C. D.: Maximum flood flows in the conterminous United States, US Geological Survey Water Supply Paper 1887, US Geological Survey, Washington, D.C., USA, 1977.

Dalrymple, T.: Flood-frequency analyses, Manual of Hydrology: Part 3, US Geological Survery Water Supply Paper 1543, US Geological Survey, Washington, D.C., USA, p. 80, 1960.

Dettinger, M. D.: Fifty-two years of "Pineapple-Express" storms across the west coast of North America, PIER Energy-Related Environmental Research, CEC-500-2005-004, US Geological Survey, Scripps Institution of Oceanography for the California Energy Commission, La Jolla, California, USA, 2004.

Dettinger, M. D., Ralph, F. M., Das, T., Neiman, P. J., and Cayan, D. R.: Atmospheric rivers, floods, and the water resources of California, Water, 3, 445–478, 2011.

Enzel, Y., Ely, L. L., House, P. K., Baker, V. R., and Webb, R. H.: Paleoflood evidence for a natural upper bound to flood magnitudes in the Colorado River Basin, Water Resour. Res., 29, 2287–2297, 1993.

Etheredge, D., Gutzler, D. S., and Pazzaglia, F. J.: Geomorphic response to seasonal variations in rainfall in the Southwest United States, Geol. Soc. Am. Bull., 116, 606–618, 2004.

FEMA – Federal Emergency Management Agency: Modernizing FEMA's flood hazard mapping program: Recommendations for using future-conditions hydrology for the National Flood Insurance Program, Final Report, US Department of Homeland Security, Washington, D.C., USA, 2001.

Fulton, R. A., Breidenbach, J. P., Seo, D. J., Miller, D. A., and O'Bannon, T.: The WSR-88D algorithm, Weather Forecast., 13, 377–395, 1998.

Giannoni, F., Smith, J. A., Zhang, Y., and Roth, G. Hydrologic modeling of extreme floods using radar rainfall estimates, Adv. Water Resour., 26, 195–203, 2003.

Grantz, K., Rajagopalan, B., Clark, M., and Zagona, E.: Seasonal shifts in the North American Monsoon, J. Climate, 20, 1923–1935, 2007.

Henderson, F. M.: Some properties of the unit hydrograph, J. Geophys. Res., 68, 4785–4793, 1963.

Herschy, R.: The world's maximum observed floods, Flow Meas. Instrum., 13, 231–235, 2002.

Hershfield, D. M.: Rainfall Frequency Atlast of the United States for Durations from 30 minutes to 24 hours and periods from 1 to 100 years, US Weather Bureau Technical Paper No. 40, US Weather Bureau, Washington, D.C., USA, p. 65, 1961.

Hidalgo, H. G. and Dracup, J. A.: ENSO and PDO Effects on hydroclimatic variations of the Upper Colorado River Basin, J. Hydrometeorol., 4, 5–23, 2003.

Hosking, J. R. M. and Wallis, J. R.: The effect of intersite dependence on regional flood frequency analysis, Water Resour. Res., 24, 588–600, 1988.

Hosking, J. R. M., and Wallis, J. R.: Regional Frequency Analysis, Cambridge University Press, New York, p. 283, 2005.

House, P. K. and Hirschboeck, K. K.: Hydroclimatological and paleohydrological context of extreme winter flooding in Arizona, 1993, in: Storm-Induced Geologic Hazards: Case Histories from the 1992–1993 Winter in Southern California and Arizona, vol. XI, edited by: Larson, R. A. and Slosson, J. E., Geological Society of America Reviews in Engineering Geology, Boulder, Colorado, 1–24, 1997.

Johnson, D., Smith, M., Koren, V., and Finnerty, B.: Comparing mean areal precipitation estimates from NEXRAD and rain gauge networks, J. Hydrol. Eng., 4, 117–124, 1999.

Kang, K. and Merwade, V.: Development and application of a storage-release based distributed hydrologic model using GIS, J. Hydrol., 403, 1–13, 2011.

Li, B.: Current status of weather radar data exchange, World Meteorological Organization Workshop on Radar Data Exchange, April 2013, 16.IV.2013, Exeter, UK, 2013.

Marani, M., Rinaldo, A., Rigon, R., Rodriquez-Iturbe, I.: Geomorphological width functions and the random cascade, Geophys. Res. Lett., 21, 2123–2126, 1994.

Mesa, O. J. and Mifflin, E. R.: On the relative role of hillslope and network geometry in hydrologic response, in: Scale Problems in Hydrology, edited by: Gupta, V. K., Rodriguez-Iturbe, I., and Wood, E. F., D. Reidel, Dordrecht, the Netherlands, 1–17, 1986.

Miller, A. J.: Flood hydrology and geomorphic effectiveness in the central Appalachians, Earth Surf. Proc., 15, 119–134, 1990.

Milly, P. C. D., Betancourt, J., Falkenmark, M., Hirsch, R. M., Kundzewicz, Z. W., Lettenmaier, D. P., and Stouffer, R. J.: Stationarity is dead: Whither water management?, Science, 319, 573–574, 2008.

Moody, T., Wirtanen, M., and Yard, S. N.: Regional relationships for bankfull stage in natural channels of the arid southwest, National Channel Design Inc., Flagstaff, AZ, 38 pp., 2003.

Morrison, J. E. and Smith, J. A.: Stochastic modeling of flood peaks using the generalized extreme value distribution, Water Resour. Res., 38, 1305, doi:10.1029/2001WR000502, 2002.

Moussa, R.: What controls the width function shape, and can it be used for channel network comparison and regionalization?, Water Resour. Res., 44, W08456, doi:10.1029/2007WR006118, 2008.

NOAA Atlas 14 Point Precipitation Frequency Estimates: http://hdsc.nws.noaa.gov/hdsc/pfds/pfds_map_cont.html, last access: 8 June 2016.

NOAA HDSG: http://dipper.nws.noaa.gov/hdsb/data/nexrad/cbrfc_stageiii.php, last access: 8 August 2014.

Ogden, F. L. and Julien, P. Y.: Runoff model sensitivity to radar rainfall resolution, J. Hydrol., 158, 1–18, 1994.

Parrett, C. and Johnson, D. R.: Methods for estimating flood frequency in Montana based on data through water year 1998, US Geological Survey Water-Resources Investigations Report 03-4308, US Geological Survey, Reston, Virginia, USA, 2003.

Parzen, E.: Nonparametric statistical data modeling, J. Am. Stat. Assoc., 74, 105–121, 1979.

Puente, C. E. and Sivakumar, B.: A deterministic width function model, Nonlin. Processes Geophys., 10, 525–529, doi:10.5194/npg-10-525-2003, 2003.

RadarEU: http://www.radareu.cz/, last access: 1 August 2014.

Reed, S. M. and Maidment, D. R.: Coordinate transformations for using NEXRAD data in GIS-based hydrologic modeling, J. Hydrol. Eng., 4, 174–182, 2006.

Reheis, M. C., Bright, J., Lund, S. P., Miller, D. M., Skipp, G., and Fleck, R. J.: A half-million-year record of paleoclimate from the Lake Manix Core, Mojave Desert, California, Palaeogeogr. Palaeocl., 365–366, 11–27, 2012.

Rinaldo, A., Marani, A., and Rigon, R.: Geomorphological dispersion, Water Resour. Res., 27, 513–525, 1991.

Rinaldo, A., Vogel, G. K., Rigon, R., and Rodriguez-Itrube, I.: Can one gauge the shape of a basin?, Water Resour. Res., 31, 1119–1127, 1995.

Rodriguez-Iturbe, I. and Rinaldo, A. (eds.): Fractal River Basins: Chance and Self-Organization, Cambridge University Press, New York, NY, USA, 2001.

Rodriguez-Iturbe, I. and Valdes, J. B.: The geomorphic structure of hydrologic response, Water Resour. Res., 15, 1409–1420, 1979.

Rosenberg, E. A., Clark, E. A., Steinemann, A. C., and Lettenmaier, D. P.: On the contribution of groundwater storage to interannual streamflow anomalies in the Colorado River basin, Hydrol. Earth Syst. Sci., 17, 1475–1491, doi:10.5194/hess-17-1475-2013, 2013.

Rutz, J. J. and Steenburgh, W. J.: Quantifying the role of atmospheric rivers in the interior western United States, Atmos. Sci. Lett., 13, 257–261, doi:10.1002/asl.392, 2012.

Saco, P. M. and Kumar, P.: Kinematic dispersion effects of hillslope velocities, Water Resour. Res., 40, W01301, doi:10.1029/2003WR002024, 2004.

Sankarasubramanian, A. and Vogel, R. M.: Hydroclimatology of the continental United States, Geophys. Res. Lett., 30, 1363, doi:10.1029/2002GL015937, 2003.

Shedd, R. C. and Fulton, R. A.: WSR-88D precipitation processing and its use in National Weather Service hydrologic forecasting, in: Proceedings, Engineering Hydrology: Proceedings of the Symposium Sponsored by the Hydrology Division of American Society of Civil Engineers, San Francisco, California, USA, 844–848, 1993.

Sivapalan, M. and Bloschl, G.: Transformation of point rainfall to areal rainfall: Intensity-duration-frequency curves, J. Hydrol., 204, 150–167, 1998.

Smith, J. A. and Krajewski, W. F.: A modeling study of rainfall rate–reflectivity relationships, Water Resour. Res., 29, 2505–2514, 1993.

Smith, J. A., Seo, D. J., Baeck, M. L., and Hudlow, M. D.: An intercomparison study of NEXRAD precipitation data, Water Resour. Res., 32, 2035–2045, 1996.

Soong, D. T., Ishii, A. L., Sharpe, J. B., and Avery, C. F.: Estimating flood-peak discharge magnitudes and frequencies for rural streams in Illinois, US Geological Survey Scientific Investigations Report 2004-5103, US Geological Survey, Reston, Virginia, USA, p. 162, 2004.

Stedinger, J. R., Vogel, R. M., and Foufoula-Georgiou, E.: Frequency Analysis of Extreme Events, in: Handbook of Hydrology, chap. 18, edited by: Maidment, D. R., McGraw-Hill, Inc., Washington, D.C., USA, 1993.

Troch, P. A., Smith, J. A., Wood, E. F., and de Troch, F. P.: Hydrologic controls of large floods in a small basin: central Appalachian case study, J. Hydrol., 156, 285–309, 1994.

USDA Geospatial Data Gateway: http://gdg.sc.egov.usda.gov, last access: 12 October 2016.

Veneziano, D., Moglen, G. E., Furcolo, P., and Iacobellis, V.: Stochastic model of the width function, Water Resour. Res., 36, 1143–1157, 2000.

Vivoni, E. R., Entekhabi, D., Bras, R. L., and Ivanov, V. Y.: Controls on runoff generation and scale-dependence in a distributed hydrologic model, Hydrol. Earth Syst. Sci., 11, 1683–1701, doi:10.5194/hess-11-1683-2007, 2007.

Vogel, R. M. and Fennesset, N. M.: Flow-duration curves I: New interpretation and confidence intervals, J. Water Resour. Bull., 31, 485–504, 1994.

Wallis, J. R., Schaefer, M. G., Barker, B. L., and Taylor, G. H.: Regional precipitation–frequency analysis and spatial mapping for 24-hour and 2-hour durations for Washington State, Hydrol. Earth Syst. Sci., 11, 415–442, doi:10.5194/hess-11-415-2007, 2007.

Wolman, M. G. and Costa, J. E.: Envelope curves for extreme flood events, J. Hydraul. Eng.-ASCE, 110, 77–78, 1984.

Xie, H., Zhou, X., Hendricks, J. M. H., Vivoni, E. R., Guan, H., Tian, Y. Q., and Small, E. E.: Evaluation of Nexrad Stage III precipitation data over a semiarid region, J. Am. Water Resour. As., 42, 237–256, 2006.

12

Monitoring and modeling infiltration–recharge dynamics of managed aquifer recharge with desalinated seawater

Yonatan Ganot[1,2]**, Ran Holtzman**[2]**, Noam Weisbrod**[3]**, Ido Nitzan**[1]**, Yoram Katz**[4]**, and Daniel Kurtzman**[1]

[1]Institute of Soil, Water and Environmental Sciences, The Volcani Center, Agricultural Research Organization, Rishon LeZion, 7528809, Israel

[2]Department of Soil and Water Sciences, The Hebrew University of Jerusalem, Rehovot, 7610001, Israel

[3]Department of Environmental Hydrology & Microbiology, Zuckerberg Institute for Water Research, Jacob Blaustein Institutes for Desert Research, Ben-Gurion University of the Negev, Midreshet Ben-Gurion, 8499000, Israel

[4]Mekorot, Water Company Ltd, Tel Aviv, 6713402, Israel

Correspondence to: Yonatan Ganot (yonatan.ganot@mail.huji.ac.il)

Abstract. We study the relation between surface infiltration and groundwater recharge during managed aquifer recharge (MAR) with desalinated seawater in an infiltration pond, at the Menashe site that overlies the northern part of the Israeli Coastal Aquifer. We monitor infiltration dynamics at multiple scales (up to the scale of the entire pond) by measuring the ponding depth, sediment water content and groundwater levels, using pressure sensors, single-ring infiltrometers, soil sensors, and observation wells. During a month (January 2015) of continuous intensive MAR (2.45×10^6 m^3 discharged to a 10.7 ha area), groundwater level has risen by 17 m attaining full connection with the pond, while average infiltration rates declined by almost 2 orders of magnitude (from ~ 11 to ~ 0.4 m d^{-1}). This reduction can be explained solely by the lithology of the unsaturated zone that includes relatively low-permeability sediments. Clogging processes at the pond-surface – abundant in many MAR operations – are negated by the high-quality desalinated seawater (turbidity ~ 0.2 NTU, total dissolved solids ~ 120 mg L^{-1}) or negligible compared to the low-permeability layers. Recharge during infiltration was estimated reasonably well by simple analytical models, whereas a numerical model was used for estimating groundwater recharge after the end of infiltration. It was found that a calibrated numerical model with a one-dimensional representative sediment profile is able to capture MAR dynamics, including temporal reduction of infiltration rates, drainage and groundwater recharge. Measured infiltration rates of an independent MAR event (January 2016) fitted well to those calculated by the calibrated numerical model, showing the model validity. The successful quantification methodologies of the temporal groundwater recharge are useful for MAR practitioners and can serve as an input for groundwater flow models.

1 Introduction

Managed aquifer recharge (MAR) is a common practice in water resources management in which excess water is stored in the aquifers for future consumption. Major techniques used for aquifer recharge include well injection, bank filtration, rainwater harvesting, and infiltration ponds (Dillon, 2005). In the Israeli Coastal Aquifer, MAR started in 1958 with Lake Kinneret water, surface runoff, and carbonate-aquifer groundwater as the recharge sources (Sellinger and Aberbach, 1973). Between the years 2000 and 2013 MAR in the Israeli Coastal Aquifer was mostly (88 %) from the soil aquifer treatment ponds at the Shafdan sites, where secondary effluents are delivered into infiltration ponds for tertiary treatment. The remaining MAR can be primarily attributed to storm runoff according to the seasonal rainfall (Israel Hydrological Service, 2013). Recently, desalinated seawater – a relatively new water source in Israel (Stanhill et al., 2015) – has been occasionally used as source water for MAR.

MAR using desalinated seawater (DSW) poses several scientific and operative challenges due to the unique water composition compared to natural groundwater. Yet, scientific publications on MAR with DSW are few. Field tests of MAR using DSW were performed during the 1970s and the 1990s in clastic and carbonate aquifers in Kuwait. Well clogging was identified as a major concern, especially in the clastic aquifers (Mukhopadhyay et al., 1994). These field tests were followed by laboratory studies focusing on clogging and geochemical processes using core experiments with DSW (Al-Awadi et al., 1995; Mukhopadhyay et al., 1998, 2004). A closely related study, on MAR with reverse-osmosis wastewater was conducted at the St André MAR site in Belgium. Reported work includes flow and transport modeling (Vandenbohede et al., 2008, 2009a; Vandenbohede and Van Houtte, 2012), isotope and geochemical analysis (Kloppmann et al., 2008; Vandenbohede et al., 2009b) and reactive transport modeling (Vandenbohede et al., 2013).

In this paper we focus on infiltration and recharge dynamics during MAR with DSW. This work is part of a comprehensive study involving field and laboratory investigations in order to better understand hydrological and geochemical processes during MAR with surplus of reverse-osmosis DSW in Israel (Ronen-Eliraz et al., 2017). The geochemical perspective of this field study will be reported in a future publication (Ganot et al., 2017). The results reported here are unique for several reasons. First, we monitored a month (January 2015) of continuous MAR with $2.45 \times 10^6 \, \text{m}^3$ of DSW (loading of about $23 \, \text{m} \, \text{month}^{-1}$), higher than in most other reported MAR at infiltration basins, comparable only to a few studies (Kennedy et al., 2014; Nadav et al., 2012; Racz et al., 2012). Second, we focus on the temporal pond-surface infiltration and groundwater recharge using field measurements and both simplified analytical methods as well as detailed numerical modeling. Third, for the numerical model we use measured data for variable-head boundary conditions at both top and bottom boundaries. Our data allow us to specifically address the lag between infiltration and groundwater recharge. This is of interest in cases of relatively deep unsaturated zone (initially $\sim 25 \, \text{m}$, in this study) in order to better estimate groundwater recharge. In contrast, conventional methods used in many studies estimate potential groundwater recharge from infiltration rates that do not necessarily represent recharge rates at the water table (Scanlon et al., 2002).

The purpose of this paper is to provide a detailed field-scale analysis of MAR with DSW from a hydrological perspective. Initially the monitoring system is described, the unsaturated zone is characterized, and several methods for calculating infiltration and recharge are presented. Next, infiltration dynamics (spatial and temporal) and its relation to the unsaturated zone lithology and to pond surface clogging is discussed. Finally, groundwater recharge estimations obtained from the analytical and numerical models are evaluated and compared.

2 Methods

2.1 Site description

The Menashe MAR site is located on sand dunes, $28 \, \text{m}$ above mean sea level (AMSL), overlaying the northern part of the Israeli Coastal Aquifer (Fig. 1a). Climate is Mediterranean with annual mean precipitation of $566 \, \text{mm} \, \text{yr}^{-1}$ (Gan Shmuel, 1987–2007). The annual average temperature is $20.2 \, ^\circ \text{C}$. The coldest month is January with average maximum and minimum temperatures of 17.4 and $10.5 \, ^\circ \text{C}$, respectively. The warmest month is August with average maximum and minimum temperatures of 28.5 and $24.8 \, ^\circ \text{C}$, respectively (Israel Meteorological Service, 2016). Below the Menashe MAR site the aquifer is about $80 \, \text{m}$ deep and consists mainly of calcareous sandstone (kurkar hereafter), sand, silty mud and clay lenses. Regional groundwater level is $\sim 3 \, \text{m}$ AMSL (September 2014), and pre-winter to post-winter seasonal groundwater-level fluctuations are $\sim 2 \, \text{m}$ (Israel Hydrological Service, 2014). Characteristic hydraulic properties of the aquifer are $10 \, \text{m} \, \text{d}^{-1}$, 0.25, and 0.4, for hydraulic conductivity, storativity, and porosity, respectively (Shavit and Furman, 2001). Operating since 1967, the Menashe MAR site diverts the natural ephemeral flow into a settling pond and from there to three infiltration ponds. The recharged water is recovered from the aquifer by dedicated production wells that encircle the site (Sellinger and Aberbach, 1973). In the vast majority of runoff events only the two northern infiltration ponds are used. Therefore, in the last few years, the southern infiltration pond is used for infiltration of surplus of DSW from the Hadera reverse-osmosis desalination plant, located 4 km to the west on the coast (Fig. 1b).

2.2 Monitoring

Monitoring of MAR activity was performed in the southern infiltration pond (herein referred to as "the pond") where DSW discharged occasionally according to operational considerations of Mekorot (Israel national water company) and the Israeli Water Authority. A dedicated monitoring system including observation wells, soil sensors, and infiltration rings was installed at the south part of the pond (Fig. 1c, e). The two groundwater observation wells (OA and OB) are 30 m deep, perforated at the lower part of the well (10 m from the bottom) and penetrating the saturated zone. Both were monitored by loggers (CTD-Diver, Eijkelkamp) measuring pressure head and electrical conductivity (EC). The shallow unsaturated zone includes eight soil sensors (5TE and GS3, Decagon Devices) at depths of 0.3, 0.5, 1, 1.5, 2, 2.5, 3, and 4 m below the pond surface, measuring volumetric water content (WC) and bulk EC (a measure of the electric conductivity of the bulk soil, which includes soil, water and air). The monitoring system continuously operated since October 2014 and measurements were obtained regularly every 15–30 min and at a finer resolution of 1–5 min during MAR

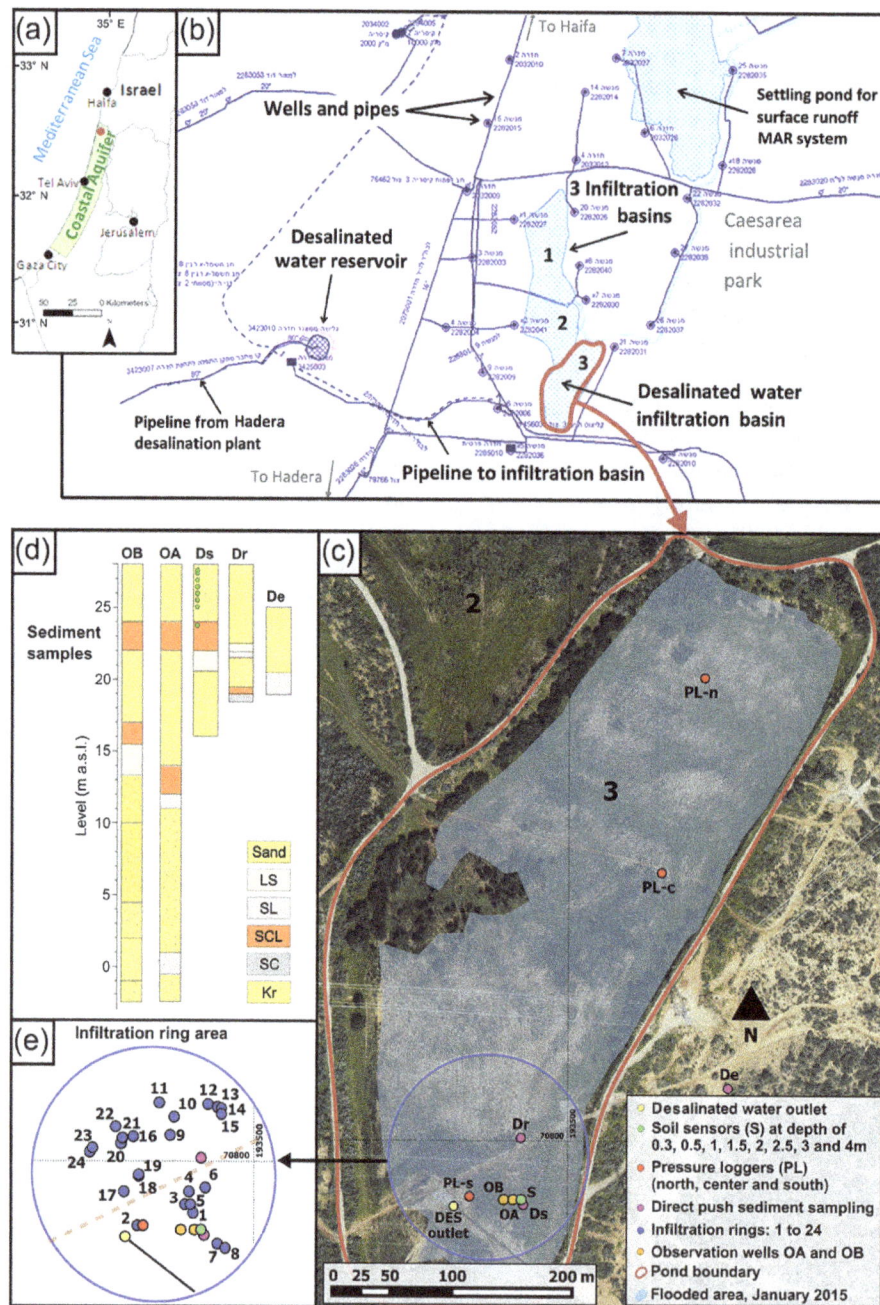

Figure 1. (a) Location of the Israeli Coastal Aquifer and the Menashe site (red circle). **(b)** Menashe MAR site. **(c)** Southern infiltration pond with the monitoring system. **(d)** Observation wells (OA and OB) and direct-push (Ds, Dr, and De) sediment profiles. The soil sensors are shown schematically on profile Ds. LS – loamy sand; SL – sandy loam; SCL – sandy clay loam; SC – sandy clay; and Kr – kurkar. **(e)** Infiltration ring locations (1–24).

or infiltration tests. In addition to the permanent monitoring system, ponding depth was monitored by three pressure loggers installed on the pond surface for the January 2015 MAR event at the north, center, and south part of the pond (Fig. 1c). All pressure head measurements were compensated by on-site logging barometer (BARO-Diver, Eijkelkamp).

2.3   Sediment sampling

Disturbed sediment samples (from auger) were taken during the drilling of the observation wells. Relatively undisturbed continuous core samples from the unsaturated zone were obtained by a direct-push rig (9700-VTR PowerProbe, AMS). Cores were taken at the following locations: next to the soil sensors (0–12 m depth), the southern road (0–9 m depth), and east of the pond (0–6 m depth, Fig. 1d). All sediment samples (both disturbed and undisturbed) were analyzed for particle size distribution by sieving for gravel (> 2 mm), sand (2–0.045 mm), and hydrometer for silt and clay. Bulk densities were calculated only for the undisturbed samples from the mass and volume of the cores and their water content.

The sediment profiles of observation wells OA and OB are similar. The top 30 m includes two repeated sequences, each consisting of a sand layer overlaying a sandy-clay-loam (SCL) layer, down to $\sim$ 15 m, of variable depth and thickness. Deeper down the profile, kurkar is dominant, alternating with layers of sand and sandy loam. The shallower direct-push profile next to the soil sensor location (Ds) is also similar to the profiles of OA and OB (due to their proximity to each other – less than 15 m apart), while the more distant profiles (Dr and De) are less similar (Fig. 1d). We cannot determine the lateral extent of the less-permeable layers based on our sediment sampling, which is spatially limited. We assume that these clayey and loamy layers are discontinuous like most of the low-permeability lenses at distances greater than 3 km from the coastline in the Israeli Coastal Aquifer (Kurtzman et al., 2012).

2.4   Calculating infiltration rates

Infiltration rates were calculated at the pond scale by pond draining rate, and at local scale by single-ring infiltrometers and wetting-front propagation (Dahan et al., 2007). Details of each method are given in what follows.

2.4.1   Pond scale

Ponding depth data were used to calculate the pond-scale infiltration rates. This method represents the infiltration rate of the whole pond, which is an average of the local infiltration rates (that may vary spatially due to sediment heterogeneity). The average pond infiltration rates were calculated by linear regression of ponding depth, which declined due to intermittent inlet discharge during the January 2015 MAR event. Each observation point of infiltration rate was calculated from a large number (tens to hundreds) of ponding-depth measurements. Two conditions must be met in order to calculate infiltration rates by this method: (1) ponding depth is declining solely due to infiltration (i.e., no other inlet/outlet source or surface flow) and (2) the time span of the descending ponding-depth data is sufficiently long (usually at least a few hours) in order to obtain regression with low-error slope (which is an estimate of the pond-scale infiltration rate).

2.4.2   Single-ring infiltrometers

In order to capture local infiltration rate variability, we used an array of 24 single-ring infiltrometers (100 cm long, 20 cm diameter) hammered 60 cm into the ground at different locations (Fig. 1e). Sediment samples taken from each infiltration ring location (outside the ring) at depth of 5 cm (undisturbed) and 60 cm (disturbed, with an auger, divided into four sections) were analyzed for bulk density (undisturbed only) and particle size distribution (both).

Infiltration tests were performed under relatively dry conditions (average WC of 0.09 $m^3\,m^{-3}$), early ponding (4 h after MAR started), and late ponding (after 1 month when discharge into the pond was ended), by continuously monitoring water level inside the rings. Under dry conditions, a fixed volume of 5.1 L was used in each ring to ensure one dimensional (1-D) flow inside the single ring. The infiltration rates were calculated by linear regression of pressure-head vs. time data, for dry conditions ($n = 24$), early ponding ($n = 11$), and late ponding ($n = 20$). Infiltration tests with relative errors higher than 15 % were omitted from the analysis. The omitted results (6 out of 61 single-ring tests) were due to insufficient measurement points during the infiltration test. We found the single-ring method more suitable than other methods we tested (double ring and Guelph permeameter), mainly because of its simplicity and permanent location that allows infiltration test repetitions under different conditions.

2.4.3   Wetting-front propagation

Monitoring of water content variation in the unsaturated zone by the soil sensors provides information on the wetting/drying-front propagation velocity. Infiltration (or drainage) rates are evident from the lag in wetting (or drying) front between different sensors at various depths. Infiltration rates were estimated from the velocity of the fronts and the difference in water content on both sides of the wetting front.

2.5   Modeling groundwater recharge

We describe the flow from the surface to the water table during a MAR operation using three different models: two simple analytical models (i.e., one using water table data, the other ponding-depth data) and one numerical model (in which both data sets were used). The simple analytical models are useful not only when there are not enough data to calibrate a numerical model; they also provide a first approximation which can be used as a preliminary test for a numerical model. In all three models we assign similar sediment profile layers and saturated hydraulic conductivity (Table 1), evaluated from pedotransfer functions (PTFs) using bulk density and particle size distribution data (Schaap et al., 2001). Only the saturated hydraulic conductivity of the top SCL layer was modified during calibration of the numerical model. Temporal and cumulative infiltration/recharge were obtained using 5 min resolution data measured during the January 2015 MAR event.

Because the pond water depth is much smaller than its horizontal dimensions, we consider 1-D vertical infiltration (perpendicular to the layers) – e.g., see Philip (1992). This assumption neglects lateral water flow, which is mainly rele-

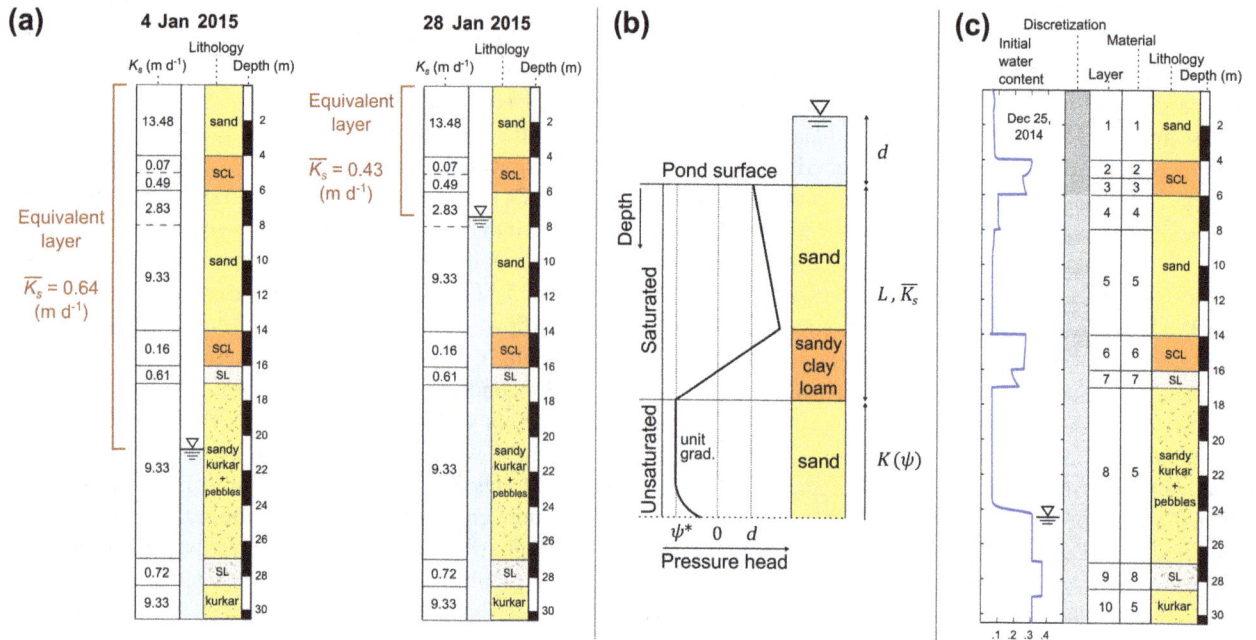

Figure 2. Recharge models. **(a)** Estimation of fluxes at early (right) and late (left) stage of the January 2015 MAR event in the lumped model. $\overline{K_S}$ is the equivalent vertical hydraulic conductivity. **(b)** The saturated portion with thickness L and $\overline{K_S}$ is used to estimate flux in the seepage model. A schematic steady-state pressure head profile shows the transition from saturated to unsaturated conditions (after Zaslavsky, 1964). **(c)** The numerical model domain.

Table 1. The material properties used for the layers in the analytical models and uncalibrated numerical model (Fig. 2): soil texture and van Genuchten–Mualem hydraulic functions parameters – residual and saturated water contents, θ_r and θ_s; fitting parameters α and n; and saturated hydraulic conductivity, K_S.

Material	Soil texture	θ_r	θ_s	α (m^{-1})	n	K_S $(m\,d^{-1})$
1	Sand	0.053	0.353	3.03	4.52	13.48
2	Sandy clay loam	0.063	0.336	2.9	1.19	0.07
3	Sandy clay loam	0.068	0.387	2.42	1.52	0.49
4	Fine sand	0.052	0.300	2.96	2.85	2.83
5	Sand/kurkar	0.050	0.310	3.09	4.05	9.33
6	Sandy clay loam	0.068	0.369	2.46	1.32	0.16
7	Sandy loam	0.064	0.365	2.32	1.71	0.61
8	Sandy loam	0.056	0.372	2.9	1.66	0.72

vant at the pond boundaries and during early and late stages of MAR (when only a portion of the pond surface is covered with water). However, during most of the January 2015 MAR event the whole pond area was covered with water and therefore the 1-D flow is a reasonable approximation. A main advantage of the 1-D model, apart from its simplicity, is that it can capture the whole-pond MAR processes in a single representative 1-D sediment profile.

2.5.1 Analytical lumped model using hydraulic conductivity and water table data

In this lumped model, we use two measured data sets – water table levels and saturated hydraulic conductivities. We consider the following transient boundary conditions: flux at the top and water table level (head) at the bottom. At the top, the recharge flux is equivalent to the flux in a saturated layered column under unit gradient flow. The lower boundary is the level of the moving water table. Calculations begin from the time that the water table starts rising, with initial conditions of fully saturated sediment profile. Assuming an initially saturated profile allows using the saturated hydraulic conductivity (K_s). We consider the flux q ($L\,T^{-1}$) to be equal to the equivalent vertical hydraulic conductivity $\overline{K_s}$ of the layers above the water table, which, for a vertical flow perpendicular to the layers, is computed from the harmonic mean of the layers' saturated hydraulic conductivities:

$$q = \overline{K_s} = L / \sum (z_i / K_{s,i}), \tag{1}$$

where z_i (L) and $K_{s,i}$ ($L\,T^{-1}$) are thickness and saturated hydraulic conductivity of layer i, respectively, and $L = \sum z_i$ is the thickness of the layers above the water table. We compute the flux every 5 min according to the measured water levels in observation well OA. Note that Eq. (1) provides temporal changes in flux because the properties (thickness and K_s) of the equivalent layer above the water table changes in time (Fig. 2a).

2.5.2 Analytical steady-state seepage model

The second analytical model assumes seepage flow through a perched water surface together with ponding depth data. We consider steady-state seepage through the topmost low-permeability SCL layer (Fig. 2a, 4–6 m depth), and that both this layer as well as the sand layer above it (0–4 m) are saturated under ponding, justified by the disparate hydraulic conductivities of the sand (high) and SCL (low) layers. By the same reasoning, the sand layer below the restrictive SCL layer remains unsaturated, maintaining steady-state flow (Fig. 2b). With the above, the 1-D steady-state flux through the saturated layers q_s can be described using Darcy's law as

$$q_s = \overline{K_s}(d + L - \psi^*)/L, \tag{2}$$

where d (L) is the ponding depth (measured every 5 min), L (L) is the thickness of the saturated layers (sand and SCL), and ψ^* (L) is the matric pressure head at the bottom of the saturated layers. The equivalent conductivity $\overline{K_s}$ for layers 1 to 3 (Table 1) was computed by Eq. (1). Assuming gravitational flow (unit gradient) in the unsaturated layer below the saturated layers (Zaslavsky, 1964), ψ^* is estimated from $q_{un} = K(\psi^*)$, where $K(\psi^*)$ is the function relating the unsaturated hydraulic conductivity to ψ^*. Mass balance implies the equality of fluxes in the saturated (q_s, Eq. 2) and unsaturated (q_{un}) layers, providing $\overline{K_s}(d + L - \psi^*)/L = K(\psi^*)$. Here, we solve iteratively for ψ^* using the van Genuchten–Mualem model (van Genuchten, 1980; Mualem, 1976), which was employed by others to estimate stream-aquifer seepage (Brunner et al., 2009; Osman and Bruen, 2002). Alternatively, to a leading order, one could take the ψ^* as the atmospheric pressure or the sediment's air entry value. The flux in Eq. (2) is mostly affected by the ponding depth (d); ψ^* is relatively insensitive to d (varying by $\pm 4\,\%$), and $\overline{K_s}$ and L are constants (unlike the lumped model where these parameters are time dependent).

2.5.3 Numerical model

Infiltration through the unsaturated zone was simulated with the HYDRUS-1-D software (version 4.16), a finite-element code for 1-D uniform water movement in variably saturated rigid porous media (for a detailed description of the governing equations see Šimůnek et al., 2009). HYDRUS-1-D was recently used to evaluate recharge in natural settings (Assefa and Woodbury, 2013; Neto et al., 2016; Turkeltaub et al., 2015). In this work HYDRUS-1-D was used to evaluate infiltration through the unsaturated zone and groundwater recharge using the 1-D Richards equation, with negligible root water uptake (sink term).

The model domain includes 10 layers within 0–30.5 m depth, with eight different compositions based on the sediment core samples, discretized into 1000 elements of thickness of 0.1–4 cm (average of 3 cm) varied according to the sediment type and location, and to the original groundwater level (Fig. 2c).

The van Genuchten–Mualem model (van Genuchten, 1980; Mualem, 1976) was used for the water retention curves and unsaturated hydraulic conductivity functions of the different sediments. The hydraulic parameters (Table 1) were calculated from the measured particle size distribution and bulk density using PTFs (ROSETTA; Schaap et al., 2001), which is incorporated in HYDRUS-1-D. An exception was made for the kurkar rock (layers 8 and 10), for which we used the hydraulic parameters of material 5 (sand) since the kurkar was crushed during drilling and its structure was destroyed. The saturated hydraulic conductivities (K_s) of wells OA and OB were estimated by single-well recovery tests interpretation (data not shown) as 3.6 and 7.5 $m\,d^{-1}$, respectively. While these values represent an effective value of several layers in which the well screen crosses rather than a specific layer, K_s values from these recovery tests are in the range of the values obtained by ROSETTA for the sand layers. In addition, previous studies (Levi, 2015; Shapira, 2012) showed that the local sediments at the Israeli Coastal aquifer show reasonable fit ($R^2 = 0.69$, $n = 53$) between the measured (in the laboratory) and estimated saturated hydraulic conductivity calculated by ROSETTA. Nevertheless, in order to test our numerical model results obtained with ROSETTA, we also preform simulations with two different USDA textural-class PTFs (Carsel and Parrish, 1988; Tóth et al., 2015; Tables S1 and S2 in the Supplement).

Boundary conditions at the soil surface (top) were the ponding depth (monitored at its surface when filled) and no flow when the pond is empty, and groundwater level (measured in well OA) at the bottom. The variable-head boundary conditions were applied at fixed locations at the top and bottom of the domain, and were updated at a 1 h resolution. Variable-head boundary conditions were selected in both boundaries in order to capture the highly dynamic behavior of the system during the January 2015 MAR event (rise of 2.2 and 17 m in ponding depth and groundwater level, re-

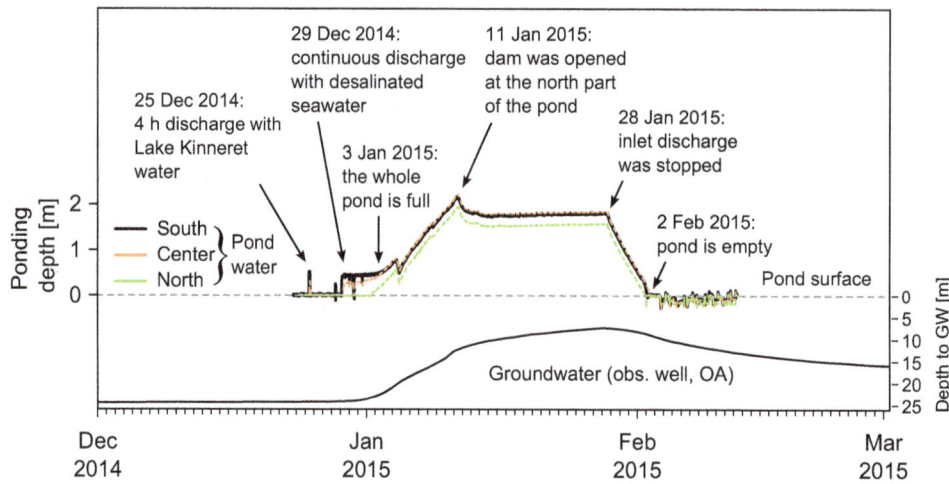

Figure 3. Water ponding depth at the south, center, and north locations inside the pond during the January 2015 MAR event. Observation well OA shows a sharp increase of the water table during MAR (17 m). Minor ticks on the *x* axis are days.

spectively). The code output is the flow at the domain boundaries: infiltration flux at the pond surface (top) and groundwater recharge (bottom). The recharge flux at the fluctuating groundwater table is similar to that at the bottom of the domain, because HYDRUS-1-D neglects the storage term in the variably saturated flow equation. Initial water content profile was obtained by field data (unsaturated at depth of 0–4 m and saturated below the water table, at 24.4 m) and by running a simulation for 88 days (September to December 2014) before the beginning of MAR that incorporates daily precipitation and evaporation.

3   Results

3.1   Monitoring pond and groundwater

The three pressure loggers monitored the local ponding depth during January 2015 MAR event, showing the filling period as water flows from the south to the north part of the pond (Fig. 3). A uniform water level was reached after the whole pond surface was covered with water and the three pressure loggers measured similar levels (beginning on 3 January 2015), with a constant difference between the pressure loggers that represents ponding depth difference due to a shallower pond surface (topography) at the northern part. Next, water ponding depth increases sharply and reaches a maximum of 2.2 m (11 January 2015). At this stage a small dam was opened at the north part of the pond allowing water to flow freely to the connecting channel and to the northern infiltration ponds (Fig. 1b) and the ponding depth was stabilized at 1.8 m for 17 days. During this period, on 21 January 2015, a flooded area of 10.7 ha was mapped with a GPS device to obtain the ponded area (Fig. 1c). After 31 days, inlet discharge was stopped and the water ponding depth de-

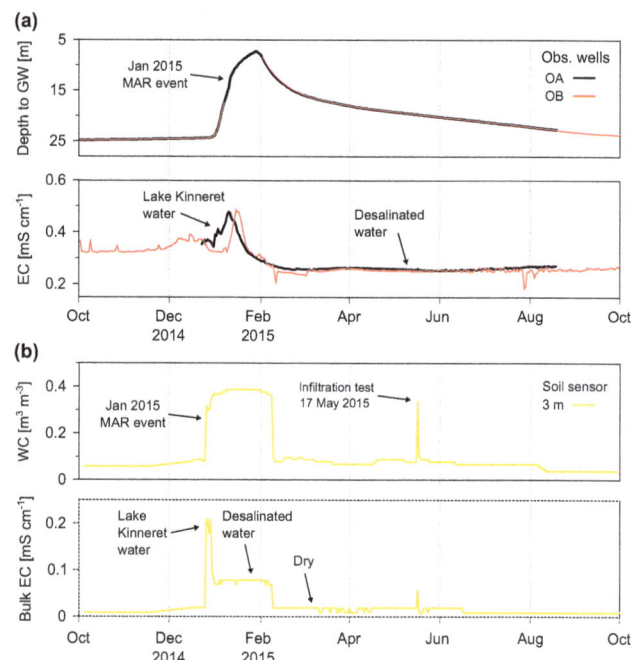

Figure 4. **(a)** Monitoring of groundwater level and EC in observation wells OA and OB during 1 year. **(b)** Volumetric water content (WC) and bulk EC at 3 m below ground surface. Note that the January 2015 MAR event and changes in water quality can be seen in both groundwater and vadose zone monitoring.

creased; finally on 2 February 2015 the pond was drained completely.

The pressure loggers in the observation wells captured a substantial rise of 17 m in groundwater level after 1 month of continuous MAR during January 2015 (Fig. 4a). Note that this rise represents the local conditions beneath the pond, while the influence on regional groundwater levels is damped farther away from the pond (e.g., a well, located 600 m to the north of the pond margins, showed a maximal groundwater-level rise of 4 m; data not shown). The lag in groundwater rise (Fig. 3) is due to the infiltration process through the unsaturated zone, which takes around 3 days until water reaches the water table, since the beginning of discharge on 29 December 2014. However, groundwater drops almost immediately (5 h) after inlet discharge was stopped and pond water level declines. This fast response implies fully connected flow (Brunner et al., 2009) between the high-level groundwater and the pond. Note that the relatively sharp decline in groundwater level during the first month after discharge was ended is followed by a gradual decline in the following months (Fig. 4a).

The EC monitoring of groundwater at the observation wells is shown in Fig. 4a to emphasize the response of groundwater to MAR with DSW (~ 0.2 mS cm^{-1}). Before the beginning of recharge with DSW on 29 December 2014, a 4 h recharge with water from Lake Kinneret was applied on 25 December 2014. The higher EC of Lake Kinneret water (~ 1.1 mS cm^{-1}) compared to DSW is related with the high EC readings (up to 0.5 mS cm^{-1}) in the observation wells, which decrease as DSW reaches the groundwater and finally the groundwater EC stabilized on 0.26 mS cm^{-1}.

Volumetric water content (WC) measurements in the vadose zone also capture the January 2015 MAR event showing constant WC during most of the ponding period and also later during the dry period. Changes in WC are most notable during wetting and drying of the upper sand layer at the beginning and end of infiltration, respectively (Fig. 4b). The bulk EC at depth of 3 m is increasing at the beginning of the MAR event and then decreases and stabilizes at 0.08 mS cm^{-1}, showing a similar trend as recorded in the observation wells and discussed above. The bulk EC (which is a function of the soil WC) decreases further to 0.01 mS cm^{-1} when the soil drains and dries after the end of MAR.

3.2 Infiltration rates

3.2.1 Pond scale

Pond infiltration rates show a general decrease during the January 2015 MAR event (Fig. 5a). Infiltration rates of 9 ± 2 (25 December 2014) and 2.9 ± 0.4 m d^{-1} (30 December 2014) were calculated before the whole pond was flooded and represents an average infiltration rate of the temporal water body, which might be biased due to surface flow. Average pond infiltration rates of 0.84 ± 0.02, 0.72 ± 0.08, and 0.36 ± 0.01 m d^{-1} were calculated after the whole pond was full. Multi-year average pan evaporation for this area in January is 0.0023 m d^{-1} (Israel Meteorological Service, 2016), which is 2 orders of magnitude smaller than the lowest infiltration rate measured for the MAR activity; hence, evaporation losses are considered negligible hereafter.

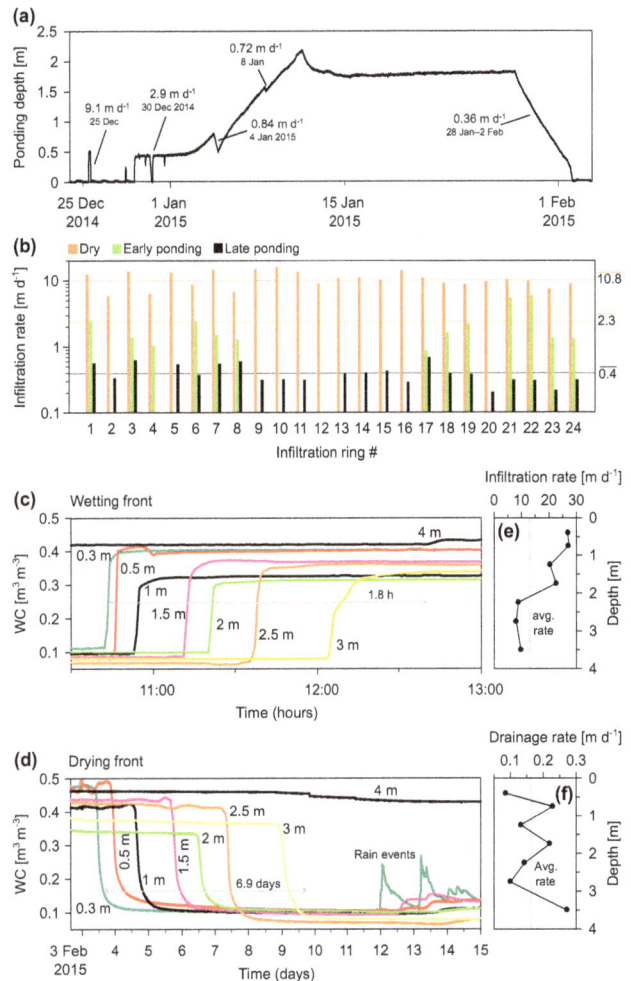

Figure 5. Infiltration rate measurements at various scales and perspectives: (**a**) pond scale (10.7 ha); (**b**) infiltration rings (0.1 m^2); note the log scale at the y axis; (**c**) wetting front propagation in the vadose zone at the beginning of infiltration; (**d**) drying front propagation in the vadose zone at the end of infiltration; (**e**) infiltration rates calculated from wetting; and (**f**) drying front.

3.2.2 Single-ring scale

Results of the single-ring infiltration tests under the different conditions show some degree of spatial and temporal variability of infiltration rates (Fig. 5b). Spatial variability was evaluated from differences in rates in different locations, and was found to be moderate (coefficient of variation, CV = 0.27) and high (CV = 0.77) for dry and early ponding conditions, respectively. Infiltration rates during late ponding were measured when pond water level decreased and the infiltration rings were gradually exposed above the water line (from north to south). This process took 5 days and therefore the variability in late ponding infiltration rates can be considered both temporal and spatial. Nevertheless, the average single-ring infiltration rates of 10.8 (standard devia-

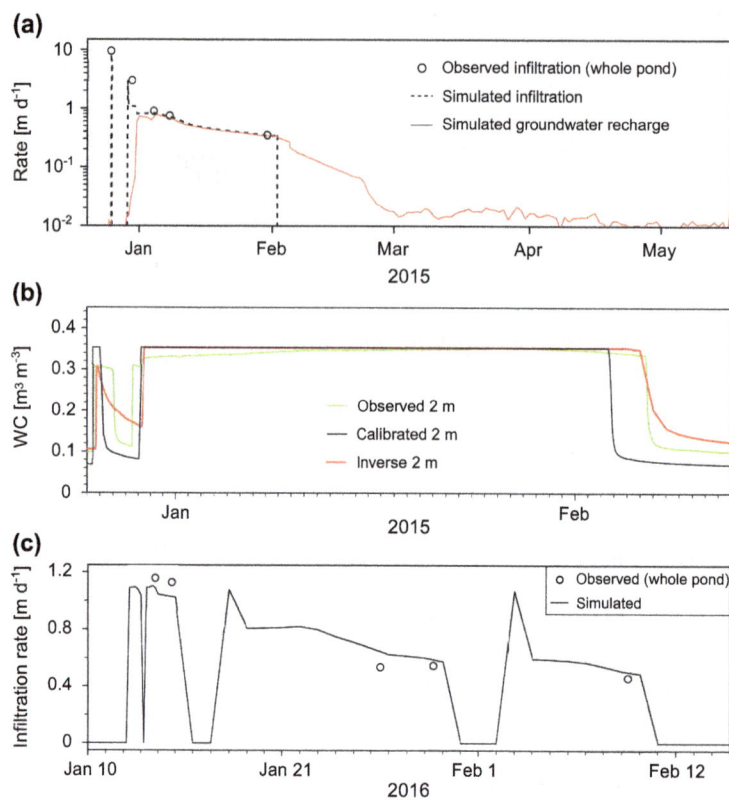

Figure 6. (a) Calibration of the infiltration rates at ground surface in the pond; note the log scale at the y axis. The simulated groundwater recharge rate is also shown for comparison. Note the large difference between simulated infiltration and recharge rates at the beginning of the MAR event versus the identity in rates that is achieved after a few days, and the continuation of recharge after the end of the infiltration. **(b)** Validation fit of the calibrated and inverse models to volumetric water content (WC) data from depth of 2 m. **(c)** Validation fit of the calibrated model to January 2016 MAR event (1.3×10^6 m^3). Minor ticks on the x axis are days in **(b)** and **(c)**.

tion, SD = 2.9), 2.3 (SD = 1.8), and 0.4 (SD = 0.15) m d^{-1} for dry conditions, and early and late ponding, respectively, show similar rates to the whole-pond rates results (Fig. 5a), indicating that the measured infiltration rates of the single rings were spatially representative.

3.2.3 Wetting-front propagation

Sharp wetting and drying (drainage) fronts, typical for sandy and coarse sediments, were observed at the beginning and end of infiltration, respectively (Fig. 5c, d). The estimated infiltration rates between couples of soil sensors (i.e., between 0.3 and 0.5 m, 0.5 and 1 m, etc.) change with time: the infiltration rate generally decreases as the wetting front advances deeper (Fig. 5e), as expected from theory (Assouline, 2013) and the impact of the SCL layer, while the drying front shows a more complex trend (Fig. 5f). In order to compare with infiltration rate measurements by other methods, the average infiltration and drainage rate of the top sand layer was estimated between the soil sensors at 0.3 and 4 m depth, as 12.8 and 0.16 m d^{-1} for dry and post-ponding conditions, respectively. The lower drainage rate of 0.16 m d^{-1} compared to the average surface infiltration rate of 0.4 m d^{-1} during late

ponding (Fig. 5a, b) is expected due to the upward capillary tension exerted during internal drainage, whereas this tension is absent during ponding.

3.3 Recharge models

3.3.1 Simplified analytical models

The lumped and seepage models provide cumulative recharge of 20.2 and 16.4 m (2.2×10^6 and 1.8×10^6 m^3 when multiplying by the active flood area), respectively. Both models assume constant flux along the profile at each time step, which means similar rates of groundwater recharge and surface infiltration. Average infiltration/recharge rates are 0.57 (SD = 0.10) and 0.46 (SD = 0.04) m d^{-1} for the lumped and seepage models, respectively. These rates are in the range of the whole-pond measured infiltration rates (Fig. 5a).

3.3.2 Numerical model

To capture temporal variations in drainage and groundwater recharge, we simulated the January 2015 MAR event from 25 December 2014 to 5 October 2015. We calibrated the numerical model to the whole-pond infiltration rate data (Fig. 5a) in order to generalize the local sediment profile into a whole-pond representative profile. Only saturated hydraulic conductivities of the top SCL section were modified during the calibration (4–6 m, layers 2 and 3, calibrated $K_s = 0.38\,\mathrm{m\,d^{-1}}$ for both layers).

The model calibration shows a good fit for 90 % of the infiltration period (4–31 January 2015) with a relative root mean square error of 4.8 % (Fig. 6a). Relatively poor fits between the model and the whole-pond infiltration data were obtained for the first two observations points (25 and 30 December 2014). These two observations were measured at early stages, when the pond was partly filled, which may overestimate infiltration rates due to surface flow. Checking the calibrated model against WC data (from the vadose zone monitoring system at 2 m depth) shows reasonable validation as the model was calibrated against whole-pond data, while the WC represents point-specific data. In terms of wetting/drying front, the model underestimates the arrival time of the front. A better fit to the WC data was achieved during the calibration process using the built-in HYDRUS-1-D inverse modeling, by fitting the van Genuchten–Mualem parameters α, n, and K_s (Fig. 6b). However, we decided not to use these calibration results (only calibrating K_s), as they underestimate the whole-pond infiltration rates, and hence the cumulative infiltration. Evaluating the calibrated model with our latest available MAR data ($1.3 \times 10^6\,\mathrm{m}^3$, recharged during 12 January to 7 February 2016) shows good validation of the whole-pond infiltration rates with a relative root mean square error of 11.4 % (Fig. 6c).

Testing the numerical model with different PTFs shows an expected variation in the results of infiltration and recharge rates. These variations clearly demonstrate the need for calibration when using PTFs to estimate deep vadose zone hydraulic parameters (Zhang et al., 2016). In this case, the simulation with the PTFs of Tóth et al. (2015) was closest to the calibrated simulation results (Fig. S1 in the Supplement).

Our numerical simulation results highlight the transient nature of groundwater recharge. At the end of the January 2015 MAR event when the pond was empty (2 February 2015), the estimated total groundwater recharge was 17.1 m, vs. 22.4 m of surface infiltration (Fig. 7). That is, during ponding $\sim 75\%$ of the infiltrated water has reached groundwater, while the remaining $\sim 25\%$ is retained in the newly saturated zone between the pre-MAR water table (24.5 m below surface) and the gradually decreasing post-MAR water table. Out of these "residual water", more than half reach the pre-MAR water table after ~ 1 month ($\sim 90\%$ of the total infiltrated water), with the remainder arriving only after ~ 6 months, as can be seen from the change in the slope of the groundwater recharge curve (Fig. 7, red line).

For a flooded pond surface of 10.7 ha the total surface infiltration gives roughly a total water volume of $2.4 \times 10^6\,\mathrm{m}^3$ that was discharged to the pond. This is in good agreement with the $2.45 \times 10^6\,\mathrm{m}^3$ that was reported by Mekorot that

supplied the water. Comparison of the estimated recharge by the simplified models and the numerical model is shown in Fig. 7. Both simplified model underestimate the total infiltration, but the lumped model is closer (20.2 m) than the seepage model (16.4 m) to the numerical model (22.4 m).

4 Discussion

4.1 Spatial and temporal variability of infiltration rates

Spatial infiltration variability depends on the soil type and structure and its spatial distribution in the pond. Single-ring dry infiltration rates showed significant correlation with bulk density ($r = -0.57$, $p = 0.003$) sampled at the soil surface (5 cm deep), but for the same samples no significant correlation was found with water content or with clay and silt fraction. Also, for the 60 cm deep samples, no significant correlation was found with clay and silt fraction. Because the upper soil in all the infiltration rings is classified as sand (at least 97 % sand) it is likely that very minor differences in the soil structure and particle size distribution are responsible for the difference between rings under dry infiltration rates (Fig. 5b). These minor changes are probably below the resolution of the particle size analysis that was conducted on all the sediment samples in this study. During the early ponding infiltration rate measurements (starting 4 h after ponding started), the wetting front was advancing further downward into the profile after it passed the top sandy layer, as evident by the soil sensors readings (Fig. 5c). At this stage, the spatial variability of infiltration rates is probably controlled by the lithology of the deep layered soil profile.

Temporal infiltration variability is evident from the single-ring tests as infiltration rates decrease from 6–16 m d^{-1} before MAR (dry conditions) to 0.1–0.7 m d^{-1} at the end of the MAR operation (late ponding, Fig. 5b). This temporal variability is similar to the variability from pond scale (Fig. 5a) and vadose zone infiltration rates (Fig. 5c–f). The main reason for this 2 orders of magnitude decrease of infiltration rate is the sharp contrast between the hydraulic conductivity of the top sand layer ($K_s = 13.5\,\mathrm{m\,d^{-1}}$) and the SCL layer underneath ($K_s = 0.07\,\mathrm{m\,d^{-1}}$, Table 1). This layer has the lowest hydraulic conductivity along the unsaturated profile, so it serves as the limiting layer for pond infiltration. Thus, the fast infiltration rate at early ponding continues as long as water flows to the north part of the pond during the pond-filling process and simultaneously the wetting front has not reached the SCL layer.

High spatial and temporal variability of infiltration rates, measured with thermal and pressure probes, was reported by Racz et al. (2012) during several months of MAR to an infiltration pond with an area of 3 ha. They postulated that small differences in the percentage of fine material in the relatively homogeneous shallow soil, clogging of the pond surface, and deeper unsaturated zone processes can explain this variabil-

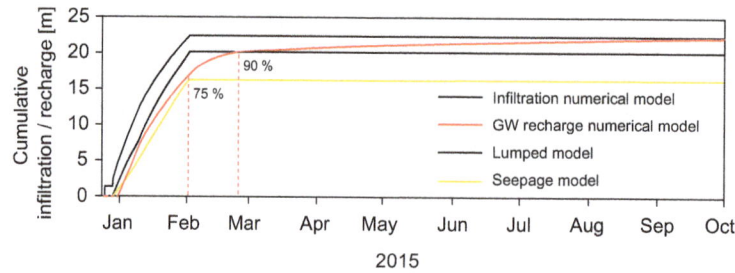

Figure 7. Cumulative infiltration and recharge of the various models during 2015. According to the numerical model, most of the infiltrated water (~ 75 %) reached the pre-MAR water table after 1 month (end of ponding), and ~ 90 % reached about 2 months after the onset of infiltration.

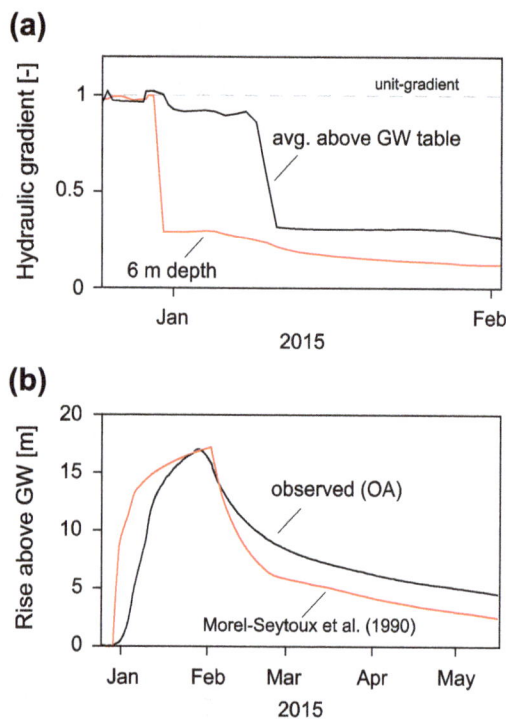

Figure 8. Testing models assumptions. **(a)** Unit-gradient assumption: the hydraulic gradients (HG) were calculated from the numerical model to test the HG of the lumped model (average HG above the fluctuating groundwater table) and of the seepage model (HG at 6 m depth). **(b)** One-dimensional flow assumption along the variably saturated zone: measured groundwater rise above the original groundwater table compared to the analytical solution of Morel-Seytoux et al. (1990) calculated using the infiltration and recharge fluxes of the numerical model.

ity. Mawer et al. (2016) used fiber optic distributed temperature sensing to monitor infiltration rates with high spatial resolution during MAR to an infiltration pond. They concluded that 80 % of the recharged water infiltrated through the most permeable 50 % surface area of the pond which was explained by heterogeneous clogging. In our study the relatively deep unsaturated zone sampling and infiltration

rate data show that the spatial and temporal variability of infiltration rates is suppressed (and controlled) by the low-permeability layers. Probably for the same reason, together with the high-quality source water (DSW), there was no field evidence in our study for clogging of the top sand layer.

4.2 Clogging

Clogging of the infiltration surface is the major operational concern in most MAR systems (Bouwer, 2002; Martin, 2013). The extent of clogging during MAR with DSW is questionable due to the low turbidity, organic matter, and total dissolved solids (TDS) of the source water (in this case, the DSW turbidity ~ 0.2 NTU and TDS ~ 120 mg L^{-1}). Vandenbohede et al. (2009b) reported on pond clogging during MAR with reverse-osmosis desalinated wastewater (TDS = 50 mg L^{-1}). It was explained by the accumulation of algae on the pond bottom, but the authors stated that "further research is needed to explain the reasons for the clogging". In laboratory experiments, Mukhopadhyay et al. (2004) reported on permeability reduction following injection of filtered (< 0.5 µm) DSW into cores initially saturated with groundwater. The authors explained this reduction by clogging with fines originating from dissolution of carbonate and gypsiferous matrix, commenting that further research is needed. In this study we did not find evidence for biological clogging, while dissolution of carbonate is a minor concern as the DSW that was used here is enriched with calcium during post-treatment of the desalination process, and therefore the DSW is saturated with respect to calcium carbonate (Ronen-Eliraz et al., 2017).

To further examine the impact of DSW on clogging, we performed preliminary infiltration column experiments in the laboratory with DSW and sand taken from the pond surface (top 0.4 m). Results showed a reduction by a factor of 1.5 compared to the initial infiltration rate, probably due to compaction clogging (see Sect. S2 in the Supplement). Similar results were obtained by Lado and Ben-Hur (2010) in a column experiment with sandy soil leached with reverse-osmosis effluent. They suggested that the relatively large average pore size in the sandy soil prevented pore clogging and

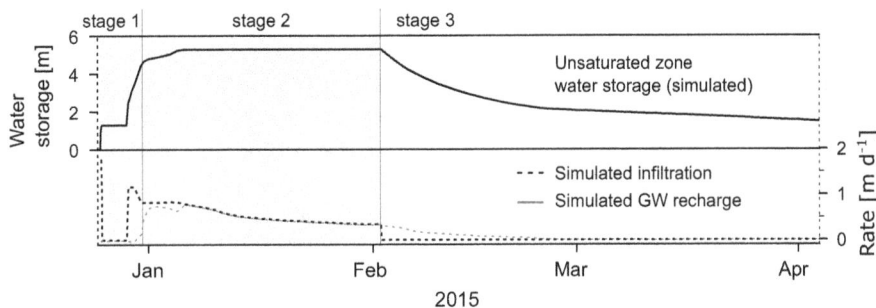

Figure 9. Change in water storage in the unsaturated zone (i.e., above the original water table) during ponding (stage 1 and 2) and drainage (stage 3). Simulated rates of surface infiltration and groundwater recharge are also shown as a reference.

K_s reduction. These findings support our field-scale assumption that infiltration-rate reduction due to clogging processes at the top sand layer is absent or negligible, compared to the impact of the low-permeability layers. This is also supported by the numerical model validation (Fig. 6c) as infiltration dynamics is captured nicely without the need to incorporate a clogging parameter in the numerical model.

It is worth noting that infiltration of low-salinity water into natural settings may cause clogging due to clay swelling, dispersion and colloidal release and deposition, which can lead to K_s reduction of up to 2 orders of magnitude (Blume et al., 2002; Lado and Ben-Hur, 2010; Mohan et al., 1993; Shainberg and Letey, 1984). The practical importance of these clogging mechanisms during MAR with DSW is unclear. Further research is needed in order to determine the long-term impact of MAR with DSW on field-scale clogging at the Menashe site.

4.3 Groundwater recharge

Clearly, the simplified models cannot capture groundwater recharge dynamics as the numerical model does, but they can serve as a first approximation for recharge when no other data are available or as complementary recharge estimation when other methods are used. The lumped model suffers from practical and theoretical limitations compared to the seepage model. The need for drilling a monitoring well inside (or very close to) the infiltration pond and in addition continuously monitoring groundwater level is the main operational limitation. Moreover, there is no field evident that supports the assumption of a saturated profile with a unit gradient along the heterogeneous sediment profile. Yet, a main advantage of the lumped model is its ability to predict groundwater recharge using only pedotransfer-based K_s values when water table data are available. Noticeably, even when water table data are unavailable, using the equivalent hydraulic conductivity ($\overline{K_s}$) of the sediments above the regional groundwater level provides an excellent estimate of the total cumulative recharge (in this study, $\overline{K_s} = 0.73 \, \mathrm{m \, d^{-1}}$ gives a total of 22.6 m during 31 days of MAR). The seepage model is practically simpler, as it does not require deep drilling (more relevant to settings with thick unsaturated zone) and only continuous monitoring of the pond water level is needed. While the seepage model requires the occurrence of a shallow low-permeability layer beneath the pond, it is not considered a major limitation because a clogging layer is usually found in most MAR systems (Bouwer, 2002).

The unit-gradient assumption in the simplified models was tested using the results of the calibrated numerical model. Checking the hydraulic gradients as calculated from the calibrated numerical model, for the lumped model (between the pond surface and the groundwater table) and for the seepage model (at 6 m depth, below the upper SCL layer), shows that the unit-gradient assumption is not always valid (Fig. 8a). This is due to the significant water table rise, the layered sediment profile, and the variably saturated conditions. These factors, together with the lack of calibration of the simplified models, provide a possible explanation for the differences between the simplified models and the calibrated numerical model.

The calibrated 1-D numerical model is a more complex tool compared to the simplified models that were presented or to other approximated methods, yet it is still simpler compared to 2-D or 3-D variably saturated models. Sumner et al. (1999) and Morel-Seytoux (2000) discussed the validity of 1-D flow along the unsaturated zone for estimating groundwater mounding during MAR. We tested our numerical model results using the analytical solution of Morel-Seytoux et al. (1990), which assumes 1-D vertical infiltration along the unsaturated zone and radial flow along the saturated zone. The calculated groundwater level below the pond (Eq. 27 in Morel-Seytoux et al., 1990), using the infiltration and recharge rate results from the calibrated 1-D numerical model, shows a reasonable fit with the observed groundwater levels, supporting our assumption that flow along the unsaturated zone is mainly vertical (Fig. 8b). The differences between the calculated and observed groundwater levels can be attributed to the analytical model assumptions and to errors associated with the estimated model parameters (0.25, 185, and 70 m for specific yield, equivalent pond radius, and saturated aquifer thickness, respectively).

The main advantage of our 1-D numerical model is its ability to capture infiltration and recharge dynamics of the MAR system based on only one representative sediment profile. The obvious main drawback of the 1-D model is its inability to capture lateral flows, both at the unsaturated and saturated zones. This limitation, to some extent, is compensated by the use of data-based variable-head boundary conditions which were employed to better estimate surface infiltration and groundwater recharge (for comparison, applying a constant-head lower boundary condition as an alternative will overestimate recharge). However, when using these boundary conditions the model is inadequate for predicting water table or ponding depth evolution during future MAR events. This limitation can be overcome by changing the model boundaries, as discussed, for example, by Neto et al. (2016).

Predicting the recharge dynamics during MAR by numerical simulations is a valuable tool for planning successive MAR events, and as an input for regional groundwater models. Groundwater recharge is governed by the boundaries of the system and the unsaturated zone hydraulic properties. In MAR sites with unsaturated zone of intermediate depth (normally ~ 25 m at the Menashe MAR site), the water storage of the unsaturated zone affects infiltration and recharge dynamics. This is shown in Fig. 9 and can be divided into three stages: (1) high infiltration and low recharge rates during the saturation process of the vadose zone; (2) full water capacity (or close to) is attained – infiltration and recharge rates are similar and finally converge and decreases due to groundwater level rise (hydraulic gradient decreases); and (3) end of ponding (infiltration ends) – recharge rate and water storage decrease further during drainage of the vadose zone. In MAR sites with shallow or no unsaturated zone (e.g., Vandenbohede et al., 2008), stages 1 and 3 are minor (if any) and the system will persist at stage 2 during the MAR operation. On the other hand, MAR sites with very deep unsaturated zone (e.g., Flint et al., 2012) may skip stage 2, crossing from stage 1 to 3 without reaching the potential water storage. The optimal extent of each stage during MAR operation is site-specific, depending on the MAR site requirements and constraints.

ter or negligible compared to the low-permeability layers. While sediment sampling and analysis is a routine procedure in hydrology science, we emphasize its crucial role in MAR projects. Careful consideration of the hydrological properties of the deep unsaturated zone is needed in order to quantify the contribution of the local sediments to infiltration rate dynamics, compared to the contribution of the MAR-related clogging layer. To date, literature on clogging during MAR with desalinated seawater is limited and the extent of field-scale clogging is unclear and probably site-specific. For this reason, the long-term impact of MAR with desalinated seawater on clogging processes at the Menashe site should be addressed in future studies. Groundwater recharge was estimated by analytical and numerical models that include ponding and groundwater head data. The simple analytical models can estimate reasonably well cumulative groundwater recharge using field data, but predicting the late recharge after pond infiltration terminates, requires a detailed unsaturated flow model. A 1-D numerical model with a whole-pond representative soil profile can capture groundwater recharge dynamics, especially when it constrained by measured variable-head boundary conditions. Validation of our numerical model in an independent MAR event shows the model robustness. The dynamic groundwater recharge described by the numerical model is useful for future MAR operation planning, and also as input for regional-scale groundwater modeling.

Competing interests. The authors declare that they have no conflict of interest.

Acknowledgements. The research leading to these results received funding from the European Union Seventh Framework Programme (FP7/2007-2013) under grant agreement no. 619120 (Demonstrating Managed Aquifer Recharge as a Solution to Water Scarcity and Drought – MARSOL). We thank Amos Russak and Raz Amir for their technical assistance and Eline Futerman-Hartog for conducting the dry infiltration experiments.

Edited by: Bill Hu

5 Summary and conclusions

Groundwater level under a sandy infiltration pond in the Israeli Coastal Aquifer rose by 17 m during 1 month of continues MAR with surplus desalinated seawater. Measured infiltration rates were relatively uniform spatially, however highly variable in time: during continuous discharge of 2.45×10^6 m^3, rates decreased by almost 2 orders of magnitude. This reduction can be explained solely by the lithology of the unsaturated zone that includes relatively low-permeability sediments, whereas clogging processes at pond-surface are negated by the high-quality desalinated seawa-

References

Al-Awadi, E., Mukhopadhyay, A., and Al-Haddad, A. J.: Compatibility Of Desalinated Water With The Dammam Formation At The Northwest Shigaya Water-Well Field, Kuwait – A Preliminary Study, Hydrogeol. J., 3, 56–73, https://doi.org/10.1007/s100400050068, 1995.

Assefa, K. A. and Woodbury, A. D.: Transient, spatially varied groundwater recharge modeling, Water Resour. Res., 49, 4593–4606, https://doi.org/10.1002/wrcr.20332, 2013.

Assouline, S.: Infiltration into soils: Conceptual approaches and solutions, Water Resour. Res., 49, 1755–1772, https://doi.org/10.1002/wrcr.20155, 2013.

Blume, T., Weisbrod, N., and Selker, J. S.: Permeability Changes in Layered Sediments: Impact of Particle Release, Ground Water, 40, 466–474, https://doi.org/10.1111/j.1745-6584.2002.tb02530.x, 2002.

Bouwer, H.: Artificial recharge of groundwater: hydrogeology and engineering, Hydrogeol. J., 10, 121–142, https://doi.org/10.1007/s10040-001-0182-4, 2002.

Brunner, P., Cook, P. G., and Simmons, C. T.: Hydrogeologic controls on disconnection between surface water and groundwater, Water Resour. Res., 45, W01422, https://doi.org/10.1029/2008WR006953, 2009.

Carsel, R. F. and Parrish, R. S.: Developing joint probability distributions of soil water retention characteristics, Water Resour. Res., 24, 755–769, https://doi.org/10.1029/WR024i005p00755, 1988.

Dahan, O., Shani, Y., Enzel, Y., Yechieli, Y., and Yakirevich, A.: Direct measurements of floodwater infiltration into shallow alluvial aquifers, J. Hydrol., 344, 157–170, https://doi.org/10.1016/j.jhydrol.2007.06.033, 2007.

Dillon, P.: Future management of aquifer recharge, Hydrogeol. J., 13, 313–316, https://doi.org/10.1007/s10040-004-0413-6, 2005.

Flint, A. L., Ellett, K. M., Christensen, A. H., and Martin, P.: Modeling a Thick Unsaturated Zone at San Gorgonio Pass, California: Lessons Learned after Five Years of Artificial Recharge, Vadose Zone J., 11, https://doi.org/10.2136/vzj2012.0043, 2012.

Ganot, Y., Russak, A., Siebner, H., Bernstein, A., Katz, Y., Guttman, J., and Kurtzman, D.: Geochemical processes in a calcareous sandstone aquifer during managed aquifer recharge with desalinated seawater, EGU General Assembly, Vienna, Austria, 23–28 April 2017, Geophys. Res. Abstr., 19, EGU2017-1633-1, 2017.

Israel Hydrological Service: http://www.water.gov.il/Hebrew/ProfessionalInfoAndData/Data-Hidrologeime/2013/hof-2013.pdf (last access: 4 November 2016), 2013.

Israel Hydrological Service: http://www.water.gov.il/Hebrew/ProfessionalInfoAndData/DataHidrologeime/DocLib2/hydrological-report-sep14.pdf (last access: 4 November 2016), 2014. Israel Meteorological Service: http://www.ims.gov.il/IMS/CLIMATE/ClimaticAtlas, last access: 16 June 2016.

Kennedy, J., Ferré, T. P. A., Güntner, A., Abe, M., and Creutzfeldt, B.: Direct measurement of subsurface mass change using the variable baseline gravity gradient method: Kennedy et al.: Variable-baseline gradients, Geophys. Res. Lett., 41, 2827–2834, https://doi.org/10.1002/2014GL059673, 2014.

Kloppmann, W., Van Houtte, E., Picot, G., Vandenbohede, A., Lebbe, L., Guerrot, C., Millot, R., Gaus, I., and Wintgens, T.: Monitoring Reverse Osmosis Treated Wastewater Recharge into a Coastal Aquifer by Environmental Isotopes (B, Li, O, H), Environ. Sci. Technol., 42, 8759–8765, https://doi.org/10.1021/es8011222, 2008.

Kurtzman, D., Netzer, L., Weisbrod, N., Nasser, A., Graber, E. R., and Ronen, D.: Characterization of deep aquifer dynamics using principal component analysis of sequential multilevel data, Hydrol. Earth Syst. Sci., 16, 761–771, https://doi.org/10.5194/hess-16-761-2012, 2012.

Levi, Y: Observations and Modeling of Nitrate Fluxes to Groundwater under Diverse Agricultural Land-Uses: From the Fields to the Pumping Wells, M.Sc. thesis, The program of Hydrology and Water Resources, The Hebrew University of Jerusalem, Israel, 2015.

Lado, M. and Ben-Hur, M.: Effects of Irrigation with Different Effluents on Saturated Hydraulic Conductivity of Arid and semiarid Soils, Soil Sci. Soc. Am. J., 74, 23–32, https://doi.org/10.2136/sssaj2009.0114, 2010.

Martin, R. (Ed.): Clogging issues associated with managed aquifer recharge methods, IAH Commission on Managing Aquifer Recharge, available at: http://recharge.iah.org/files/2015/03/Clogging_Monograph.pdf (last access: 6 September 2017), Australia, 2013.

Mawer, C., Parsekian, A., Pidlisecky, A., and Knight, R.: Characterizing Heterogeneity in Infiltration Rates During Managed Aquifer Recharge, Groundwater, 54, 818–829, https://doi.org/10.1111/gwat.12423, 2016.

Mohan, K. K., Vaidya, R. N., Reed, M. G., and Fogler, H. S.: Water sensitivity of sandstones containing swelling and non-swelling clays, Colloids and Surfaces A, 73, 237–254, https://doi.org/10.1016/0927-7757(93)80019-B, 1993.

Morel-Seytoux, H. J.: Effects of Unsaturated Zone on Ground-Water Mounding, J. Hydrol. Eng., 5, 435–436, 2000.

Morel-Seytoux, H. J., Miracapillo, C., and Abdulrazzak, M. J.: A reductionist physical approach to unsaturated aquifer recharge from a circular spreading basin, Water Resour. Res., 26, 771–777, https://doi.org/10.1029/WR026i004p00771, 1990.

Mualem, Y.: A new model for predicting the hydraulic conductivity of unsaturated porous media, Water Resour. Res., 12, 513–522, https://doi.org/10.1029/WR012i003p00513, 1976.

Mukhopadhyay, A., Szekely, F., and Senay, Y.: Artificial Ground Water Recharge Experiments in Carbonate and Clastic Aquifers of Kuwait, JAWRA J. Am. Water Resour. Assoc., 30, 1091–1107, https://doi.org/10.1111/j.1752-1688.1994.tb03355.x, 1994.

Mukhopadhyay, A., Al-Awadi, E., AlSenafy, M. N., and Smith, P. C.: Laboratory investigations of compatibility of the Dammam Formation Aquifer with desalinated freshwater at a pilot recharge site in Kuwait, J. Arid Environ., 40, 27–42, https://doi.org/10.1006/jare.1998.0428, 1998.

Mukhopadhyay, A., Al-Awadi, E., Oskui, R., Hadi, K., Al-Ruwaih, F., Turner, M., and Akber, A.: Laboratory investigations of compatibility of the Kuwait Group aquifer, Kuwait, with possible injection waters, J. Hydrol., 285, 158–176, https://doi.org/10.1016/j.jhydrol.2003.08.017, 2004.

Nadav, I., Tarchitzky, J., and Chen, Y.: Soil cultivation for enhanced wastewater infiltration in soil aquifer treatment (SAT), J. Hydrol., 470/471, 75–81, https://doi.org/10.1016/j.jhydrol.2012.08.013, 2012.

Neto, D. C., Chang, H. K., and van Genuchten, M. T.: A Mathematical View of Water Table Fluctuations in a Shallow Aquifer in Brazil, Groundwater, 54, 82–91, https://doi.org/10.1111/gwat.12329, 2015.

Osman, Y. Z. and Bruen, M. P.: Modelling stream–aquifer seepage in an alluvial aquifer: an improved loosing-stream package for MODFLOW, J. Hydrol., 264, 69–86, https://doi.org/10.1016/S0022-1694(02)00067-7, 2002.

Philip, J. R.: Falling head ponded infiltration, Water Resour. Res., 28, 2147–2148, https://doi.org/10.1029/92WR00704, 1992.

Racz, A. J., Fisher, A. T., Schmidt, C. M., Lockwood, B. S., and Huertos, M. L.: Spatial and Temporal Infiltration Dynamics During Managed Aquifer Recharge, Groundwater, 50, 562–570, https://doi.org/10.1111/j.1745-6584.2011.00875.x, 2012.

Ronen-Eliraz, G., Russak, A., Nitzan, I., Guttman, J., and Kurtz-man, D.: Investigating geochemical aspects of managed aquifer recharge by column experiments with alternating desalinated water and groundwater, Sci. Total Environ., 574, 1174–1181, https://doi.org/10.1016/j.scitotenv.2016.09.075, 2017.

Scanlon, B. R., Healy, R. W., and Cook, P. G.: Choosing appropriate techniques for quantifying groundwater recharge, Hydrogeol. J., 10, 18–39, https://doi.org/10.1007/s10040-001-0176-2, 2002.

Schaap, M. G., Leij, F. J., and van Genuchten, M. T.: rosetta: a computer program for estimating soil hydraulic parameters with hierarchical pedotransfer functions, J. Hydrol., 251, 163–176, https://doi.org/10.1016/S0022-1694(01)00466-8, 2001.

Sellinger, A. and Aberbach, S. H.: Artificial recharge of coastal-plain aquifer in Israel, IAHS Publ., 111, 701–714, 1973.

Shainberg, I. and Letey, J.: Response of soils to sodic and saline conditions, Hilgardia, 52, 1–57, https://doi.org/10.3733/hilg.v52n02p057, 1984.

Shapira, R. H.: Nitrate Flux to Groundwater Under Citrus Orchards: Observations, Modeling and Simulating Different Nitrogen Application Rates, M.Sc. thesis, The program of Hydrology and Water Resources, The Hebrew University of Jerusalem, Israel, 2012.

Shavit, U. and Furman, A.: The location of deep salinity sources in the Israeli Coastal aquifer, J. Hydrol., 250, 63–77, https://doi.org/10.1016/S0022-1694(01)00406-1, 2001.

Šimůnek, J., van Genuchten, M. T., Sejna, M., Saito, H., and Sakai, M.: HYDRUS-1-D technical manual, available from: http://www.pc-progress.com/Downloads/Pgm_hydrus1D/HYDRUS1D-4.08.pdf (last access: 1 November 2014), 2009.

Stanhill, G., Kurtzman, D., and Rosa, R.: Estimating desalination requirements in semi-arid climates: A Mediterranean case study, Desalination, 355, 118–123, https://doi.org/10.1016/j.desal.2014.10.035, 2015.

Sumner, D. M., Rolston, D. E., and Mariño, M. A.: Effects of unsaturated zone on ground-water mounding, J. Hydrol. Eng., 4, 65–69, 1999.

Tóth, B., Weynants, M., Nemes, A., Makó, A., Bilas, G., and Tóth, G.: New generation of hydraulic pedotransfer functions for Europe: New hydraulic pedotransfer functions for Europe, Eur. J. Soil Sci., 66, 226–238, https://doi.org/10.1111/ejss.12192, 2015.

Turkeltaub, T., Kurtzman, D., Bel, G., and Dahan, O.: Examination of groundwater recharge with a calibrated/validated flow model of the deep vadose zone, J. Hydrol., 522, 618–627, https://doi.org/10.1016/j.jhydrol.2015.01.026, 2015.

Vandenbohede, A. and Van Houtte, E.: Heat transport and temperature distribution during managed artificial recharge with surface ponds, J. Hydrol., 472/473, 77–89, https://doi.org/10.1016/j.jhydrol.2012.09.028, 2012.

Vandenbohede, A., Houtte, E. V., and Lebbe, L.: Groundwater flow in the vicinity of two artificial recharge ponds in the Belgian coastal dunes, Hydrogeol. J., 16, 1669–1681, https://doi.org/10.1007/s10040-008-0326-x, 2008.

Vandenbohede, A., Houtte, E., and Lebbe, L.: Sustainable groundwater extraction in coastal areas: a Belgian example, Environ. Geol., 57, 735–747, https://doi.org/10.1007/s00254-008-1351-8, 2009a.

Vandenbohede, A., Houtte, E. V., and Lebbe, L.: Water quality changes in the dunes of the western Belgian coastal plain due to artificial recharge of tertiary treated wastewater, Appl. Geochem., 24, 370–382, https://doi.org/10.1016/j.apgeochem.2008.11.023, 2009b.

Vandenbohede, A., Wallis, I., Houtte, E. V., and Ranst, E. V.: Hydrogeochemical transport modeling of the infiltration of tertiary treated wastewater in a dune area, Belgium, Hydrogeol. J., 21, 1307–1321, https://doi.org/10.1007/s10040-013-1008-x, 2013.

van Genuchten, M. T.: A Closed-form Equation for Predicting the Hydraulic Conductivity of Unsaturated Soils, Soil Sci. Soc. Am. J., 44, 892–898, https://doi.org/10.2136/sssaj1980.03615995004400050002x, 1980.

Zaslavsky, D.: Theory of unsaturated flow into a non-uniform soil profile, Soil Sci., 97, 400–410, 1964.

Zhang, Y., Schaap, M. G., Guadagnini, A., and Neuman, S. P.: Inverse modeling of unsaturated flow using clusters of soil texture and pedotransfer functions, Water Resour. Res., 52, 7631–7644, https://doi.org/10.1002/2016WR019016, 2016.

13

Convective rainfall in a dry climate: relations with synoptic systems and flash-flood generation in the Dead Sea region

Idit Belachsen[1,2], Francesco Marra[2], Nadav Peleg[3], and Efrat Morin[2]

[1]Hydrology and Water Resources Program, Hebrew University of Jerusalem, 91904, Israel
[2]Institute of Earth Sciences, Hebrew University of Jerusalem, 91904, Israel
[3]Institute of Environmental Engineering, Hydrology and Water Resources Management, ETH Zurich, Switzerland

Correspondence to: Idit Belachsen (idit.belachsen@mail.huji.ac.il) and Efrat Morin (efrat.morin@mail.huji.ac.il)

Abstract. Spatiotemporal patterns of rainfall are important characteristics that influence runoff generation and flash-flood magnitude and require high-resolution measurements to be adequately represented. This need is further emphasized in arid climates, where rainfall is scarce and highly variable. In this study, 24 years of corrected and gauge-adjusted radar rainfall estimates are used to (i) identify the spatial structure and dynamics of convective rain cells in a dry climate region in the Eastern Mediterranean, (ii) to determine their climatology, and (iii) to understand their relation with the governing synoptic systems and with flash-flood generation. Rain cells are extracted using a segmentation method and a tracking algorithm, and are clustered into three synoptic patterns according to atmospheric variables from the ERA-Interim reanalysis. On average, the cells are about $90\,km^2$ in size, move $13\,m\,s^{-1}$ from west to east, and live for 18 min. The Cyprus low accounts for 30 % of the events, the low to the east of the study region for 44 %, and the Active Red Sea Trough for 26 %. The Active Red Sea Trough produces shorter rain events composed of rain cells with higher rain intensities, longer lifetime, smaller area, and lower velocities. The area of rain cells is positively correlated with topographic height. The number of cells is negatively correlated with the distance from the shoreline. Rain-cell intensity is negatively correlated with mean annual precipitation. Flash-flood-related events are dominated by rain cells of large size, low velocity, and long lifetime that move downstream with the main axis of the catchments. These results can be further used for stochastic simulations of convective rain storms and serve as input for hydrological models and for flash-flood nowcasting systems.

1 Introduction

A flash flood is a rapid runoff response of a catchment to intense precipitation. Owing to their short response time and high intensity, flash floods are difficult to predict and result in economic damages and casualties (Borga et al., 2011). In fact, they are among the most dangerous meteorological hazards affecting the Mediterranean countries (Llasat et al., 2010; Tarolli et al., 2012). Many factors contribute to flash-flood generation, such as rainfall conditions (e.g., amount, intensity, and spatial and temporal distribution), catchment morphological properties (e.g., slope and surface cover), and hydrological preconditions (e.g., soil saturation). The magnitude of a flash flood is determined by the interactions between these factors (Borga et al., 2014; Nied et al., 2014; Smith et al., 2002; Wright et al., 2014).

In particular, rainfall spatial and temporal variability is a key factor in runoff response prediction (Bahat et al., 2009; Faurés et al., 1995; Morin et al., 2006; Morin and Yakir, 2014; Rozalis et al., 2010; Yakir and Morin, 2011; Yang et al., 2016a). Rozalis et al. (2010) found great sensitivity in flash-flood generation and magnitude to the intra-storm rain intensity distribution. Andréassian et al. (2004) and Zoccatelli et al. (2010) reported that neglecting rainfall spatial variability resulted in a considerable degradation of the modeling results. Yakir and Morin (2011) observed high sensitivity in the response of an arid catchment to location, direction, and velocity of the convective storm.

The need to account for rainfall variability is accentuated in arid and semi-arid regions, where rainfall is often of a convective nature and characterized by extremely variable, high-intensity, short-duration events (Goodrich et al., 1995;

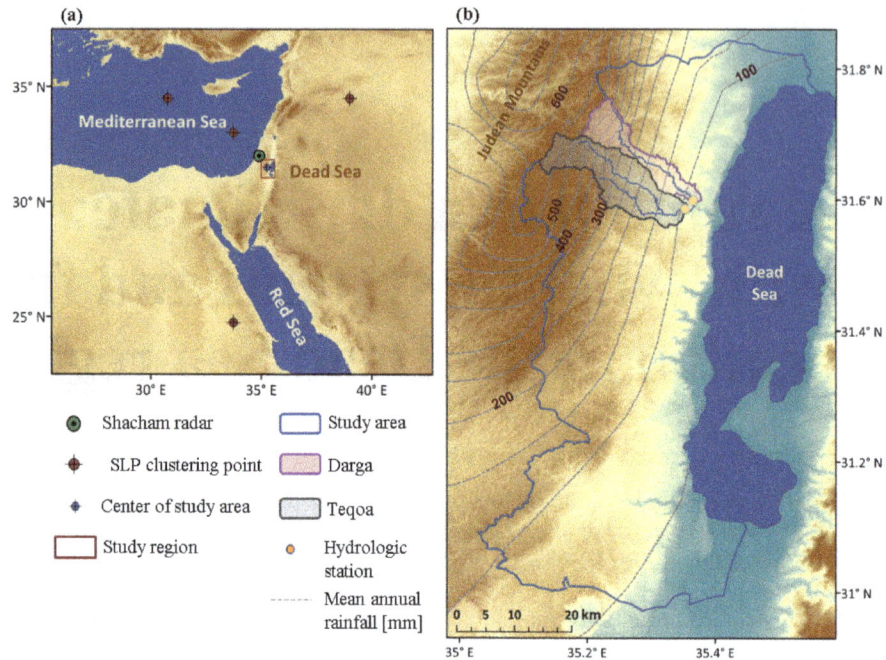

Figure 1. Map of the Eastern Mediterranean area presenting radar location and coordinates used to derive sea level pressure (SLP) differences for synoptic classification **(a)**. A map of the study area with the Darga and Teqoa catchments **(b)**. Isohyets over the study region represent long-term (30 years, 1980–2010) mean annual rainfall (mm).

Syed et al., 2003; Segond et al., 2007). Although antecedent soil-moisture conditions are known to play a role in runoff generation, studies conducted in semi-arid and arid areas ascribed them only a minor influence on flood response, due to low infiltration capacities of the ground, high evaporation rates (Ries et al., 2017; Syed et al., 2003; Yair and Lavee, 1985), and long dry spells between rainfall events (Saaroni et al., 2014). Hence, high-resolutions in space and time, and over large areas, are required to adequately represent rainfall spatio-temporal distributions. These can be best achieved by remote sensing tools such as weather radars (e.g., Barnolas et al., 2010; Berne and Krajewski, 2013; Karklinsky and Morin, 2006; Krajewski and Smith, 2002; Peleg and Morin, 2012).

The spatial distribution of rainfall in convective environments is often examined by focusing on the properties of the convective rain cells (abbreviated hereafter as rain cells), which can be directly derived by exploiting the full three-dimensional structure of the cells (Dixon and Wiener, 1993; Johnson et al., 1998; Steiner et al., 1995) or, more commonly, extracting the convective two-dimensional segments from radar data (Barnolas et al., 2010; Cox and Isham, 1988; Féral, 2003; Féral et al., 2000; von Hardenberg, 2003; Karklinsky and Morin, 2006; Northrop, 1997). A widely used approach requiring only two-dimensional information is to define them as areas in which the rain intensity exceeds a certain threshold. This simplified representation of the rain field allows focusing on the high flash-flood generating potential portion of

the storm, and is used in synthetic rainfall generators and hydrological models (e.g., Morin et al., 2006; Peleg and Morin, 2012; Wheater et al., 2000; Yakir and Morin, 2011).

Rain cells can be represented by fitting an ellipse around the local rain maxima in a radar image (Féral et al., 2000; Karklinsky and Morin, 2006) and geometrical properties of the cells such as area, axes length, orientation angle, and maximal intensity can then be derived. Some studies accounted for rain-cell dynamics by monitoring their progress over time with tracking algorithms (e.g., Dixon and Wiener, 1993; Johnson et al., 1998; Kyznarová and Novák, 2009; Peleg and Morin, 2012; Rinehart and Garvey, 1978). This allowed the derivation of additional parameters such as rain-cell lifetime, velocity, and direction of movement.

The atmospheric conditions generating a rainfall event are expected to influence the properties of rain cells and, consequently, the rainfall–catchment interactions and the runoff response. The objective of this study is to quantify the properties of rain cells originated within different synoptic systems in an arid climate region and to understand the rain cell–catchment interactions with the generation of flash floods. Specific questions motivating this study include the following. (i) What are the property distributions of convective rain cells in dry environments? (ii) How do they vary between different synoptic systems? (iii) How do rain-cell characteristics change within the study region? (iv) What are the cell properties that dominate the formation and magnitude of flash floods? These questions are examined through a statistical

Table 1. Morphological and hydrological characteristics of the Darga and Teqoa catchments.

Catchment property	Darga	Teqoa
Area (km^2)	73	140
Height range (m above sea level)	-19 to $+813$	-20 to $+992$
Mean channel gradient (–)	0.027	0.029
Mean hillslope gradient (–)	0.114	0.135
Percentage of desert soils (%)	42	37
Maximal observed peak discharge ($m^3\,s^{-1}$)*	61.2	158.5
Average number of flow events per year*	1.96	2.13
Threshold discharge value (according to Shamir et al., 2013) ($m^3\,s^{-1}$)	0.25	2

* Data record of hydrological years 1990/1991–2014/2015.

analysis of rain cells derived from 24 years of weather radar data over the western tributaries of the Dead Sea and flash flood data from two catchments within this region (Fig. 1).

The paper is organized as follows. The study area and data are described in Sect. 2. Section 3 analyzes the relations between the properties of the rain cells and the governing synoptic system. Section 4 presents the impact of rain-cell properties on flash-flood generation. The results of the study are discussed in Sect. 5. Section 6 reports the concluding remarks.

2 Regional background and data

This study focuses on the western tributaries of the Dead Sea in the Eastern Mediterranean (EM, Fig. 1) that drain from the Judean Mountains water divide (600–1000 m a.s.l.) towards the Dead Sea (currently 430 m below sea level). The study area is of $3315\,km^2$ (~ 50 km west to east and ~ 80 km north to south).

2.1 Climate

The area is dominated by semi-arid and arid climates except for the northwestern part that is governed by a Mediterranean climate (Greenbaum et al., 2006). Mean annual precipitation shows a steep gradient from over 500 mm in the northwestern portion of the area to about 150 mm and even less than 50 mm in the northeastern and southern parts, respectively (Fig. 1b). The west-to-east gradient is due to the rain shadow effect caused by the Judean Mountains and by the low topography of the Dead Sea valley. The north-to-south gradient is related to the distance from the shoreline and from the main tracks of Mediterranean storms. Rainfall occurs from October to May, with no rain during summer (Goldreich, 2003). Intensities and duration of extreme events differ dramatically within the study area, with the relative frequency of high rainfall intensities increasing as the mean annual precipitation decreases (Marra et al., 2017; Marra and Morin, 2015; Morin et al., 2009; Sharon and Kutiel, 1986).

Most intense rainfall episodes over the study area, and the EM in general, are associated with the cold fronts of mid-latitude lows: the Cyprus low – a Mediterranean low located around Cyprus, and the Syrian low – a well-developed Mediterranean low located over Syria (Dayan et al., 2015; Dayan and Morin, 2006; Goldreich, 2003; Kahana et al., 2002). The region is also affected by more localized convective showers associated with the Active Red Sea trough (ARST), a surface low-pressure trough extending from eastern Africa along the Red Sea towards the Middle East in its active phase (Ashbel, 1938). The ARST is more frequent during the transition seasons and its contribution to rainfall and flash floods in the EM decreases going north (Kahana et al., 2002; Dayan and Morin, 2006). According to Kahana et al. (2002), the ARST accounts for most of the major floods over the arid catchments in the south of the study area, followed by the Syrian low. Some rare events of relatively widespread rainfall leading to flash floods in the region are associated with the subtropical jet and the conveying of air moisture of tropical origin over Africa to the Eastern Mediterranean; a system often referred to as tropical plume (Dayan and Morin, 2006; Kahana et al., 2002; Tubi and Dayan, 2014; Ziv, 2001)

2.2 Hydrology

Two side-by-side gauged catchments, located in the northern part of the study area (Fig. 1b), are chosen for the analysis due to the availability of long and concurrent records of water discharge and radar data; these are the Darga ($73\,km^2$) and Teqoa ($140\,km^2$). These catchments have ephemeral dry channels, and flash-flood events occur on average twice a year (Table 1). The surface is characterized by large areas of bare rock, shallow soils of low permeability, and sparse vegetation.

Water discharge data were obtained from the Israel Hydrological Service for the 24 hydrological years (October to September) 1990/1991–2013/2014. In order to avoid analysis of low flows, events with peak discharges lower than $0.25\,m^3\,s^{-1}$ for Darga and $2\,m^3\,s^{-1}$ for Teqoa were excluded from the analysis. These thresholds were based on Shamir

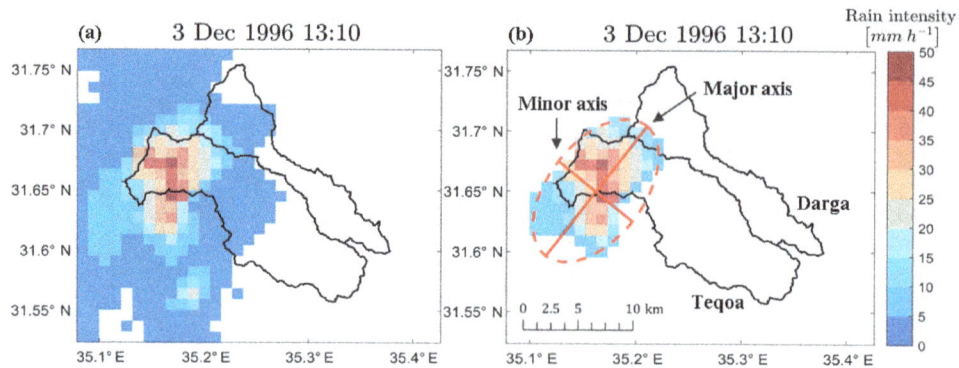

Figure 2. An example of a rain-cell derivation for 3 December 1996, 13:10 (UTC) over the Darga and Teqoa catchments. Radar rainfall for the examined time (**a**), a segment of an identified rain cell and its fitted ellipse (**b**). The spatial properties of the presented cell are the following: area, 90 km^2; orientation, 51°; major axis length, 15 km; minor axis length, 9 km; ellipticity, 0.58; max rainfall intensity, 47 mm h^{-1}; and mean rainfall intensity, 20 mm h^{-1}.

et al. (2013) who developed a method for determining minimal flash-flood thresholds for arid regions using geomorphic indexes.

2.3 Weather radar

Rainfall data used for the research are based on the Shacham weather radar, a C-band non-Doppler instrument, located within the Ben-Gurion international airport (Israel), 50–125 km northwest of the study area (Fig. 1). The observation geometry of the radar is characterized by a spatial polar resolution of 1.4° × 1 km and a temporal resolution of about 5 min per volume scan. Its archived data record is of 24 hydrological years 1990/1991–2013/2014. Such a long record represents a clear advantage for climatological and hydrological studies in an arid region and data from this radar were fruitfully used for research in many studies so far (Karklinsky and Morin, 2006; Marra et al., 2017; Marra and Morin, 2015; Morin et al., 2001; Morin and Gabella, 2007; Morin and Yakir, 2014; Peleg et al., 2016, 2013, 2015a, b; Peleg and Morin, 2012; Rozalis et al., 2010). An extensive description of the quantitative radar precipitation estimation and of its assessment is provided in Marra and Morin (2015). The distance from the radar increases going south and east in the study area, so that the instrument samples higher atmospheric levels (sampling elevation between 1000 and 4500 m). The radar data are corrected taking into account the vertical profile of reflectivity, but overshooting of precipitation in the study area is possible. However, this problem is expected to be negligible for this study, in which vertically developed systems, such as the convective cells, are examined.

3 Rain cell characterization

3.1 Rain cell identification and tracking

The identification of rain cells is done by partitioning each radar image into segments, and is followed by a cell tracking algorithm (Peleg and Morin, 2012). Rain cells are defined as connected radar pixels with the following: (i) more than 5 mm h^{-1} rain intensity, (ii) at least one peak exceeding 10 mm h^{-1}, and (iii) an area larger than 9 km^2. These thresholds were suggested and used in previous studies (e.g., von Hardenberg, 2003; Morin et al., 2006; Syed et al., 2003) and allow focusing on the convective part of the rain without excluding moderate rainfall intensities that could be relevant for the flood generation. It should be noted that the selection of the thresholds can affect some of the derived properties (e.g., cell area and mean areal rain intensity), but it should not affect the comparison of these properties between different groups of cells.

The spatial properties extracted from each segment are the following: area (km^2), length of the major and minor axes of the ellipse fitted to the segment (km), ellipse center location, ellipticity (minor-to-major axis ratio), orientation (angle of the major axis in degrees relative to the west–east orientation, positive values are counter-clockwise), maximum rain intensity (mm h^{-1}), and mean areal intensity (i.e., the mean intensity over the area of the segment, in mm h^{-1}). Figure 2 presents an example of one radar image and a derived rain cell with its spatial properties.

The cell tracking algorithm links rain cells in consecutive images and allows characterizing of the rain-cell lifetime and average advection (velocity and direction). The algorithm, developed by Kyznarová and Novák (2009) and modified by Peleg and Morin (2012), is based on Pearson's correlation between shifted successive images. The term "lifetime" relates to the length of the individual cell's life, while "duration" to the length of the rain event. Frequencies of different tracking

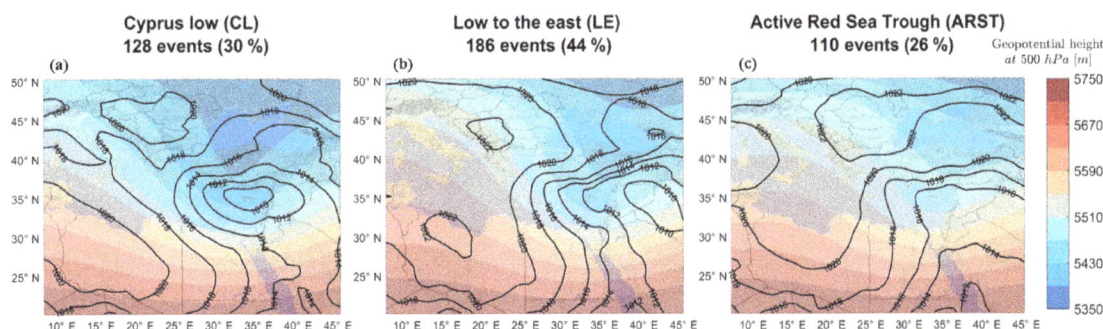

Figure 3. Average SLP (hPa, black contour lines) and geopotential height at 500 hPa (m, in color) for each cluster; **(a)** Cyprus low, **(b)** low to the east, and **(c)** Active Red Sea Trough (ARST).

categories during the cell's full life cycle (i.e., frequency of splits and merges) were left out of the analysis as no added value to the presented results was given. A total of 10 447 rain cells (composing 2632 tracks) were derived. The rain record was divided into 424 rain events, defined as separated by dry spells longer than 6 h. This allowed associating each rain cell to the governing synoptic system of the rain event. Rain events for which less than 80 % of the radar scans were available have been removed.

3.2 Synoptic classification

Rain events were classified into synoptic types using cluster analysis. The clustering was aimed at relying mainly on the sea level pressure (SLP) map, the most commonly used map for synoptic classification in the EM (Alpert et al., 2004; Dayan et al., 2012; Kahana et al., 2002; Zangvil et al., 2003) and in other places in the world (Cannon et al., 2002; Hewitson and Crane, 2002). In addition, surface temperature was used to distinguish between an ARST and a cold Mediterranean low, since the former is usually initiated by thermal instability caused by differential heating between the surface (where a warm advection from the southeast takes place) and the upper atmospheric levels (Dayan et al., 2001). Other atmospheric variables (e.g., specific humidity at 700 hPa and temperature at 850 hPa) were tested and found to have a negligible influence on the clustering. Each rain event was linked to the ERA-Interim global reanalysis atmospheric variables (Dee et al., 2011) – SLP and near-surface (1000 hPa) air temperature, obtained for the time closest to the rain event's center of mass (chosen out of Era-Interim 4 times daily available times – 00:00, 06:00, 12:00, 18:00 UTC).

A hierarchical agglomerative clustering technique using Ward's criterion was applied based on the following: (i) location of the minimum SLP within the EM region (25.5–42.75° E and 22.5–37.5° N); (ii) north–south SLP difference between two points – (33.75° E, 33° N) and (33.75° E, 24.75° N); (iii) west–east SLP difference between two points – (30.75° E, 34.5° N) and (39° E, 34.5° N); and (iv) near-surface temperature at the grid point closest to the center of

the study area (35.25° E, 31.5° N). All the mentioned points are presented in Fig. 1a.

The rain events were found to be best described by three clusters (Fig. 3). The first cluster (128 events, Fig. 3a) describes a Mediterranean low located west of the shoreline, and is associated with a Cyprus low (CL). The second cluster (186 events, Fig. 3b) describes a low to the east (LE) and could be associated with a Syrian low, or with any other Mediterranean low settled east of the study area. The third cluster (110 events, Fig. 3c) describes a surface trough extending from the south and is associated with an ARST. The radar rain record did not include tropical plume events and therefore this synoptic type is not considered in the present analysis.

On average, both Mediterranean lows (Fig. 3a and b) are accompanied by a pronounced 500 hPa trough extending from eastern Europe towards the EM, with an axis orientation of north–south to northeast–southwest. Besides the known effect of an upper level trough on the intensification of the low on the surface (Ahrens, 2003), under such atmospheric conditions the northwesterly flow is enriched with moisture from the sea, increasing the probability of rainfall in the southern EM (Zangvil and Druian, 1990; Ziv et al., 2006). The upper trough of the ARST (Fig. 3c) is shallower and has a similar axis orientation.

Validation of the clustering results was done according to expert examinations. Maps displaying contour lines of SLP and wind directions at 850 hPa for 30 randomly chosen events were given to three experts to evaluate. The experts were asked to choose the synoptic system that best describes a given map out of four possible options: CL, LE, ARST, and "none of the above". Mismatches between the three experts' classifications and the automated clustering were then counted in relation to the extent of disagreement between the experts. For 19 maps there was an agreement between the experts and the automated procedure. Three out of 30 maps were agreed on between all experts but were classified differently from the automated procedure, resulting in a 10 % classification error. For 8 maps there was no agreement among the experts, resulting in 7 maps with matches between only

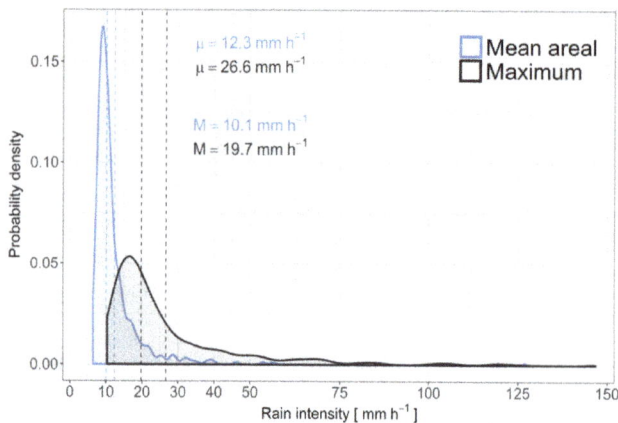

Figure 4. Empirical probability density functions of mean areal (blue) and maximum (black) rain intensities of rain cells over the Dead Sea (smoothed). μ represents the mean and M represents the median.

some of the experts and the automated procedure and 1 map with no matches.

3.3 Spatial and temporal rain-cell characteristics

In this section, the differences between properties of rain cells originated by different synoptic systems are analyzed. Event properties and average spatial and temporal characteristics of rain cells are presented in Table 2. The use of averaged cell characteristics allows for neutralizing the dependency between individual rain cells of the same event and is crucial for the statistical comparison between the properties of cells generated by different synoptic systems. From this point on, unless stated otherwise, all values of rain-cell properties mentioned are the average value during the event. Advection direction is defined following the meteorological standard, i.e., it represents the direction of the origin (e.g., direction of 270° represents a movement from west to east).

The average duration of all rain events is 5.4 h. On average, the rain cells are 92 km² in area, advecting from west to east (274°) at a velocity of 12.8 m s⁻¹ and living 18.1 min. Ellipticity of cells is 0.57 (minor to major axis length ratio of 3 : 5). The major axis is close to alignment (18°) with their direction of movement. Similar values of ellipticity and orientation were found in previous studies conducted close to the study area (Karklinsky and Morin, 2006; Morin et al., 2006; Peleg and Morin, 2012; Yakir and Morin, 2011), and in other regions such as Catalonia and France (Barnolas et al., 2010; Féral et al., 2000). The empirical probability density functions of the cells' rain intensities are shown in Fig. 4. The mean areal and maximal rain intensities are 12.3 and 26.6 mm h⁻¹, respectively, and both functions are positively skewed (skew coefficients of 7.4 and 2.7, respectively) as a result of extreme rainfall events.

Table 2. Mean and standard deviation (in parentheses) of spatial and temporal properties of events and derived convective rain cells.

	Number of events	Total number of rain cells	Number of cells in a rain event	Event duration (h)	Cell area (km²)	Major axis length (km)	Minor axis length (km)	Maximum rain intensity (mm h⁻¹)	Mean areal rain intensity (mm h⁻¹)	Ellipticity (-)	Orientation (°)	Average lifetime (min)	Mean velocity (m s⁻¹)	Mean direction (°)
All events	424	10447	24.6 (31.4)	5.4 (6.5)	92.1 (102.7)	13.6 (7.4)	6.9 (3.6)	26.6 (19.2)	12.3 (8.4)	0.57 (0.09)	14.4 (25.9)	18.1 (11.4)	12.8 (5.7)	273.9 (29.9)
Synoptic system														
CL	128	3042	23.7 (29.5)	5.5 (6.3)	110.4 (110.6)	15.6 (8.2)	7.7 (3.7)	22.8 (14.6)	10.6 (4.2)	0.54 (0.09)	16.1 (25.5)	18.1 (10.5)	14.5 (6.2)	267.6 (24.9)
LE	186	4964	26.7 (36.1)	6.3 (6.9)	88.2 (98.6)	13.0 (6.9)	6.8 (3.4)	26.4 (18.2)	12.2 (7)	0.58 (0.1)	14.8 (26.4)	17.7 (8.7)	12.3 (5.5)	277 (30.3)
ARST	110	2441	22.19 (24.1)	3.8 (4.6)	77.2 (97.6)	12.3 (6.9)	6.3 (3.5)	31.5 (24.3)	14.5 (12.7)	0.57 (0.1)	11.8 (25.6)	20.8 (14.6)	11.4 (5)	276.5 (33.7)
P value[a]			0.46	<0.01	0.03	<0.01	<0.01	0.01	<0.01	0.03	0.42	0.03	<0.01	<0.01
Significance differences[b]			2	1	1	1,3	1	1	1	3		2	1,3	1,3

a P value of the ANOVA test applied for the different rain-cell properties between the three synoptic systems. All data sets were tested first for variance heterogeneity using Levene's test (with squared deviations). The Box-Cox transformation technique was applied to properties of unequal variances to obtain normality. In these cases, Welch's test followed by a multiple comparison using the Bonferroni method was used.
b Pair of groups of significance difference at 0.05 level: (1) ARST-CL, (2) ARST-LE, (3) CL-LE.

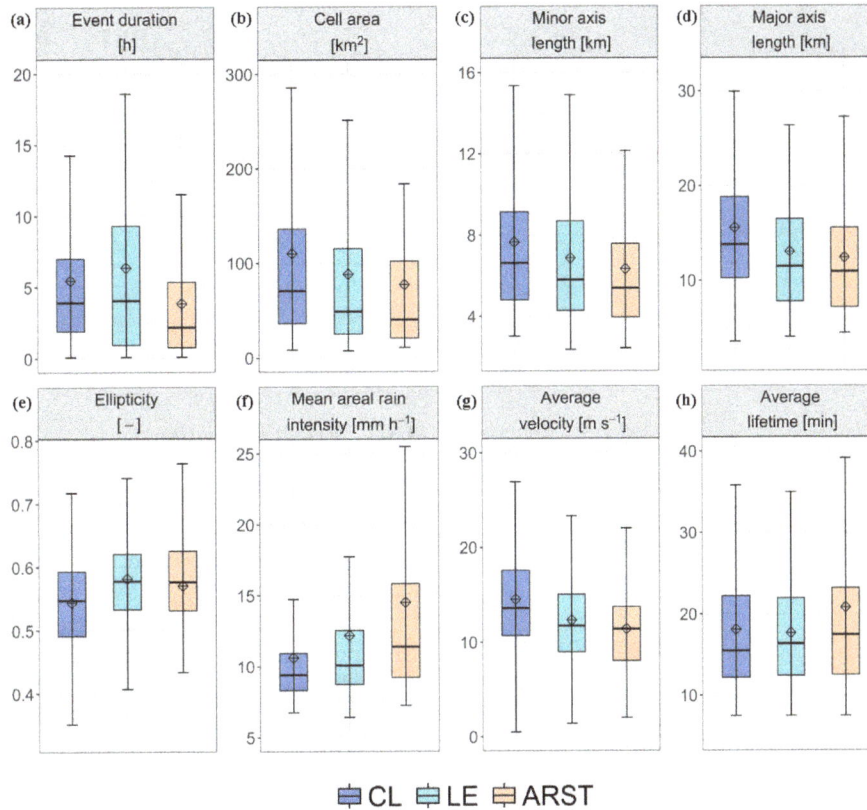

Figure 5. Comparison of rain-cell properties (averaged for each rain event) of the three synoptic systems: CL (dark blue), LE (light blue) and ARST (orange); **(a)** average event duration (h), **(b)** cell area (km^2), **(c)** minor axis length (km), **(d)** major axis length (km), **(e)** ellipticity (–), **(f)** mean areal rain intensity (mm h^{-1}), **(g)** average velocity (m s^{-1}), and **(h)** average lifetime (min). The black line in each boxplot marks the median, black diamond the mean, boxes lower and upper borders mark the 25 and 75 % quartiles, respectively, and the whiskers mark minimum and maximum values unless these values exceed $1.5 \cdot$ IQR (inter quartile range – the distance between lower and upper quartiles). Outliers are not shown. See Table 2 for numerical values of the mean and standard deviation.

3.3.1　Effect of synoptic system

Spatial and temporal properties of rain cells originated by different synoptic systems are compared (Table 2 and Figs. 5 and 6) using one-way analysis of variance (ANOVA) followed by a multiple pairwise comparison of the three groups' means using Tukey's honest significant difference criterion. Statistically significant differences have been found and are highlighted in Table 2.

The LE rain events are characterized by the highest average duration of all synoptic systems (6.3 h compared to 5.5 and 3.8 h of the CL and ARST events, respectively, Fig. 5a). The area of the ARST rain cells was found to be smaller than the area of the Mediterranean lows (76 km^2 compared to 110 and 87 km^2 of the CL and LE, respectively, Fig. 5b), but cells were found to live longer than cells in other synoptic systems (20.8 min compared to 18.1 and 17.7 min of the CL and LE, respectively, Fig. 5h). Moreover, ARST rain cells' mean areal rain intensity (14.5 mm h^{-1}) and maximal rain intensity (31.5 mm h^{-1}, Table 2) were found to be higher than both CL intensities (10.6 and 22.8 mm h^{-1}) and LE in-

tensities (12.2 and 26.4 mm h^{-1}) and these events have the highest variability in mean areal rain intensities (Fig. 5f).

The rain cells generally preserve the same orientation for all three synoptic types, but their shape is different: CL cells are characterized by lower ellipticity (0.54 compared to 0.58 and 0.57 of the LE and ARST, respectively). General mean orientation is west to east with a 12–16° counterclockwise tilt from the west–east axis (Table 2). The CL events are characterized by rain cells generally moving from west to east (268°, Fig. 6a), whereas the LE and ARST events are characterized by a slightly stronger northwestern component (about 277°, Fig. 6b and c). In the ARST case, the direction distribution is bimodal, with cells originating from west-southwest and west-northwest (Fig. 6c). ARST events are characterized by lower average velocities (11.4 m s^{-1} compared to 14.5 and 12.3 m s^{-1} of the CL and LE, respectively, Fig. 5g).

Significant differences (P value < 0.05) are found between the following: (i) ARST and CL cell areas, minor axis length, maximum rain intensities, and mean areal rain intensities; (ii) ARST and LE cells average lifetime and event duration; (iii) CL and both ARST and LE major axis length,

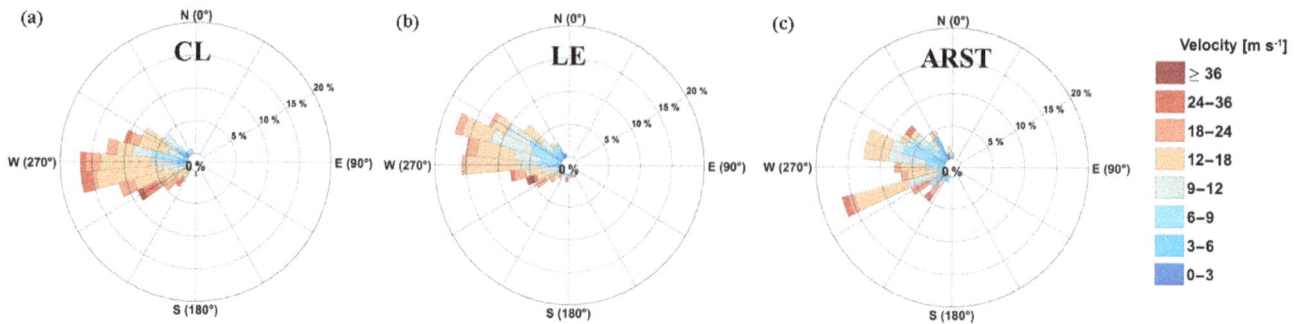

Figure 6. Advection (velocity and direction) distributions of CL (**a**), LE (**b**), and ARST (**c**) events.

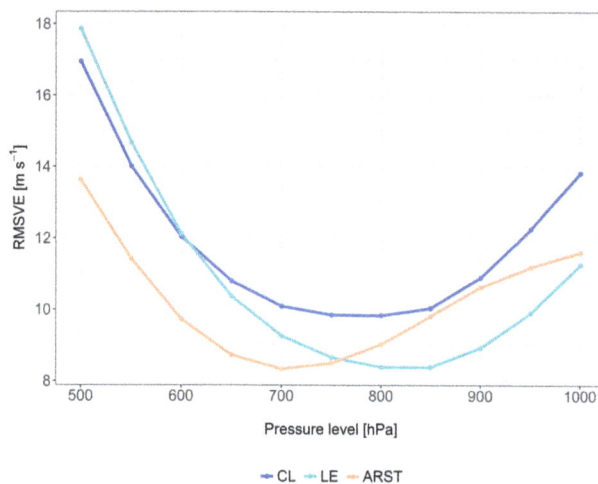

Figure 7. Calculated root mean square vector error (RMSVE) between average advection components and wind zonal and meridional components obtained from Era-Interim reanalysis for pressure levels of 500–1000 hPa for each synoptic system.

average direction, and average velocity of rain cells; and (iv) CL and LE cells' ellipticity. Rain cells are broadly advecting with the mean wind through some deep tropospheric layer, in which the cloud is embedded (Chappell, 1986; Doswell et al., 1996). To identify the layers that rain cells are commonly moving in over the study area, the advection vector components were compared with zonal and meridional wind components at pressure levels 500–1000 hPa (increments of 50 hPa), extracted from Era-Interim reanalysis for the times closest to each rain event's mass center at a grid point closest to the center of the study area (35.25° E, 31.5° N). A root mean square vector error (RMSVE) was calculated for each pressure level and synoptic system (Fig. 7). The pressure levels with minimum RMSVE are in general 700 to 850 hPa, and in particular 700–850 hPa for CL, 800–850 hPa for LE, and around 700 hPa for ARST.

3.3.2 Effect of location

The variations in rain-cell characteristics along the north–south latitudinal gradient were examined (Fig. 8). The total number and the average cell area of the two Mediterranean lows is decreasing from north to south (Fig. 8a and b), whereas their velocity is increasing (Fig. 8c). The mean areal rain intensity of CL cells is increasing from north to south, but LE cells show an increasing trend only between latitudes 31.6 and 31.1° N (Fig. 8d).

A positive correlation between topographic height and cells area (Fig. 8e and b) and a negative correlation between distance from shoreline and number of cells (Fig. 8f and a) is seen in both Mediterranean lows in the northern part of the study area, and especially for the LE system.

The ARST rain cells follow different trends: (i) the total number of rain cells shows a smaller variation with latitude (Fig. 8a), (ii) the cells moderately increase in size along the north–south axis reaching a peak around latitudes 31.3–31.5° N (Fig. 8b), (iii) the cells have a moderate decrease in their velocity with latitude (Fig. 8c), and (iv) no clear trend in mean areal rain intensity is seen (Fig. 8d).

The region with higher mean annual rainfall (Fig. 8g) overlaps the regions of maximal number of cells (Fig. 8a), maximal cell area (Fig. 8b), and low velocities (Fig. 8c) of both Mediterranean lows. The region of maximal mean areal rain intensity (Fig. 8d), however, is not collocated with maximal rainfall amounts. This fits previous findings that in dryer regions rainfall is generally more intense over short durations (Marra et al., 2017; Marra and Morin, 2015).

The moderate increasing trend in velocity of Mediterranean lows' cells along the north–south axis may have resulted from a bias in favor of stronger storms in the southern part, i.e., regions that are most distant from the sea and from the Mediterranean storm tracks. Mediterranean storms that produce rainfall over those regions were most likely deep lows of stronger winds that had managed to transport clouds from the Mediterranean Sea far inland (Saaroni et al., 2010).

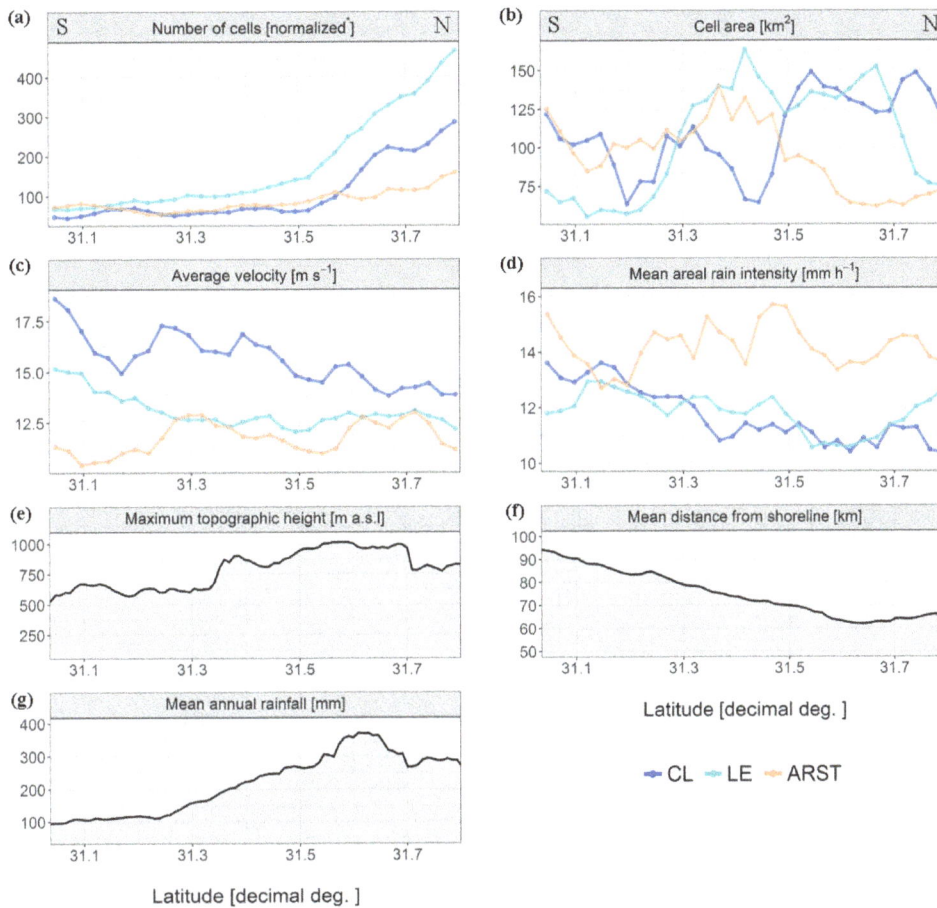

Figure 8. Mean rain-cell properties along a north–south (right-to-left axes) orientation of the different synoptic systems: (**a**) number of cells[1], (**b**) cell area (km^2), (**c**) average velocity (m s^{-1}), (**d**) mean areal rain intensity (mm h^{-1}). Potentially related variables along the same axis: (**e**) maximal topographic height (m a.s.l.), (**f**) distance from shoreline (km), and (**g**) mean annual rainfall (mm). Each point represents the mean or maximal value in a west to east strip (in panels **a–d**, the strip is 2.5 km wide and a running average of 7.5 km is applied).

4 Relations between rain-cell properties and flash floods

The relationship between properties of rain cells and the occurrence and magnitude of flash floods in the Darga and Teqoa catchments (Fig. 1b) was explored. A flash flood was defined according to the criteria specified in Sect. 2.2. No distinction between the two catchments was made due to their similar morphology, and their small and narrow shape relative to an average size rain cell (Fig. 2). Out of the 424 detected rain events, 173 events had rain cells tracked above the catchments, 29 of which (532 rain cells) were associated with flash-flood events. Since the same rain event can potentially lead to a flash flood in both catchments, the 29 rain events corresponded to 41 measured flash floods (21 in Darga and 20 in Teqoa). The remaining 144 rain events (988 rain cells) were classified as "non-flash-flood" events.

Examining the event duration and total rain depth over the catchments (Fig. 9) reveal that the recorded flash floods were produced from events of a few minutes to 2 days long and areal rain depth from 1 to 100 mm. In general, long duration and high rain depth are conditions favoring the occurrence of flash floods. A number of rain events with similar duration and depth did not lead to flash floods, confirming that short duration intensities are also important. In fact, ARST events leading to flash floods generate high peak discharges, in spite of their general shorter durations and lower rain depths than both Mediterranean lows.

Flash floods in the desert are often triggered by one or two rain cells (David-Novak et al., 2004; Yakir and Morin, 2011). In this study, the "dominant rain cell" of each event, i.e., the rain cell that contributed the largest amount of rainfall over the two catchments, was identified. The mean, median, and interquartile ranges of the properties characterizing the dominant cells of flash-flood and non-flash-flood events during their lifetime over the catchments were compared, and are presented in Fig. 10. If not stated otherwise, the results presented below are significant at 0.05 level.

---

[1]Number of cells is normalized to the relative strip area.

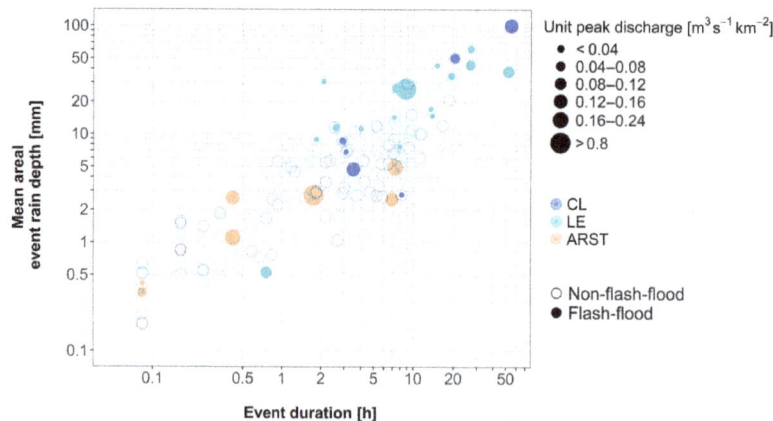

Figure 9. Scatter plot of rain events' mean areal rain depth over the catchments vs. duration, for flash-flood-associated (filled circles) and non-flash-flood events (empty circles), with respect to different synoptic types and different unit peak discharges of the flash-flood-related events. In case of flow in both catchments, maximum unit peak discharge is presented. Over 90 % of the watershed area is covered by rainfall during flash-flood-related events (79 % for the non-flash-flood-related events). Event duration refers to the part of the event where rain cells were found over the catchments.

Results show that dominant rain cells associated with flash floods are (i) larger than dominant cells not associated with flash floods ($247 \, km^2$ compared to $159 \, km^2$; Fig. 10a) and, accordingly, cover a larger portion of the catchments area ($77 \, km^2$ compared to $55 \, km^2$; Fig. 10b); (ii) have lower average velocities ($11.6 \, m \, s^{-1}$ compared to $14 \, m \, s^{-1}$, significance level 0.1; Fig. 10c); and (iii) persist longer (41 min compared to 20 min; Fig. 10d). The mean areal and maximum rain intensities of flash-flood-related dominant cells are higher than non-flash-flood dominant cells (Fig. 10g and h), though this difference was found to be not statistically significant.

Figure 11a and b shows the distribution of the advection of dominant cells for non-flash-flood and flash-flood events. Non-flash-flood dominant cells are generally more westerly and characterized by higher velocities ($14 \, m \, s^{-1}$ and 275°, on average) than flash-flood dominant cells ($11.6 \, m \, s^{-1}$ and 286°). Flash-flood dominant cells are characterized by a bimodal distribution with low velocity, north-northwesterly cells (generally, $< 12 \, m \, s^{-1}$, 300–360°) and higher velocities, westerly cells (generally, $> 12 \, m \, s^{-1}$, 240–300°). Considering only high magnitude flash floods (with peak discharge larger than the median) it is found that dominant rain cells are related to low velocities and north-northwesterly directions ($9.8 \, m \, s^{-1}$ and 301° on average; Fig. 11c) that match the main drainage axis of the two studied catchments (Fig. 1b).

5 Discussion

The properties of convective rain cells in the arid area of the Dead Sea western tributaries are discussed in relation to the governing synoptic system, location, and to flash-flood generation.

5.1 Variation between different synoptic systems

Rain-cell properties are distinctly associated with the characteristics of three synoptic systems governing rain events in the region: the Cyprus low, the low to the east, and the Active Red Sea Trough (CL, LE and ARST, respectively; see Sect. 3.2). ARST events have shorter duration and their rain cells are characterized by higher intensities, smaller areas, longer lifetimes, and lower velocities compared to the two Mediterranean lows. The low cell velocities are likely due to the more continental nature of this system (Goldreich et al., 2004) and to the smaller pressure gradients (Dayan et al., 2012), while the higher intensities could be related to higher surface temperatures (observed in the clustering results, not shown) leading to greater atmospheric instability (Ahrens, 2003).

The average directions of rain cells are southwest to west for CL events and west to northwest for LE events. In ARST events both modes are common. These results are explained by the location of the surface low in CL and LE systems and by the cyclonic geostrophic wind (Fig. 3); a CL located west of the study area is usually identified with southwesterly to westerly winds at low levels, while northwesterly wind directions are more dominant when the low is located at the East.

In both Mediterranean lows, southwestern cell directions are associated with higher cell velocities, while northwestern directions with lower cell velocities (Fig. 6). The lower cell velocities could be related to the slower movement of the Mediterranean low when located above land and to the larger distance of the center of the low from the study area (Fig. 3a and b). The lower average cell velocities of the LE events is therefore explained by a larger portion of northwestern cell directions. Furthermore, average cell velocity

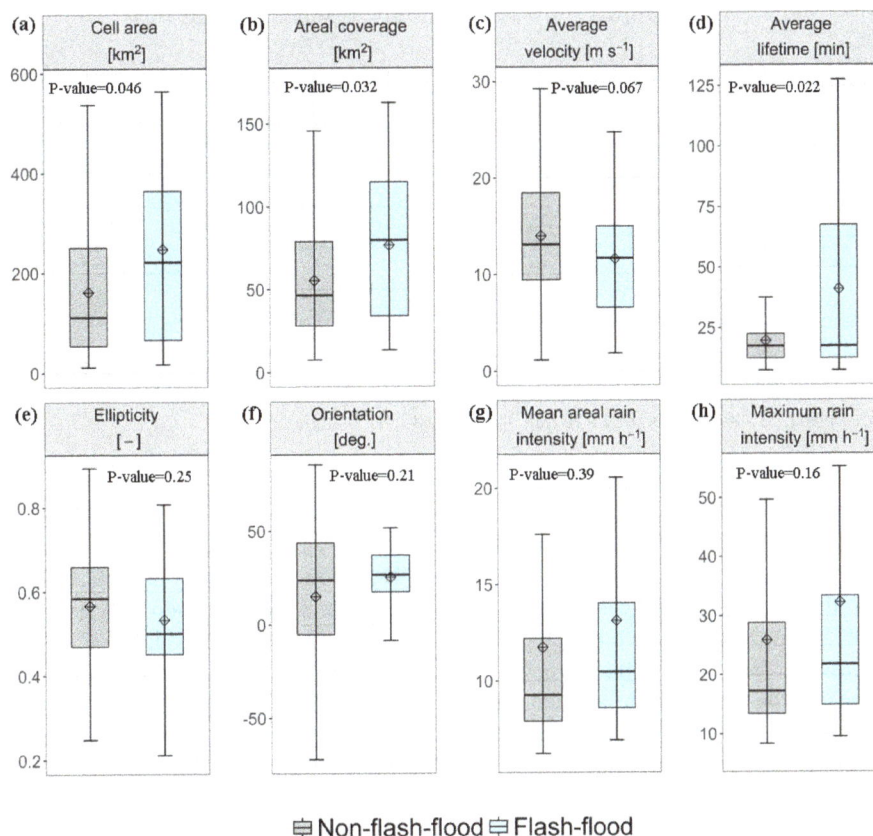

Figure 10. Comparison of dominant flash-flood-related cells (blue, $N = 29$) and non-flash-flood-related cell (grey, $N = 144$) properties: (a) cell area (km^2), (b) areal coverage (km^2), (c) average velocity ($m\,s^{-1}$), (d) average lifetime (min), (e) ellipticity, (f) orientation ($^\circ$), (g) mean areal rain intensity ($mm\,h^{-1}$), and (h) maximum rain intensity ($mm\,h^{-1}$). Values are averaged over the dominant cell's lifetime. Boxplot properties are as specified in Fig. 5. Reported P values are of the ANOVA test applied between dominant flash-flood-related cells and non-flash-flood-related cells. All data sets were tested first for variance heterogeneity using Levene's test (with squared deviations). The Box-Cox transformation technique was applied to properties of unequal variances to obtain normality. In these cases, Welch's test was used.

components of Mediterranean low events are in better agreement with low level (750–850 hPa) wind components, while ARST with higher levels (700 hPa; Fig. 7). These findings agree with the synoptic understanding that ARST events are usually identified with medium-level clouds and Mediterranean lows with low-level clouds. Unlike the Mediterranean lows, which have the Mediterranean Sea as their major moisture supplier, the moisture essential for the development of convective rain cells in ARST events must be transported at the medium levels from remote southern origins, since a dry easterly wind flow is found at the lower levels (Dayan et al., 2001; Kahana et al., 2004; Krichak and Alpert, 1998).

5.2   Variation within the study region

Rain-cell properties vary in space. Variations along the north–south axis of the study area, characterized by a sharp decrease in mean annual rainfall (Fig. 8g), an increase in distance from the Mediterranean Sea's shoreline (Fig. 8f), and a change of topography (Fig. 8e) were examined. The rel-

ative frequency of high-intensity rainfall increases with the reduction of annual rainfall amounts (Goodrich et al., 1995; Marra et al., 2017; Marra and Morin, 2015; Sharon and Kutiel, 1986), and orographic effects lead to enhanced rainfall generation (Houze, 2012; Sharon and Kutiel, 1986; Warner, 2004; Wheater et al., 1991). Both phenomena are reflected in the characteristics of the rain cells of Mediterranean lows: (i) mean rain intensities are generally increasing with the degradation of mean annual rainfall southward towards drier regions. Though the increase in rainfall intensities southward is seen explicitly in the Mediterranean lows, the increasing dominance of the ARST rain cells towards the south is likely to play a significant role in the latitudinal increase in rainfall intensities (Sharon and Kutiel, 1986); (ii) rain cells are larger where topography is higher. As stated by Goldreich et al. (2004), clouds distancing themselves from the shoreline and ascending the mountains, tend to become more uniform and continuous than over the coastal plain, thus increasing their size. When the terrain features are low enough, preexisting clouds that move over them produce maximum precipitation

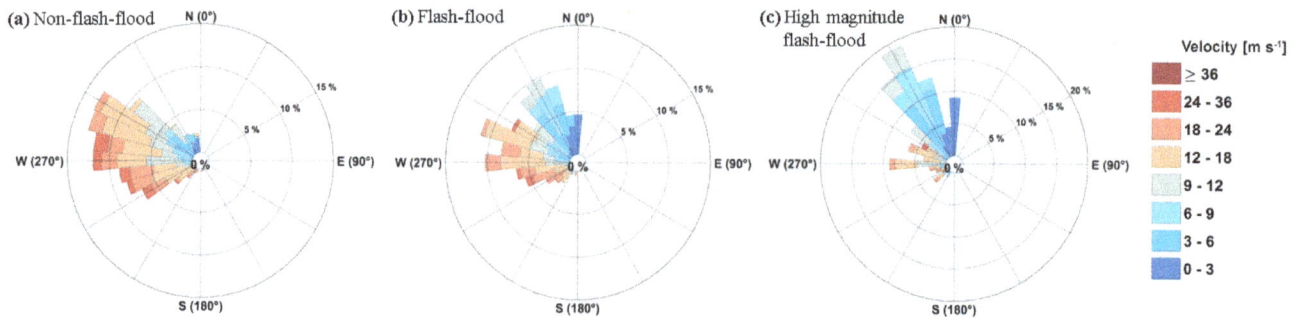

Figure 11. Advection (velocity and direction) distributions of dominant rain cells for **(a)** non-flash-flood events (490 cells, 144 events), **(b)** flash-flood events (220 cells, 29 events), and **(c)** high-magnitude flash-flood events (111 cells, 15 events). High magnitude is defined based on median values of measured flash floods in Darga and Teqoa used in this analysis.

on the upwind side of the barrier. As the precipitating cloud is advected to the lee side, the precipitating capacity is weakened by the downslope air motion (Houze, 2012). Our analyses included rain cells with centroids located east of the water divide, i.e., on the lee side of the mountain range (Fig. 1). Nevertheless, in many cases the cell area includes precipitation also on the west side of the water divide and thus the total effect obtained is the cell area increase along with the rain intensity weakening over the mountainous areas.

5.3   Flash-flood-related characteristics

A few studies had focused on the contribution of the spatial and temporal characteristics of rain cells to flash-flood generation in arid environments. Doswell et al. (1996) reported that slow storm velocities contribute to flash-flood generation and Syed et al. (2003) underlined the importance of the areal coverage of the storm core for runoff generation in a semi-arid catchment. Bracken et al. (2008) suggested that floods, though mainly related to the total rainfall, were eventually triggered by intense bursts of rain. Our results support these previous findings, showing that flash-flood-related and non-flash-flood-related rain cells differ in size, areal coverage, velocity, lifetime over the catchment, and rain intensity.

Other studies wished to analyze the impact of the spatial and temporal characteristics of rain cells on flash-flood magnitude by using case studies of real storms (Smith et al., 2010; Yang et al., 2016a) or model simulations (Morin et al., 2006; Rozalis et al., 2010; Wright et al., 2014; Yakir and Morin, 2011; Yang et al., 2016b). While some argued about the importance of rainfall intensity distribution (Chang, 2007; Rozalis et al., 2010; Yang et al., 2016b), a common conclusion concerned the movement of the storm: slower storms directed downstream of the catchment seem to produce flash floods of higher magnitudes (Doswell et al., 1996; Singh, 1997; Smith et al., 2000; Yakir and Morin, 2011). Our results confirm this effect: rain cells moving downstream with a direction close to the orientation of the

principal axis of the catchments (Fig. 1) at low average velocities ($< 12\,\mathrm{m\,s^{-1}}$) are related to higher peak discharges.

Due to the small sample of flash-flood events (29 events), the influence of the synoptic type on flash-flood generation was not taken into consideration. Nevertheless, results presented in Sect. 3 suggest that ARST rain cells should have larger flash-flooding potential, due to the lower velocities, longer lifetime, and higher intensities; for example, 55 % of the ARST rain cells analyzed had velocities lower than $12\,\mathrm{m\,s^{-1}}$) against only 33 and 46 % for CL and LE. Out of them, the fraction with mean areal rain intensities higher than $10\,\mathrm{mm\,h^{-1}}$) are 63, 39, and 46 % for ARST, CL, and LE cells, respectively. In fact, 6 out of the 7 flash floods generated by ARST events were characterized by high magnitude. Other studies underlined the localized and intense nature of ARST storms and their high flash-flooding potential (Dayan et al., 2001; Kahana et al., 2002; Ziv et al., 2004).

Some of the rain-cell properties associated with flash floods are tied with the catchment properties. For example, large catchments might be less influenced by ARST rain cells due to their smaller size. The more northwesterly cell directions associated with LE events and part of the ARST events might present better flooding conditions relative to other cell directions, due to the northwest to southeast orientation ($\sim 315°$) of the Darga and Teqoa catchments (Fig. 1), but might not have any advantage in the case of a different catchment orientation. This seems to be reflected in the increased representation of LE in flash-flood-related events in Darga and Teqoa (55 %) in comparison with the general frequency of LE events in the entire study area (44 %). Furthermore, Kahana et al. (2002) found that the frequency of flash floods in the Negev (south of our study area) generated by ARST events is slightly higher (38 % in comparison with 24 % found in our study) than Syrian low events (33 %), which are most likely the equivalent to our LE events. These differences may arise from the different sample sets analyzed by Kahana et al. (2002) (only floods > 5 years recurrence interval were taken) and from the more northeastern location of the Dead Sea region than the Negev, and may indicate the importance of the LE events in generating flash floods in the Dead Sea region, especially in the northern parts of our study area.

6 Conclusions

This study provides a climatology of spatial and temporal properties of radar-derived convective rain cells over the dry area of the Dead Sea (Eastern Mediterranean region). These properties are examined in relation to the governing synoptic system and to flash-flood generation and magnitude. The study offers a statistical approach to relate rainfall properties (e.g., the properties of the most contributing cell) to catchment response. The main findings of the study are as follows:

- Convective rain cells are on average $92 \, \text{km}^2$ in size, move at a velocity of $12.8 \, \text{m s}^{-1}$ from west to east, and live for 18.1 min.

- ARST events are characterized by the shortest event duration, highest cell mean areal rain intensity, smallest cell area, longest cell lifetime, and lowest cell velocity.

- The area of rain cells generated by Mediterranean lows is positively correlated with the topographic height in the northern part of the study area, the number of cells is negatively correlated with the distance from the shoreline, and the mean rain intensities are negatively correlated with mean annual rainfall.

- High mean annual rainfall in the northern mountainous part of the study area results from a large number of rain cells with low velocities and large area rather than cells of high rain intensities. Rain cells related to flash-flood events are characterized by larger area, lower velocity, and longer lifetime over the catchment.

- Rain cells with lower velocities (generally $< 12 \, \text{m s}^{-1}$) and of north to northwestern origins, directed downstream with the main catchment axis, lead to high magnitude flash floods.

Results from this study add insights and quantitative information to previous studies in the Dead Sea region and in other arid regions worldwide. This advocates the robustness of the methods applied and the adequacy of the radar data used to represent rainfall over the study area. The distributions of the convective rain-cell characteristics extracted in this work can be used for stochastic simulations of convective rain storms and serve as input for hydrological models and for flash-flood nowcasting systems.

Competing interests. The authors declare that they have no conflict of interest.

Special issue statement. This article is part of the special issue "Environmental changes and hazards in the Dead Sea region (NHESS/ACP/HESS/SE inter-journal SI)". It is not associated with a conference.

Acknowledgements. The study was partially funded by the Dead Sea Drainage Authority, the Israel Water Authority, the Israel Science Foundation (grant no. 1007/15), the NSF-BSF grant (BSF 2016953) and the Lady Davis Fellowship Trust (project: RainFreq). This work is a contribution to the HyMeX program. The authors thank Maya Bartov, Moshe Armon and Uri Dayan for their assistance in validating the automatic classification of the synoptic systems.

Edited by: Florian Pappenberger

References

Ahrens, C. D.: Meteorology today: an introduction to weather, climate, and the environment, Brooks/Cole-Thomson Learning, 7th Edn., 2003.

Alpert, P., Osetinsky, I., Ziv, B., and Shafir, H.: Semi-objective classification for daily synoptic systems: application to the eastern Mediterranean climate change, Int. J. Climatol., 24, 1001–1011, https://doi.org/10.1002/joc.1036, 2004.

Andréassian, V., Oddos, A., Michel, C., Anctil, F., Perrin, C., and Loumagne, C.: Impact of spatial aggregation of inputs and parameters on the efficiency of rainfall–runoff models: A theoretical study using chimera watersheds, Water Resour. Res., 40, W05209, https://doi.org/10.1029/2003WR002854, 2004.

Ashbel, D.: Great floods in Sinai Peninsula, Palestine, Syria and the Syrian Desert, and the influence of the red sea on their formation, Q. J. Roy. Meteor. Soc., 64, 635–639, https://doi.org/10.1002/qj.49706427716, 1938.

Bahat, Y., Grodek, T., Lekach, J., and Morin, E.: Rainfall-runoff modeling in a small hyper-arid catchment, J. Hydrol., 373, 204–217, https://doi.org/10.1016/j.jhydrol.2009.04.026, 2009.

Barnolas, M., Rigo, T., and Llasat, M. C.: Characteristics of 2-D convective structures in Catalonia (NE Spain): an analysis using radar data and GIS, Hydrol. Earth Syst. Sci., 14, 129–139, https://doi.org/10.5194/hess-14-129-2010, 2010.

Berne, A. and Krajewski, W. F.: Radar for hydrology: Unfulfilled promise or unrecognized potential?, Adv. Water Resour., 51, 357–366, https://doi.org/10.1016/j.advwatres.2012.05.005, 2013.

Borga, M., Anagnostou, E. N., Blöschl, G., and Creutin, J. D.: Flash flood forecasting, warning and risk management: The HYDRATE project, Environ. Sci. Pol., 14, 834–844, https://doi.org/10.1016/j.envsci.2011.05.017, 2011.

Borga, M., Stoffel, M., Marchi, L., Marra, F., and Jakob, M.: Hydrogeomorphic response to extreme rainfall in headwater systems: Flash floods and debris flows, J. Hydrol., 518, 194–205, https://doi.org/10.1016/j.jhydrol.2014.05.022, 2014.

Bracken, L. J., Cox, N. J., and Shannon, J.: The relationship between rainfall inputs and flood generation in south-east Spain, Hydrol. Process., 22, 683–696, https://doi.org/10.1002/hyp.6641, 2008.

Cannon, A. J., Whitfield, P. H., and Lord, E. R.: Synoptic map-pattern classification using recursive partitioning and principal component analysis, Mon. Weather Rev., 130, 1187–1206, https://doi.org/10.1175/1520-0493(2002)130<1187:SMPCUR>2.0.CO;2, 2002.

Chang, C.-L.: Influence of Moving Rainstorms on Watershed Responses, Environ. Eng. Sci., 24, 1353–1360, https://doi.org/10.1089/ees.2006.0220, 2007.

Chappell, C. F.: Quasi-Stationary Convective Events, 289–310, American Meteorological Society, Boston, MA, https://doi.org/10.1007/978-1-935704-20-1_13, 1986.

Cox, D. R. and Isham, V.: A Simple Spatial-Temporal Model of Rainfall, P. R. Soc. Lond. A Mat., 415, 317–328, 1988.

David-Novak, H. B., Morin, E., and Enzel, Y.: Modern extreme storms and the rainfall thresholds for initiating debris flow on the hyperarid western escarpment of the Dead Sea, Israel, B. Geol. Soc. Am., 116, 718–728, https://doi.org/10.1130/B25403.2, 2004.

Dayan, U. and Morin, E.: Flash flood–producing rainstorms over the Dead Sea: A review, Geol. Soc. Am. S., 401, 53–62, https://doi.org/10.1130/2006.2401(04), 2006.

Dayan, U., Ziv, B., Margalit, A., Morin, E., and Sharon, D.: A severe autumn storm over the Middle-East: Synoptic and mesoscale convection analysis, Theor. Appl. Climatol., 69, 103–122, https://doi.org/10.1007/s007040170038, 2001.

Dayan, U., Tubi, A., and Levy, I.: On the importance of synoptic classification methods with respect to environmental phenomena, Int. J. Climatol., 32, 681–694, https://doi.org/10.1002/joc.2297, 2012.

Dayan, U., Nissen, K., and Ulbrich, U.: Review Article: Atmospheric conditions inducing extreme precipitation over the eastern and western Mediterranean, Nat. Hazards Earth Syst. Sci., 15, 2525–2544, https://doi.org/10.5194/nhess-15-2525-2015, 2015.

Dee, D. P., Uppala, S. M., Simmons, A. J., Berrisford, P., Poli, P., Kobayashi, S., Andrae, U., Balmaseda, M. A., Balsamo, G., and Bauer, P.: The ERA-Interim reanalysis: Configuration and performance of the data assimilation system, Q. J. Roy. Meteor. Soc., 137, 553–597, https://doi.org/10.1002/qj.828, 2011.

Dixon, M. and Wiener, G.: TITAN: Thunderstorm Identification, Tracking, Analysis, and Nowcasting – A Radar-based Methodology, 10, 785–797, https://doi.org/10.1175/1520-0426(1993)010<0785:TTITAA>2.0.CO;2, 1993.

Doswell, C. A., Brooks, H. E., and Maddox, R. A.: Flash Flood Forecasting: An Ingredients-Based Methodology, Weather Forecast., 11, 560–581, https://doi.org/10.1175/1520-0434(1996)011<0560:FFFAIB>2.0.CO;2, 1996.

Faurés, J.-M., Goodrich, D. C., Woolhiser, D. A., and Sorooshian, S.: Impact of small-scale spatial rainfall variability on runoff modeling, J. Hydrol., 173, 309–326, https://doi.org/10.1016/0022-1694(95)02704-S, 1995.

Féral, L.: HYCELL – A new hybrid model of the rain horizontal distribution for propagation studies: 2. Statistical modeling of the rain rate field, Radio Sci., 38, 1–18, https://doi.org/10.1029/2002RS002803, 2003.

Féral, L., Mesnard, F., Sauvageot, H., Castanet, L., and Lemorton, J.: Rain Cells Shape and Orientation Distribution in South-West of France, 25, 1073–1078, 2000.

Goldreich, Y.: The Climate of Israel: Observation, Research and Application, Kluwer Acad., New York, 2003.

Goldreich, Y., Mozes, H., and Rosenfeld, D.: Radar Analysis of Cloud Systems and Their Rainfall Yield in Israel, Isr.

J. Earth Sci., 53, 63–76, https://doi.org/10.1560/G68K-30MN-D5V0-KUHU, 2004.

Goodrich, D. C., Faurés, J.-M., Woolhiser, D. A., Lane, L. J., and Sorooshian, S.: Measurement and analysis of small-scale convective storm rainfall variability, J. Hydrol., 173, 283–308, https://doi.org/10.1016/0022-1694(95)02703-R, 1995.

Greenbaum, N., Ben-Zvi, A., Haviv, I., and Enzel, Y.: The hydrology and paleohydrology of the Dead Sea tributaries, Geol. Soc. Am. S., 401, 63–93, https://doi.org/10.1130/2006.2401(05), 2006.

Hewitson, B. C. and Crane, R. G.: Self-orginizing maps, Application to synoptic climatology, Clim. Res., 22, 13–26, https://doi.org/10.3354/cr022013, 2002.

Houze, R.: Orographic Effects on Precipitating Clouds, Rev. Geophys., 50, RG1001, https://doi.org/10.1029/2011RG000365, 2012.

Johnson, J. T., MacKeen, P. L., Witt, A., Mitchell, E. D. W., Stumpf, G. J., Eilts, M. D., and Thomas, K. W.: The Storm Cell Identification and Tracking Algorithm: An Enhanced WSR-88D Algorithm, Weather Forecast., 13, 263–276, https://doi.org/10.1175/1520-0434(1998)013<0263:TSCIAT>2.0.CO;2, 1998.

Kahana, R., Ziv, B., Enzel, Y., and Dayan, U.: Synoptic climatology of major floods in the Negev Desert, Israel, Int. J. Climatol., 22, 867–882, https://doi.org/10.1002/joc.766, 2002.

Kahana, R., Ziv, B., Dayan, U., and Enzel, Y.: Atmospheric predictors for major floods in the Negev Desert, Israel, Int. J. Climatol., 24, 1137–1147, https://doi.org/10.1002/joc.1056, 2004.

Karklinsky, M. and Morin, E.: Spatial characteristics of radar-derived convective rain cells over southern Israel, Meteorol. Z., 15, 513–520, https://doi.org/10.1127/0941-2948/2006/0153, 2006.

Krajewski, W. and Smith, J.: Radar hydrology: rainfall estimation, Adv. Water Resour., 25, 1387–1394, https://doi.org/10.1016/S0309-1708(02)00062-3, 2002.

Krichak, S. O. and Alpert, P.: Role of large scale moist dynamics in November 1–5, 1994, hazardous Mediterranean weather, J. Geophys. Res., 103, 19453–19468, 1998.

Kyznarová, H. and Novák, P.: CELLTRACK – Convective cell tracking algorithm and its use for deriving life cycle characteristics, Atmos. Res., 93, 317–327, https://doi.org/10.1016/j.atmosres.2008.09.019, 2009.

Llasat, M. C., Llasat-Botija, M., Prat, M. A., Porcú, F., Price, C., Mugnai, A., Lagouvardos, K., Kotroni, V., Katsanos, D., Michaelides, S., Yair, Y., Savvidou, K., and Nicolaides, K.: High-impact floods and flash floods in Mediterranean countries: the FLASH preliminary database, Adv. Geosci., 23, 47–55, https://doi.org/10.5194/adgeo-23-47-2010, 2010.

Marra, F. and Morin, E.: Use of radar QPE for the derivation of Intensity-Duration-Frequency curves in a range of climatic regimes, J. Hydrol., 531, 427–440, https://doi.org/10.1016/j.jhydrol.2015.08.064, 2015.

Marra, F., Morin, E., Peleg, N., Mei, Y., and Anagnostou, E. N.: Intensity-duration-frequency curves from remote sensing rainfall estimates: comparing satellite and weather radar over the eastern Mediterranean, Hydrol. Earth Syst. Sci., 21, 2389–2404, https://doi.org/10.5194/hess-21-2389-2017, 2017.

Morin, E. and Gabella, M.: Radar-based quantitative precipitation estimation over Mediterranean and dry climate regimes, J. Geophys. Res., 112, D20108, https://doi.org/10.1029/2006JD008206, 2007.

Morin, E. and Yakir, H.: Hydrological impact and potential flooding of convective rain cells in a semiarid environment, Hydrolog. Sci. J., 59, 1353–1362, https://doi.org/10.1080/02626667.2013.841315, 2014.

Morin, E., Enzel, Y., Shamir, U., and Garti, R.: The characteristic time scale for basin hydrological response using radar data, J. Hydrol., 252, 85–99, https://doi.org/10.1016/S0022-1694(01)00451-6, 2001.

Morin, E., Goodrich, D. C., Maddox, R. a., Gao, X., Gupta, H. V., and Sorooshian, S.: Spatial patterns in thunderstorm rainfall events and their coupling with watershed hydrological response, Adv. Water Resour., 29, 843–860, https://doi.org/10.1016/j.advwatres.2005.07.014, 2006.

Morin, E., Jacoby, Y., Navon, S., and Bet-Halachmi, E.: Towards flash-flood prediction in the dry Dead Sea region utilizing radar rainfall information, Adv. Water Resour., 32, 1066–1076, https://doi.org/10.1016/j.advwatres.2008.11.011, 2009.

Nied, M., Pardowitz, T., Nissen, K., Ulbrich, U., Hundecha, Y., and Merz, B.: On the relationship between hydrometeorological patterns and flood types, J. Hydrol., 519, 3249–3262, https://doi.org/10.1016/j.jhydrol.2014.09.089, 2014.

Northrop, P.: A clustered spatial-temporal model of rainfall, 454, 1875–1888, https://doi.org/10.1098/rspa.1998.0238, 1997.

Peleg, N. and Morin, E.: Convective rain cells: Radar-derived spatiotemporal characteristics and synoptic patterns over the eastern Mediterranean, J. Geophys. Res., 117, D15116, https://doi.org/10.1029/2011JD017353, 2012.

Peleg, N., Ben-Asher, M., and Morin, E.: Radar subpixel-scale rainfall variability and uncertainty: lessons learned from observations of a dense rain-gauge network, Hydrol. Earth Syst. Sci., 17, 2195–2208, https://doi.org/10.5194/hess-17-2195-2013, 2013.

Peleg, N., Bartov, M., and Morin, E.: CMIP5-predicted climate shifts over the East Mediterranean: implications for the transition region between Mediterranean and semi-arid climates, Int. J. Climatol., 35, 2144–2153, https://doi.org/10.1002/joc.4114, 2015a.

Peleg, N., Shamir, E., Georgakakos, K. P., and Morin, E.: A framework for assessing hydrological regime sensitivity to climate change in a convective rainfall environment: a case study of two medium-sized eastern Mediterranean catchments, Israel, Hydrol. Earth Syst. Sci., 19, 567–581, https://doi.org/10.5194/hess-19-567-2015, 2015b.

Peleg, N., Marra, F., Fatichi, S., Paschalis, A., Molnar, P., and Burlando, P.: Spatial variability of extreme rainfall at radar subpixel scale, J. Hydrol., https://doi.org/10.1016/j.jhydrol.2016.05.033, online first, 2016.

Ries, F., Schmidt, S., Sauter, M., and Lange, J.: Controls on runoff generation along a steep climatic gradient in the Eastern Mediterranean, J. Hydrol.: Regional Studies, 9, 18–33, https://doi.org/10.1016/j.ejrh.2016.11.001, 2017.

Rinehart, R. E. and Garvey, E. T.: Three-dimensional storm motion detection by conventional weather radar, 273, 287–289, https://doi.org/10.1038/273287a0, 1978.

Rozalis, S., Morin, E., Yair, Y., and Price, C.: Flash flood prediction using an uncalibrated hydrological model and radar rainfall data in a Mediterranean watershed under changing hydrological conditions, J. Hydrol., 394, 245–255,

https://doi.org/10.1016/j.jhydrol.2010.03.021, 2010.

Saaroni, H., Halfon, N., Ziv, B., Alpert, P., and Kutiel, H.: Links between the rainfall regime in Israel and location and intensity of Cyprus lows, Int. J. Climatol., 30, 1014–1025, https://doi.org/10.1002/joc.1912, 2010.

Saaroni, H., Ziv, B., Lempert, J., Gazit, Y., and Morin, E.: Prolonged dry spells in the Levant region: Climatologic-synoptic analysis, International J. Climatol., 2236, 2223–2236, https://doi.org/10.1002/joc.4143, 2014.

Segond, M. L., Wheater, H. S., and Onof, C.: The significance of spatial rainfall representation for flood runoff estimation: A numerical evaluation based on the Lee catchment, UK, J. Hydrol., 347, 116–131, https://doi.org/10.1016/j.jhydrol.2007.09.040, 2007.

Shamir, E., Ben-Moshe, L., Ronen, A., Grodek, T., Enzel, Y., Georgakakos, K. P., and Morin, E.: Geomorphology-based index for detecting minimal flood stages in arid alluvial streams, Hydrol. Earth Syst. Sci., 17, 1021–1034, https://doi.org/10.5194/hess-17-1021-2013, 2013.

Sharon, D. and Kutiel, H.: The distribution of rainfall intensity in Israel, its regional and seasonal variations and its climatological evaluation, J. Climatol., 6, 277–291, https://doi.org/10.1002/joc.3370060304, 1986.

Singh, V. P.: Effect of spatial and temporal variability in rainfall and watershed characteristics on stream flow hydrograph, Hydrol. Process., 11, 1649–1669, https://doi.org/10.1002/(SICI)1099-1085(19971015)11:12<1649::AID-HYP495>3.0.CO;2-1, 1997.

Smith, J. a., Baeck, M. L., Morrison, J. E., and Sturdevant-Rees, P.: Catastrophic Rainfall and Flooding in Texas, J. Hydrometeorol., 1, 5–25, https://doi.org/10.1175/1525-7541(2000)001<0005:CRAFIT>2.0.CO;2, 2000.

Smith, J. A., Baeck, M. L., Morrison, J. E., Sturdevant-Rees, P., Turner-Gillespie, D. F., and Bates, P. D.: The Regional Hydrology of Extreme Floods in an Urbanizing Drainage Basin, J. Hydrometeorol., 3, 267–282, https://doi.org/10.1175/1525-7541(2002)003<0267:TRHOEF>2.0.CO;2, 2002.

Smith, J. a., Baeck, M. L., Villarini, G., and Krajewski, W. F.: The Hydrology and Hydrometeorology of Flooding in the Delaware River Basin, J. Hydrometeorol., 11, 841–859, https://doi.org/10.1175/2010JHM1236.1, 2010.

Steiner, M., Houze, R. A., and Yuter, S. E.: Climatological Characterization of Three-Dimensional Storm Structure from Operational Radar and Rain Gauge Data, 34, 1978–2007, https://doi.org/10.1175/1520-0450(1995)034<1978:CCOTDS>2.0.CO;2, 1995.

Syed, K. H., Goodrich, D. C., Myers, D. E., and Sorooshian, S.: Spatial characteristics of thunderstorm rainfall fields and their relation to runoff, J. Hydrol., 271, 1–21, https://doi.org/10.1016/S0022-1694(02)00311-6, 2003.

Tarolli, P., Borga, M., Morin, E., and Delrieu, G.: Analysis of flash flood regimes in the North-Western and South-Eastern Mediterranean regions, Nat. Hazards Earth Syst. Sci., 12, 1255–1265, https://doi.org/10.5194/nhess-12-1255-2012, 2012.

Tubi, A. and Dayan, U.: Tropical Plumes over the Middle East: Climatology and synoptic conditions, Atmos. Res., 145–146, 168–181, https://doi.org/10.1016/j.atmosres.2014.03.028, 2014.

von Hardenberg, J.: The shape of convective rain cells, Geophys.

Res. Lett., 30, 4–7, https://doi.org/10.1029/2003GL018539, 2003.

Warner, T. T.: Desert Meteorology, 61, 89–90, https://doi.org/10.1256/wea.201.04, 2004.

Wheater, H. S., Butler, A. P., Stewart, E. J., and Hamilton, G. S.: A multivariate spatial-temporal model of rainfall in southwest Saudi Arabia. I. Spatial rainfall characteristics and model formulation, J. Hydrol., 125, 175–199, https://doi.org/10.1016/0022-1694(91)90028-G, 1991.

Wheater, H. S., Isham, V. S., Cox, D. R., Chandler, R. E., Kakou, A., Northrop, P. J., Oh, L., Onof, C., and Rodriguez-Iturbe, I.: Spatial-temporal rainfall fields: modelling and statistical aspects, Hydrol. Earth Syst. Sci., 4, 581–601, https://doi.org/10.5194/hess-4-581-2000, 2000.

Wright, D., Smith, J., and Baeck, M.: Flood frequency analysis using radar rainfall fields and stochastic storm transposition, Water Resour. Res., 50, 1592–1615, https://doi.org/10.1002/2013WR014224, 2014.

Yair, A. and Lavee H.: Runoff Generation in arid and semi-arid zones, in: Hydrological Forecasting, edited by: Anderson, M. G. and Burt, T. P., Wiley, Chichester, UK, 183–220, 1985.

Yakir, H. and Morin, E.: Hydrologic response of a semi-arid watershed to spatial and temporal characteristics of convective rain cells, Hydrol. Earth Syst. Sci., 15, 393–404, https://doi.org/10.5194/hess-15-393-2011, 2011.

Yang, L., Smith, J., Baeck, M. L., Smith, B., Tian, F., and Niyogi, D.: Structure and evolution of flash flood producing storms in a small urban watershed, J. Geophys. Res.-Atmos., 121, 3139–3152, https://doi.org/10.1002/2015JD024478, 2016a.

Yang, L., Smith, J., Baeck, M. L., and Zhang, Y.: Flash flooding in small urban watersheds: Stormevent hydrologic response, Water Resour. Res., 52, 4571–4589, https://doi.org/10.1002/2015WR018326, 2016b.

Zangvil, A. and Druian, P.: Upper air trough axis orientation and the spatial distribution of rainfall over Israel, Int. J. Climatol., 10, 57–62, 1990.

Zangvil, A., Karas, S., and Sasson, A.: Connection between Eastern Mediterranean seasonal mean 500 hPa height and sea-level pressure patterns and the spatial rainfall distribution over Israel, Int. J. Climatol., 23, 1567–1576, https://doi.org/10.1002/joc.955, 2003.

Ziv, B.: A subtropical rainstorm associated with a tropical plume over Africa and the Middle-East, Theor. Appl. Climatol., 69, 91–102, https://doi.org/10.1007/s007040170037, 2001.

Ziv, B., Dayan, U., and Sharon, D.: A mid-winter, tropical extreme flood-producing storm in southern Israel: Synoptic scale analysis, Meteorol. Atmos. Phys., 88, 53–63, https://doi.org/10.1007/s00703-003-0054-7, 2004.

Ziv, B., Dayan, U., Kushnir, Y., Roth, C., and Enzel, Y.: Regional and global atmospheric patterns governing rainfall in the southern Levant, Int. J. Climatol., 26, 55–73, https://doi.org/10.1002/joc.1238, 2006.

Zoccatelli, D., Borga, M., Zanon, F., Antonescu, B., and Stancalie, G.: Which rainfall spatial information for flash flood response modelling? A numerical investigation based on data from the Carpathian range, Romania, J. Hydrol., 394, 148–161, https://doi.org/10.1016/j.jhydrol.2010.07.019, 2010.

Permissions

List of Contributors

Stephen T. Casey, Matthew J. Cohen and Subodh Acharya
School of Forest Resources and Conservation, University of Florida, Gainesville, FL, USA

David A. Kaplan
Engineering School of Sustainable Infrastructure and Environment, Environmental Engineering Sciences Department, University of Florida, Gainesville, FL, USA

James W. Jawitz
Soil and Water Science Department, University of Florida, Gainesville, FL, USA

Gabriëlle J. M. De Lannoy
KU Leuven, Department of Earth and Environmental Sciences, Heverlee, Belgium

Rolf H. Reichle
NASA Goddard Space Flight Center, Global Modeling and Assimilation Office, Greenbelt, Maryland, USA

Luis Samaniego, Rohini Kumar, Stephan Thober, Oldrich Rakovec, Matthias Zink and Sabine Attinger
Department of Computational Hydrosystems, UFZ-Helmholtz Centre for Environmental Research, Leipzig, Germany

Niko Wanders
Department of Civil and Environmental Engineering, Princeton University, Princeton, NJ 08544, USA
Universiteit Utrecht, Department of Physical Geography, Utrecht, the Netherlands

Stephanie Eisner
Center for Environmental Systems Research, University of Kassel, Kassel, Germany
Division for Forestry and Forest Resources, Norwegian Institute of Bioeconomy Research, Ås, Norway

Hannes Müller Schmied
Institute of Physical Geography, Goethe-University Frankfurt, Frankfurt, Germany
Senckenberg Biodiversity and Climate Research Centre (BiK-F), Frankfurt, Germany

Edwin H. Sutanudjaja
Universiteit Utrecht, Department of Physical Geography, Utrecht, the Netherlands

Kirsten Warrach-Sagi
Institute of Physics and Meteorology, University of Hohenheim, Stuttgart, Germany

Shirley Echendu, Yunqing Xuan, Mike Webster and Ian Cluckie
College of Engineering, Swansea University Bay Campus, Swansea, SA1 8EN, UK

Dehua Zhu
College of Engineering, Swansea University Bay Campus, Swansea, SA1 8EN, UK
School of Hydrometeorology, Nanjing University of Information Science and Technology. Nanjing, 210044, China

Søren Thorndahl
Department of Civil Engineering, Aalborg University, Aalborg, 9220, Denmark

Aske Korup Andersen and Anders Badsberg Larsen
Niras A/S, Aalborg, 9000, Denmark

Steven L. Markstrom and Lauren E. Hay
US Geological Survey, P.O. Box 25046, MS 412, Denver Federal Center, Denver, Colorado, 80225, USA

Martyn P. Clark
National Center for Atmospheric Research, P.O. Box 3000, Boulder, Colorado, 80307, USA

Erin Coughlan de Perez
Red Cross Red Crescent Climate Centre, The Hague, 2521 CV, the Netherlands
Institute for Environmental Studies, VU University Amsterdam, 1081 HV, the Netherlands
International Research Institute for Climate and Society, Columbia University, New York, 10964, USA

Maarten van Aalst
Red Cross Red Crescent Climate Centre, The Hague, 2521 CV, the Netherlands
International Research Institute for Climate and Society, Columbia University, New York, 10964, USA

Elisabeth Stephens
School of Archaeology, Geography and Environmental Science, University of Reading, Reading, RG6 6AH, UK

Konstantinos Bischiniotis
Institute for Environmental Studies, VU University Amsterdam, 1081 HV, the Netherlands

Simon Mason and Hannah Nissan
International Research Institute for Climate and Society, Columbia University, New York, 10964, USA

Bart van den Hurk
Royal Netherlands Meteorological Institute (KNMI), De Bilt, 3731 GA, the Netherlands

Florian Pappenberger
European Centre for Medium-Range Weather Forecasts, Reading, RG2 9AX, UK

Chao Deng, Pan Liu, Shenglian Guo and Zejun Li
State Key Laboratory of Water Resources and Hydropower Engineering Science, Wuhan University, Wuhan, China
Hubei Provincial Collaborative Innovation Center for Water Resources Security, Wuhan, China

Dingbao Wang
Department of Civil, Environmental & Construction Engineering, University of Central Florida, Orlando, FL, USA

Alejandra Stehr and Mauricio Aguayo
Centre for Environmental Sciences EULA-CHILE, University of Concepción, Concepción, Chile
Faculty of Environmental Sciences, University of Concepción, Concepción, Chile

Lamprini V. Papadimitriou, Aristeidis G. Koutroulis and Manolis G. Grillakis
Technical University of Crete, School of Environmental Engineering, Chania, Greece

Ioannis K. Tsanis
Technical University of Crete, School of Environmental Engineering, Chania, Greece
McMaster University, Department of Civil Engineering, Hamilton, ON, Canada

Caitlin A. Orem and Jon D. Pelletier
Department of Geosciences, The University of Arizona, 1040 E. 4th Street, Tucson, AZ 85721, USA

Ido Nitzan and Daniel Kurtzman
Institute of Soil, Water and Environmental Sciences, The Volcani Center, Agricultural Research Organization, Rishon LeZion, 7528809, Israel

Yonatan Ganot
Institute of Soil, Water and Environmental Sciences, The Volcani Center, Agricultural Research Organization, Rishon LeZion, 7528809, Israel
Department of Soil and Water Sciences, The Hebrew University of Jerusalem, Rehovot, 7610001, Israel

Ran Holtzman
Department of Soil and Water Sciences, The Hebrew University of Jerusalem, Rehovot, 7610001, Israel

Noam Weisbrod
Department of Environmental Hydrology & Microbiology, Zuckerberg Institute for Water Research, Jacob Blaustein Institutes for Desert Research, Ben-Gurion University of the Negev, Midreshet Ben Gurion, 8499000, Israel

Yoram Katz
Mekorot, Water Company Ltd, Tel Aviv, 6713402, Israel

Idit Belachsen
Hydrology and Water Resources Program, Hebrew University of Jerusalem, 91904, Israel
Institute of Earth Sciences, Hebrew University of Jerusalem, 91904, Israel

Francesco Marra and Efrat Morin
Institute of Earth Sciences, Hebrew University of Jerusalem, 91904, Israel

Nadav Peleg
Institute of Environmental Engineering, Hydrology and Water Resources Management, ETH Zurich, Switzerland

Index